Studies in Applied Mechanics 48

Stability of Nonlinear Shells
On the example of spherical shells

Dov Shilkrut
Department of Mechanical Engineering
Ben-Gurion University of the Negev
Beer-Sheva, Israel

Scientific Editor
Eduard Riks
Department of Aerospace Engineering
University of Delft, The Netherlands

2002

ELSEVIER
Amsterdam - London - New York - Oxford - Paris - Shannon - Tokyo

ELSEVIER SCIENCE Ltd
The Boulevard, Langford Lane
Kidlington, Oxford OX5 1GB, U.K.

First edition 2002

Library of Congress Cataloging in Publication Data
A catalog record from the Library of Congress has been applied for.

British Library Cataloguing in Publication Data
A catalogue record from the British Library has been applied for.

ISBN: 0 08 044085 1

♾The paper used in this publication meets the requirements of ANSI/NISO Z39.48-1992 (Permanence of Paper).
Printed in The Netherlands.

Dov Shilkrut, 1925 – 1998

IN MEMORIAM

Dov Shilkrut, 1925-1998

Loving husband, father, and grandfather

Father lived a difficult but interesting life. He was born in Romania in 1925, and received his education in Russia — in part, in the Asian part of Russia during the hard times of the Second World War.

His whole life was directed by immense talent and ability. His thirst for knowledge, his love of science and unlimited capacity for work, and, of course, his unique talents, were the factors that defined his natural state — his constant concentration and devotion to solving numerous scientific problems.

His creative energy was inexhaustible and, in the last period of his life, this energy had not diminished, but actually grew stronger. "Give me 500 doctoral students and I will give each one of them a topic for their thesis," he used to say.

His interests were not limited to science, but were unusually wide-ranging. He was fascinated by philosophy, the history of science, general history, geography, and philately. In his leisure time, he collected stamps, greatly enjoyed the visual arts and music, and was a devoted opera lover. Because of his youthful soul, his deep pleasure in being with younger people, and the readiness to help, our house was always an open house, full of students — undergraduates as well as doctoral candidates. He also kept in contact with a large number of scientists all over the world, many of whom had been his students or were colleagues, past and present. His unassuming manner, fairness in dealing with others, and sense of humor made him an interesting conversationalist and a sought-after guest in many different social circles.

He loved his family very much, and, being a perceptive person, showed deep concern for our problems and welfare. His positivity, optimism, and decisiveness all helped us to overcome whatever adversity we faced.

Our memories of him fill our hearts with love, pride, and respect.

The present book is the crown of his life-long scientific contribution and the many, many publications from past years of meticulous research. It brings together the sum of his scientific activity for the last 15 years. It was his last and greatest creative effort, and we offer it in his memory as his legacy to the scientific community.

It was not intended to be the end of his special contribution to his field. Before his last tragic trip to Amsterdam, which was in connection with the publication of *Stability of Nonlinear Shells*, Father told us about a theory that had already ripened into completeness in his mind — a theory that would extend the subject of this book, and which, in his opinion, would surpass all that he had created till then. He was only waiting for the finalization of the present book to begin writing.

But, he never came back. He died on his way from the Schiphol Airport to Delft.

The Family would like to express our gratitude to those who did everything possible to see this book published—Professor Eliahu Zahavi of Ben-Gurion University, Father's close friend, who took upon himself to coordinate the completion of the manuscript; Dr. Eduard Riks of Delft, a colleague and long-time associate of Father, who agreed to act as scientific editor; and Ms. Catherine Logan, also from Israel and Father's copyeditor for over 6 years, who produced this book from its earliest stages to its final camera-ready form. They labored hard and had to use their initiative continuously, not having the author to consult, all the while making sure that Father's original manuscript was unchanged.

The Family is also indebted to Ben-Gurion University of the Negev for their financial support of this publication, and, especially, to Professor Nachum Finger, the Rector of Ben-Gurion University, and to Professor Mordechai Perl, Dean of the Faculty of Engineering Sciences. Both, in recognition of their friendship with Father, took a personal interest to ensure that the book was published.

Special thanks are due to the publishers, Elsevier Science Ltd., and to the many people there—especially Ms. Lorna Canderton, Administrative Editor, and the editor in charge, Ms. Nicola Garvey—whose understanding, patience, and goodwill supported the team in their difficult undertaking.

The Shilkrut Family

ACKNOWLEDGEMENT

Starting to prepare the manuscript for publication right after the loss of the author was, foremost, a step that helped me cope with the grief over the abrupt end of my close friendship with Professor Shilkrut. Since I was aware of the work's substantial contribution to Applied Mechanics, it was only natural for me to approach those already acquainted with the work, Dr. Eduard Riks and Ms Catherine Logan. I would like to express my deep acknowledgement for their enthusiastic consent and their devotion in carrying out the difficult job. However, Ms Catherine Logan, who cooperated with Professor Shilkrut on many of his prior publications, must be credited for guarding the authenticity of the manuscript.

Professor Eliahu Zahavi
Ben-Gurion University of the Negev

A PERSONAL NOTE

On November 10, 1998, Dov Shilkrut left his home in Beer-Sheva, Israel to visit me in Delft, the Netherlands. This was after we had agreed to speed up the scientific part of the editing of his manuscript by going through it together and clearing up some points I had raised concerning some technical matters. I expected a pleasant visit with my colleague, since we were basically in agreement on the outstanding issues.

While I was waiting for his telephone call to tell me that he had arrived, the phone rang and the voice of our secretary came through with the message that shook me rather gravely. Dov had passed away while boarding the train at Schiphol airport. The blow was devastating; and it left a numbing feeling that is very difficult to describe even today.

It was several weeks later, while I was still trying to cope with the thought that Dov was no longer with us, that I received the request from the family Shilkrut — conveyed to me by Alec Zahavi — that I do the scientific part of the editing and checking of the manuscript and help Alec and Catherine Logan bring the book "into life." I felt very moved by this request, not only because of the trust the family granted me in being able to help finalize the work for publication and the challenge it entailed, but also because this task gave me the opportunity to pay tribute to a man who had been a dear and close friend of mine for many years.

Dov and I met each other in Stuttgart in 1980, as I recall, at a conference devoted to shell stability and related topics. We immediately developed a mutual understanding that went far beyond the technical interests we shared as professionals. During the 20 years that passed we saw each other at many occasions, among others, the usual conferences and the extended periods that Dov stayed and worked at the Faculty of Aerospace Engineering of Delft University as visiting professor.

I knew Dov Shilkrut as a man who had strong views that he expressed with great intensity. This intensity was not always understood by people who knew him superficially. However, for those closer to him, like myself, he was a very generous and sensitive man, a man who was intensely loyal and who radiated a lot of warmth to those he considered his friends.

It is my wish and hope that this book will stand as a lasting monument of his achievements in the field of shell stability and that it will help to keep his memory alive for all of us who had the privilege to know him personally.

Eduard Riks

PREFACE

Ben-Gurion University of the Negev was established in the first half of the 1960s with a strong commitment to serve not only as an institution of higher learning, dedicated to the pursuit of scholarship and education, but also to stimulate the economic, social, and cultural development of the entire Negev region. Soon after opening its doors, the university became a magnet for talented scientists from all over the world, and especially for those arriving in Israel with the first wave of immigration from the Soviet Union. Ben-Gurion University was particularly fortunate that, among these scientists, was Professor Dov Shilkrut, whose reputation preceded him. Easily identifying with the pioneering spirit of the university, Professor Shilkrut joined the Faculty of Engineering, soon becoming one of the pillars of the Faculty and of the Department of Mechanical Engineering. He was instrumental in establishing a research group in Mechanics, which is very active in the Department to this very day.

I recently came across a report of Professor Shilkrut's activities during his first year at the university alone, listing a number of publications that had appeared in print, an additional number submitted for publication, as well as an impressive list of papers delivered at scientific conferences and seminars. What a report such as this cannot illuminate, of course, was Dov's personality: his ebullience, his drive, his vivacious and yet no-nonsense approach to everything with which he dealt.

Dov considered this book the culmination of his professional contribution to knowledge to the moment of his death; it is my personal feeling that it is also most fitting for it to serve as a partial legacy. A large part of this legacy, however, does not lie in the written word, be it book, journal, or conference abstract. Dov imparted elements of his intellectual spirit to several generations of students, both in the former Soviet Union and in Israel, some of whom, as scientists in their own right, actively pursue his ideas at our university today, as well as in universities, research institutes, and private companies worldwide. Dov dedicated his life to passing on the torch of his ideology, his devotion to science, and his dynamic approach to daily challenges — qualities that are much less tangible than his extensive body of published and unpublished works, but nonetheless genuine.

It is an honor to write the preface to this book, although it inevitably saddens me that it is posthumous. I doubt whether it is possible to assess the extent of creative energy lost to the scientific world with Dov Shilkrut's untimely passing.

Prof. Nachum Finger
Rector,
Ben-Gurion University of the Negev
Beer-Sheva, Israel

INTRODUCTION

Geometrically nonlinear (flexible) shells, one of the most effective thin-walled structures, are meeting with steadily broadening applications in almost all branches of modern industry, from aircraft, space craft, and ship structures to chemical apparatus and building constructions. With the combination of light weight with high strength, the flexible shell is frequently an indispensable structural form. Its use is continuously growing for various purposes and under very complicated conditions.

Being flexible, the behavior of these shells under loading is accompanied by essentially nonlinear effects, which cannot, in principle, be described by linear theories. Thus the prob-lem of deformation and stability of geometrically nonlinear shells are very diverse and complex, and demands serious consideration. The understanding of the nonlinear effects is absolutely necessary for the design of the modern, greatly diverse, structures and machines that contain these shells. An optimized design demands the most complete information on the behavior of the structure from the beginning of loading up to its limit state, and the designer must possess "nonlinear intuition" and "nonlinear psychology." The latter may be acquired only by mastering the vast body of up-to-date information on the behavior of nonlinear objects. It is necessary to stress that, until now, well-founded recommendations for the design and control of the strength and stability of flexible shells and other similar structures have not been available.

On the other hand, the development of the theory of nonlinear shells and plates is of the greatest importance from a purely theoretical point of view, since it belongs to the field of modern nonlinear mechanics and applied mathematics. This is precisely the reason why this theory has been attracting so much attention.

As is well known, there are two basic models of the nonlinear theory of shells — the spherical and the cylindrical, representing the class of nondevelopable and developable surfaces, respectively. The main properties of these models are characteristic for all nonlinear shells with the exception of those having non-negative Gaussian curvature; that is, they are characteristic for the majority of the shell structures used in present day engineering. *The Stability of Nonlinear Shells* deals with the study of the behavior of spherical caps and circular plates subjected to loadings of different types under various conditions of their applications.

There are many works which investigate the deformation and stability of geometrically nonlinear spherical shells and caps. However, the important information about these results is scattered among a number of publications, and, therefore, a clear and comprehensive analy-tical survey of this subject has not as yet been presented, despite several surveys of general character [e.g., Bushnell, 1989; Kaplan, 1974; Volmir, 1963].

The goal of this book is to compile such an analytical survey and to present a summary of the available published results and their analysis in order to present as complete a picture as possible of the behavior of nonlinear shells based on the most up-to-date information. In particular, this survey will include a compilation of investigations by the author and his students. A survey of this nature can never be completely exhaustive, and therefore the author offers his apologies for inadvertent omissions of contributions he may have overlooked.

It is clear that such a book serves both designers and scientists in the corresponding fields, and also, of course, students of various fields of engineering, mathematics, and physics.

The first chapter of *The Stability of Nonlinear Shells* deals with the basic equations which describe the geometrically nonlinear flexible shells of revolution in different cases. Various versions of these are presented. In addition to the well-known basic equations of the von Karman-Marguerre type for shallow shells, more exact basic equations for non-shallow shells are given. For the latter, in contradistinction to the conventional approach, the components ε of the tensor of deformation (which are small in the elastic range) are not rejected in the expression of type $1+\varepsilon$, which offers the possibility of keeping the derivative ε' in the respective terms of the basic equations after they have been differentiated. This is very important for flexible non-shallow shells, which can possess a large number of geometrically complicated equilibrium forms for which ε' is not negligible, despite the smallness of ε.

The second chapter contains some qualitative properties of the solutions of the basic equations, which enable one to establish a number of important qualitative features of the behavior of shells under loading. The author obtained these results by direct use of the basic equations that govern system behavior, but without actually solving them, which is an approach similar to the qualitative theory of differential equations (dynamic systems). In particular, a class of so-called reciprocal systems is revealed which has a series of interesting properties, including, especially, an exact analytical relation between the upper and lower critical values of the loading.

The generalization, given in the second chapter, of the features of reciprocal systems extended to those that are nonreciprocal essentially enlarges the qualitative picture of the behavior of shells. Another important result of this qualitative approach is the elucidation of the crucial role of membrane stresses (their distribution, magnitudes, and signs) on all types of phenomena (processes) of loss of stability. At the end of Chapter 2 there is a discussion of an interesting paradox: when the equation of oscillations of one degree of freedom cannot be expressed in an explicit form, relative to the inertial term (the normal form). In actuality, this is a case of oscillations of a mass attached to a geometrically nonlinear spherical cap, where, as is usual, the small mass of the cap is neglected in comparison to the added mass. Poincaré's phase plane of this system reveals a series of new qualitative features which do not exist for a normal conservative system. Thus, we have here a case whereby the usual approach of the idealization of the mass distribution cannot bring about a proper solution.

The third chapter deals with the results of numerical solutions for axisymmetrically loaded, geometrically nonlinear, shallow spherical shells. The considerations deal with both axisymmetric deformations and bifurcation phenomena, when axisymmetrically loaded shells can exhibit symmetric equilibrium states in addition to axisymmetric states. Included are also some cases of the influence of geometrical imperfections on the critical values of loading parameters. It is of interest to note that geometric imperfections are sometimes able to increase the critical values of the loading, and thus do not always lead to a decrease of the critical load, as is usually assumed.

These numerous solutions, distributed over in a multitude of different publications, are gathered, analyzed, and criticized here. This provides the possibility of revealing the whole complexity of nonlinear effects that already arise in such relatively simple cases as deformations of shallow flexible shells of revolution under axisymmetric single-parametric loading. For the first time, practically all the available numerically obtained information about the behavior of shallow, geometrically nonlinear, spherical caps subjected to a single-parametric loading is collected and analyzed in one place.

Chapter 4, which consists almost entirely of new results obtained by the author and his students, is devoted to the behavior of a cap under a system of multiparametric, independently acting, external loading systems. These investigations are of special interest from many points of view, taking into consideration the absence of the principle of superposition because of the nonlinearity of the problems. The main tool of the investigation of multiparametric problems is the so-called "deformation map," which is considered in detail in this chapter. This map is a global storage of information of the behavior of caps under two- and three-parametric excitations. Chapter 4 contains a number of features of the behavior of circular plates and shells, which were obtained by means of the deformation map. Among these is the chaotic-like behavior of an orthotropic circular plate under three-parametric static loading systems (two forces and a temperature field). Thus, there may exist chaos in statics, which, in a physical sense, means that the behavior of the system is, de facto, unpredictable, despite the existence of only a fixed, limited number of possible stable equilibrium states. It is likely that this number is limited to a maximum three, as is brought out by the map (see also Chapter 6). The application of multiparametric loading systems of course requires the determination of the influence of the loading path on the final state of the considered deformable systems. This new problem was investigated dynamically; it could not be considered otherwise, since each loading path is, in fact, a process that passes in time even when the path is of a quasi-static character. It was revealed that two different paths either end up in the same equilibrium state, or in two different ones. Both possibilities exist, but which of these will occur under the given conditions is unknown in advance. Thus, the final result is again unpredictable. This result is also very important for numerical methods and algorithms, where an incremental (step-wise) approach is generally used. Each specific organization of the incremental multiparametric loading is in reality a selection of a corresponding loading path.

Chapter 5 first deals with a series of specific contact problems which are of interest from the theoretical (physical) or engineering points of view. Among these problems is the deformation of a nonlinear circular plate "bonded" to a linear elastic foundation. When debonding occurs, the problem is changed into a nonlinear contact problem of a plate resting on a unilateral foundation. Another contact problem concerns the deformation and loss of stability of a cap loaded by a force applied to the shell indirectly via an intermediate rigid plate. The second part of Chapter 5 is devoted to the survey and analysis of numerical results on the axisymmetric deformation of non-shallow spherical shells. The main conclusion from these results is that, in general, there is no essential qualitative difference between solutions based on the theories of shallow and non-shallow shells. Of course, the initial geometry of non-shallow spherical caps depends, contrary to the case of shallow caps, on two geometric para-meters. Because of this, all the parameters of the deformation (in particular the magnitude of the values of critical loading) depend on these two parameters.

Finally, Chapter 6 consists of a series of investigations, examples, and remarks which serve to elucidate some of the general notions concerning the stability or instability of geomet-rically nonlinear spherical caps and geometrically nonlinear systems in general. The contents of this chapter are very important for an understanding of the behavior and the first order stability of this behavior of these systems. An essential part of Chapter 6 is devoted to the discussion of methods and results of the testing of stability-in-the-small of equilibrium states of nonlinear spherical shells. It was determined that, for a single-parametric loading, there are no more than two stable states, in spite of the fact, proven in Chapter 2, that the number of possible equilibrium states generated by a given value of the loading parameters increases with decreasing shell thickness. This number tends to infinity when the shell's body thickness tends to zero. A similar result holds for two-parametric loading systems. For three-parametric loading, some cases were discovered where three-stable states exist.

The process of stability loss by snapping when only two-stable states are available can be clearly explained physically, taking into consideration that the shell is a continuous system with an infinite number of degrees of freedom. But in the presence of three-stable states, the continuity of the system is no longer sufficient to provide an explanation for the snapping process. Here a new and very important problem arises about the manifolds of attraction of different equilibrium states in the dynamics of such systems. Without the dynamic approach, the snap-through process of stability loss can therefore not be considered.

It is shown that Poincaré's classical theory of stability, which was developed primarily for a dynamic system of a single degree of freedom and thereafter generalized to a finite number of degrees of freedom, cannot always describe the stability phenomena of continuous systems in an adequate manner. Other remarks and interpretations of questions of stability aspects are also offered.

The results are illustrated by numerous tables and graphs which may be used as reference material in the design of shell design. The text is of interest for researchers, designers, and students whose work deals with the theory and practice of shells and shell constructions.

In conclusion, I have the pleasure to thank my colleagues for their help in the preparation of the book. Specifically, many thanks to my friends Professors J. Arbocz and E. Riks from the Aeronautical Department of Delft University of Technology for very useful discussions, and, in particular, for Professor Riks's expert specialist editorial comments. I am very grateful to the Rector of Ben-Gurion University of the Negev for his support. I am also deeply grateful to Ms. Catherine Logan for her inestimable handiwork during the preparation of the manuscript.

Finally, I want to especially express my gratitude and warm feelings for Professor W. Koiter for his exceptional attitude toward me and my work. Professor Koiter, who was one of the most famous scientists in the field of stability of deformable systems in the second half of the twentieth century, unfortunately died before this book could be completed.

Professor Dov Shilkrut
Ben-Gurion University of the Negev
October 1998

CONTENTS

Chapter 1

Basic Equations of Geometrically Nonlinear Shells

Chapter 2

Qualitative Investigations of Geometrically Nonlinear Shells

Chapter 3

Numerical Investigations of Axisymmetrically Loaded Geometrically Nonlinear Shallow Spherical Caps and Circular Plates (A Survey)

Chapter 4

Spherical Caps Subjected to Multi-Parameter Loading. The Deformation Map. Influence of the Loading Path on the Cap's Behavior. Thermo-Elastic Deformations

Chapter 5
Some Special Problems and the Behavior of Deep Caps

Chapter 6
The Stability of Equilibrium States of Geometrically Nonlinear Shells

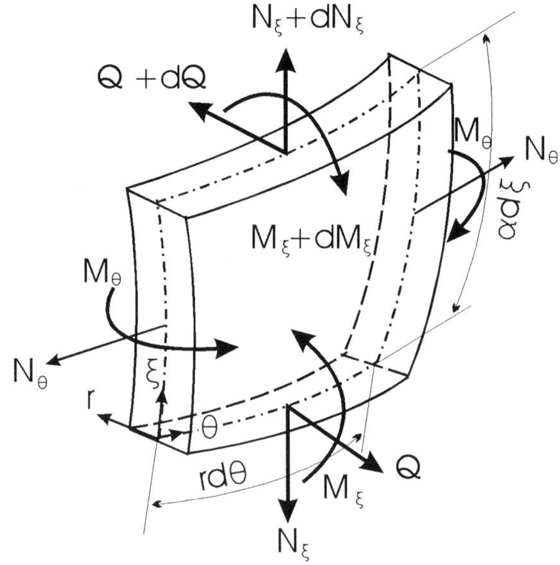

Figure 1.1 Basic conventions, force, and moment resultants.

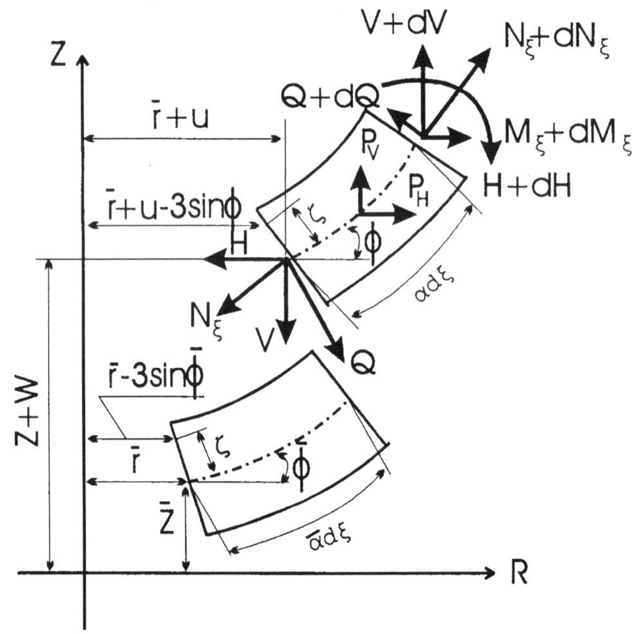

Figure 1.2 Basic conventions and undeformed and deformed meridional sections of the shell.

This hypothesis is similar to the assumption that, between longitudinal sections of a bent bar, no normal stresses are acting. As a result, the three-dimensional problem of deformation of the shell is reduced to a two-dimensional one, similar to the theory of bars, where the problem is reduced to the one-dimensional case. (The analyses of the meaning and the nature of these hypotheses were clarified in detail by Novozhilov [1953]).

By following Reissner [1950], we can write down the equations that describe the equilibrium of an infinitesimal element of the deformed shell (shown in Figs. 1.1 and 1.2). This element is cut out by letting the normal to the midsurface move along two meridians and two circles of the midsurface of the undeformed shell. The equations are

$$(rV)' + r\alpha p_V = 0. \tag{1.1}$$

$$(rH)' - \alpha N_\theta + r\alpha p_H = 0. \tag{1.2}$$

$$\left(rM_\xi\right)' - \alpha \cos\Phi M_\theta - r\alpha Q = 0. \tag{1.3}$$

where M_ξ, N_ξ and M_θ, N_θ are the (internal) bending moment stress resultants and membrane stress resultants acting in the circumferential and meridional sections perpendicular to the deformed midsurface, respectively; while Q is the (internal) transverse force acting on the circumferential normal section. Figure 1.1 shows the sign convention for the resultants M, N, and Q. The components H and V represent the decomposition of the resultant pair N_ξ, Q in the radial direction (which lies in the horizontal plane) and the direction of the vertical, respectively (see Fig. 1.2). The coordinates of a point of the deformed midsurface (r, z) are measured in these directions and this also counts for the components (p_V, p_H) of the externally applied pressure p. ξ Is the shell coordinate that determines the distance along the meridian of the midsurface of the shell. The derivative with respect to ξ is denoted by $(')$. We note here that ξ can be chosen in various ways, as is most convenient, but all functions that describe the considered problem depend upon it. The geometric factor α is defined below by expressions (1.8) and (1.9).

It is interesting to note that these equations of equilibrium are formally identical to the corresponding equations of the classical linear theory of shells, which are written with respect to the initial (undeformed) state. Figure 1.2 shows the undeformed element drawn below the deformed infinitesimal element of a meridional section. All the geometric parameters that correspond to the undeformed (initial) state are denoted by an additional we shall assume that the parameters E_ξ, α_ξ $(\xi \Leftrightarrow \theta)$ in (1.15) can depend on ξ, but not on ζ and they are also not dependent on the changes in the temperature field $T(\xi, \zeta)$.

Hooke-Neumann's law (1.15) is written for the plane stress problem only, because the stress σ_ζ is negligible compared to σ_ξ or σ_θ under the Kirchhoff-Love hypotheses. The first relation of (1.16) is a consequence of the symmetry of the matrix of coefficients of the general linear strain-stress relations [Lekhnitskii, 1963]. The inequality in (1.16) is a consequence of the negative sign of the change in volume of an infinitesimal element, which undergoes hydrostatic compression [see Lekhnitskii, 1963; Shilkrut, 1983b]. By inverting equations (1.15), we obtain relations (1.17)

$$\sigma_\xi = \frac{E_\xi}{1-\nu_\xi\nu_\theta}\left[\left(\varepsilon_\xi - \alpha_\xi T\right) + \nu_\theta\left(\varepsilon_\theta - \alpha_\theta T\right)\right];\ \xi \Leftrightarrow \theta. \tag{1.17}$$

Thus, if $h/R << 1$, the bending moments and membrane forces are

$$N_\xi = \int_{-h/2}^{h/2} \sigma_\xi d\zeta = \frac{E_\xi h}{1-\nu_\xi\nu_\theta}\left(\varepsilon_\xi + \nu_\theta\varepsilon_\theta\right) + N_{\xi T};\ \xi \Leftrightarrow \theta. \tag{1.18}$$

$$M_\xi = \int_{-h/2}^{h/2} \sigma_\xi \zeta d\zeta = \frac{E_\xi h^3}{12\left(1-\nu_\xi\nu_\theta\right)}\left(K_\xi + \nu_\theta K_\theta\right) + M_{\xi T};\ \xi \Leftrightarrow \theta, \tag{1.19}$$

where

$$N_{\xi T} = -\frac{E_\xi}{1-\nu_\xi\nu_\theta}\left(\alpha_\xi + \nu_\theta\alpha_\theta\right)\int_{-h/2}^{h/2} T d\zeta;$$

$$M_{\xi T} = -\frac{E_\xi}{1-\nu_\xi\nu_\theta}\left(\alpha_\xi + \nu_\theta\alpha_\theta\right)\int_{-h/2}^{h/2} T\zeta d\zeta;\ \xi \Leftrightarrow \theta. \tag{1.20}$$

As is well known [e.g., Reissner, 1950], the problem can be reduced to a system of two differential equations. For this, it is necessary to select two independent, unknown functions which are the generating functions for all the parameters that govern the problem. Such a selection can be made in different ways. Reissner chose the two basic unknowns, the stress resultant $\Psi = \bar{r}\ H$ and the angle Φ. In order to derive the corresponding basic equations, he simplified the system of equations of equilibrium (1.1), (1.2) and (1.3) by replacing r by $\bar{r}$ and α by $\bar{\alpha}$, using the assumption that the deformations are small. In that case, one can set

$$1+\varepsilon_{\xi 0}\approx 1;\quad \xi\Leftrightarrow\theta. \tag{1.21}$$

Billington [1967] goes further, suggesting an additional simplification by replacing $\cos\Phi$ by $\cos\overline{\Phi}$ in the compatibility equation (1.13), but this step will not be used here.

Reissner's equations, which are based on the simplification given above, and which we shall call "simplified," can be obtained by the following elaboration. The force V is found from (1.1) by integration. By means of (1.2), N_θ is then expressed via the two basic unknown functions Ψ and Φ. The same can be done for N_ξ and Q using (1.4) and (1.5), respectively. Furthermore, the resultant Q obtained in this way is substituted into (1.3), where the moments M_ξ and M_θ are expressed as functions of Φ via the curvature changes K_ξ and K_θ using (1.19) and finally (1.12). In this manner, the first simplified basic equation, which contains Φ'', is obtained.

The second equation can easily be derived from the equation of compatibility (1.13). In order to do this, it is necessary to express $\varepsilon_{\xi 0}$ and $\varepsilon_{\theta 0}$ as functions of Ψ and Φ using (1.18), and the formulae for N_ξ and N_θ that were obtained before via Ψ and Φ. After this elaboration, the second simplified equation with Ψ'' is derived. This system of equations in its entirety is given below, but, before writing them down, it is extremely important to understand when the simplifications (1.21) are, and when they are not, justified.

The terms $1+\varepsilon_{\xi 0}$ ($\xi\Leftrightarrow\theta$), which are present in the expressions for α and r (1.19) are differentiated before entering the equations of equilibrium (1.1)-(1.3). In this connection, it is necessary to pay attention to the fact that the assumption (1.21) can only be applied after differentiation of (1.19), but not before. This follows because differentiation of assumption (1.21) would lead to $\varepsilon'_{\xi 0}=0$; ($\xi\Leftrightarrow\theta$), which is an outcome that is not acceptable in general. For example, for plates (where $\xi=\bar{r}$ and $\overline{\Phi}\equiv 0$), differentiation of the second equation of (1.9), together with $\varepsilon'_{\theta 0}\approx 0$, gives

$$\varepsilon_{\theta 0}\approx u'. \tag{1.22}$$

Then (1.10) yields

$$1+\varepsilon_{\xi 0}=\frac{1+u'}{\cos\Phi}\approx\frac{1+\varepsilon_{\theta 0}}{\cos\Phi}.$$

Consequently, by taking $\varepsilon'_{\xi 0}\approx 0$ in unison with the assumption that the strains $\varepsilon_{\xi 0}$ and $\varepsilon_{\theta 0}$ remain small, we imply that Φ is small; i.e., that the deformed shape of the shell remains close to the undeformed state. This outcome puts a limit on the magnitude of the allowable rotations and contradicts the main goal of the theory, which is to describe the deformation of non-shallow shells with unrestricted displacements and rotations.

A similar difficulty also occurs for spherical caps of radius R_0, so that $\overline{\alpha} = R_0$, and consequently, $\overline{\alpha}' = 0$. In that case, it would follow from (1.21) that $\alpha' \approx 0$. But because $\alpha' = R_0\varepsilon'_{\xi 0}$, this would mean that $\varepsilon'_{\xi 00} \approx 0$, which, together with $\varepsilon_{\xi 0} \approx 0$, cannot be accepted as a general rule. Novozhilov [1953] remarks that the smallness of the derivatives do not follow from the smallness of the components of deformations, because the former are of the same order of magnitude as that of the slopes Φ. These clarifications about the proper use of assumption (1.21) were given by Shilkrut [1969b].

We now present three sets of equations: simplified, exact (no simplifications of type (1.21)), and intermediate. The simplified equations are:

$$\Phi'' - \overline{\Phi}'' + \frac{1}{\overline{r}}\left(\Phi' - \overline{\Phi}'\right)\left(\overline{r}' - \frac{\overline{r}\overline{\alpha}'}{\overline{\alpha}} - \nu_\xi \frac{D_\theta \overline{\alpha}}{D_\xi}\cos\Phi\right) - \left(\overline{r}M_{\xi T}\right)' \frac{\overline{\alpha}}{\overline{r}D_\xi}$$

$$+\frac{\overline{\alpha}D'_\xi}{D_\xi}\left(\frac{\Phi' - \overline{\Phi}'}{\overline{\alpha}} + \nu_\theta \frac{\sin\Phi - \sin\overline{\Phi}}{\overline{r}}\right) + \frac{\nu_\theta\overline{\alpha}}{\overline{r}}\left(\Phi'\cos\Phi - \overline{\Phi}'\cos\overline{\Phi}\right)$$

$$-\frac{\overline{\alpha}^2}{\overline{r}^2}\frac{D_\theta}{D_\xi}\cos\Phi\left(\sin\Phi - \sin\overline{\Phi}\right) + \frac{\overline{\alpha}^2}{\overline{r}D_\xi}M_{\theta T}\cos\Phi + \frac{\overline{\alpha}^2}{\overline{r}D_\xi}\left[-\Psi\sin\Phi + \overline{r}V\cos\Phi\right] = 0. \tag{1.23}$$

$$\Psi'' + \Psi'\left[-\frac{\overline{\alpha}'}{\overline{\alpha}} + \frac{\overline{r}'}{\overline{r}} - \frac{\left(E_\theta h\right)'}{E_\theta h}\right] - \Psi\left[-\frac{\nu_\theta\overline{\alpha}}{\overline{r}}\frac{\left(E_\theta h\right)'}{E_\theta h}\cos\Phi + \frac{E_\theta}{E_\xi}\frac{\overline{\alpha}^2}{\overline{r}^2}\cos^2\Phi\right.$$

$$\left.-\frac{\nu_\theta\overline{\alpha}}{\overline{r}}\Phi'\sin\Phi\right] - \overline{r}V\left[-\frac{\nu_\theta\overline{\alpha}}{\overline{r}}\frac{\left(E_\theta h\right)'}{E_\theta h}\sin\Phi + \frac{E_\theta\overline{\alpha}^2}{E_\xi\overline{r}^2}\cos\Phi\sin\Phi + \frac{\nu_\theta\overline{\alpha}}{\overline{r}}\Phi'\cos\Phi\right]$$

$$-\frac{\nu_\theta\overline{\alpha}}{\overline{r}}\left(\overline{r}V\right)' + E_\theta h\overline{\alpha}\left[\frac{\overline{r}'}{\overline{r}}\alpha_\theta T + \left(\alpha_\theta T\right)' - \frac{\overline{\alpha}}{\overline{r}}\alpha_\xi T\cos\Phi\right]$$

$$+\overline{r}p_H\left[\frac{\overline{\alpha}\overline{r}'}{\overline{r}} - \overline{\alpha}\frac{\left(E_\theta h\right)'}{E_\theta h} + \frac{\nu_\theta\overline{\alpha}^2}{\overline{r}}\cos\Phi\right] + \overline{\alpha}\left(rp_H\right)' = \frac{E_\theta h\overline{\alpha}^2}{\overline{r}}\left(\cos\Phi - \cos\overline{\Phi}\right), \tag{1.24}$$

where the basic unknowns are $\Phi(\xi)$ and $\Psi(\xi) = \overline{r}(\xi)H(\xi)$.

These equations were derived on the assumption that ν_ξ and ν_θ are constant, but $E_\xi h$, D_ξ (see (1.27)), and α_ξ ($\xi \Leftrightarrow \theta$) may depend on ξ. The increment $T(\xi, \zeta)$ of the temperature field is generally a function of ξ and ζ.

The exact basic equations are:

$$\Phi'' - \frac{r'^2}{D_\xi r\cos\Phi}\left\{D_\xi(1-\nu_\theta)\left(-\frac{\Phi'}{r'}\cos\Phi + \frac{\overline{\Phi}'}{\overline{\alpha}}\right) + D_\theta(1-\nu_\xi)\left(\frac{\sin\Phi}{r} - \frac{\sin\overline{\Phi}}{\overline{r}}\right)\right.$$

$$+M_{\xi T} - M_{\theta T} + \frac{rD'_\xi}{r'}\left(-\frac{\Phi'}{r'}\cos\Phi + \frac{\overline{\Phi}'}{\overline{\alpha}} - \frac{\nu_\theta \sin\Phi}{r} + \frac{\nu_\theta \sin\overline{\Phi}}{\overline{r}}\right)$$

$$+\frac{rM'_{\xi T}}{r'} + \frac{rD_\xi}{r'}\left[\frac{\Phi'}{r'^2}\left(r''\cos\Phi + r'\Phi'\sin\Phi\right) + \frac{\overline{\Phi}''}{\overline{\alpha}} - \frac{\overline{\alpha}'}{\overline{\alpha}^2}\overline{\Phi}' - \frac{\nu_\theta\Phi'\cos\Phi}{r}\right.$$

$$\left.\left.+\frac{\nu_\theta r'}{r^2}\sin\Phi + \frac{\nu_\theta\cos\overline{\Phi}\overline{\Phi}'}{\overline{r}} - \frac{\nu_\theta\sin\overline{\Phi}\overline{r}'}{\overline{r}^2}\right]\right\} - \frac{r'^2\sin\Phi}{D_\xi\cos^3\Phi}\left[m_\xi\left(\frac{r'}{\alpha\cos\Phi}\right.\right.$$

$$\left.\left.-1+\frac{\nu_\theta u}{\overline{r}}\right) + N_{\xi T}\right] + \frac{r'^2(rV)}{rD_\xi\cos^3\Phi} = 0. \tag{1.25}$$

$$\frac{ru''m_\xi}{\overline{\alpha}} + rm_\xi\left(\frac{\overline{r}''}{\overline{\alpha}} - \frac{r'\overline{\alpha}'}{\overline{\alpha}^2} + \frac{r'\Phi'\sin\Phi}{\overline{\alpha}\cos\Phi} + \frac{\nu_\theta u'\cos\Phi}{\overline{r}} - \frac{\nu_\theta u\overline{r}'\cos\Phi}{\overline{r}^2}\right) + rN'_{\xi T}\cos\Phi$$

$$+\left[m_\xi\left(\frac{r'}{\overline{\alpha}\cos\Phi} - 1 + \frac{\nu_\theta u}{\overline{r}}\right) + N_{\xi T}\right]\left(r'\cos\Phi + r\Phi'\sin\Phi\right) + rm'_\xi\left(\frac{r'}{\overline{\alpha}} - \cos\Phi + \frac{\nu_\theta u\cos\Phi}{\overline{r}}\right)$$

$$-r'\left[m_\theta\left(\frac{u\cos\Phi}{\overline{r}} + \frac{\nu_\xi r'}{\overline{\alpha}} - \nu_\xi\cos\Phi\right) + N_{\theta T}\cos\Phi\right]$$

$$-\Phi'(rV) + rr'\left(p_H\cos\Phi + p_V\sin\Phi\right) = 0 \tag{1.26}$$

where

$$D_\xi = \frac{E_\xi h^3}{12(1-\nu_\xi\nu_\theta)};\quad m_\xi = \frac{E_\xi h}{1-\nu_\xi\nu_\theta};\quad (\xi \Leftrightarrow \theta). \tag{1.27}$$

The basic unknowns are the generating functions $\Phi(\xi)$ and $\Psi(\xi)$. The expressions for r and α in terms of the basic unknowns may be found by combining (1.9) and (1.10). Using (1.1), the function (rV) can be found when the external forces p_V are known. The last set of equations is derived with the same assumptions previously introduced: ν_ξ is constant, and $E_\xi h$,

D_ξ, and α_ξ ($\xi \Leftrightarrow \theta$) may depend on ξ, but the temperature field $T(\xi, \zeta)$ may depend on both ξ and ζ. Equations (1.25) and (1.26) are obtained from the equations of equilibrium (1.2) and (1.3), where all of their terms are expressed by Φ and u using the relations (1.1.), (1.4)-(1.5), (1.10)-(1.12), and (1.18)-(1.20).

Both systems of governing equations are of the fourth order and thus demand four boundary conditions. Usually two boundary conditions are given along the edge of the shell and two at the pole when there is no central hole (i.e., the shell is a closed cap). The conditions at the pole ($\xi = 0$) are:

$$\Psi(0) = u(0) = \Phi(0) = 0. \tag{1.28}$$

The relations (1.28) follow from the axial symmetry and boundedness of the basic unknowns at the apex. For the simplified theory, the first and third conditions (1.28) are necessary, while for the exact equations (1.25) and (1.26), the second and third conditions are required. The last expression (1.28) cannot be used when the pole is a singular point on the surface of the shell (e.g., a tip of a cone), or when the unknown functions are singular in the neighborhood of point $\xi = 0$ as a result of the orthotropy (such cases are described in Chapter 4). In these cases, the basic functions in the neighborhood of $\xi = 0$ are of the type $\Phi(\xi) = A\xi^k$, where $k > 0$, and can be smaller than one. It is then necessary to exclude $\xi = 0$ and take a neighboring point $\xi = \Delta << 1$, where, instead of the last condition (1.28), a new condition (1.29) can be formulated, which follows directly from the form $A\xi^k$.

$$k\Phi(\Delta) = \Delta\Phi'(\Delta), \qquad \Delta << 1. \tag{1.29}$$

This relation implies that the cone is truncated near the tip.

At the edge of $\xi = c$, the corresponding two boundary conditions can usually be written in the linear form

$$\alpha_1 u(c) + \alpha_2 H(c) + \alpha_3 N_\xi(c) = \alpha_4. \tag{1.30}$$

$$\beta_1 M_\xi(c) + \beta_2 \Phi(c) = \beta_3. \tag{1.31}$$

Relation (1.30) characterizes the degree of mobility of the edge contour in its plane, while (1.31) characterizes the type of support affecting the bending conditions along this boundary. In the majority of cases, equations (1.30) and (1.31) are independent of each other. Parameters α_i (i = 1,2,3) and β_j (j = 1,2) depend on the way the edge is supported. Their values for the various boundary conditions are listed in Table 1.1. α_4 and β_3 determine the external membrane forces and bending moments (as line loads) along the edge, respectively.

Using Table 1.1, we can choose from the (complete) set of commonly applied boundary conditions. For the immovable clamped boundary, for example, it is necessary to take the first and fifth condition. A movable hinged support is specified by the second (or third) and fourth boundary condition, and so on. Some examples are shown on Figure 1.3.

Table 1.1

The basic boundary conditions.

No.	Kinds of supports	Values of α_i and β_i
1	Immovable in the plane of the contour.	$\alpha_1 = 1;\ \alpha_2 = \alpha_3 = \alpha_4 = 0$
2	Movable in the plane of the contour, where an inplane membrane contour force H_c is acting.	$\alpha_1 = 0;\ \alpha_2 = 1;\ \alpha_3 = 0;\ \alpha_4 = H_c$
3	Movable in the meridional direction, where the meridional membrane contour force N_c is acting.	$\alpha_1 = \alpha_2 = 0;\ \alpha_3 = 1;\ \alpha_4 = N_c$
4	Hinged, where a contour bending moment M_c is acting.	$\beta_1 = \beta_2 = 0;\ \beta_3 = M_c$
5	Clamped boundary	$\beta_1 = 0;\ \beta_2 = 1;\ \beta_3 = \overline{\Phi}(c)$.

1.2 Basic equations for circular plates and spherical caps.

We shall present here the equations for circular plates and spherical caps, which are particularizations of the general equations that were derived previously. These particular cases are, by themselves, of great theoretical and practical importance.

The first case considered is the circular plate of radius a, with coinciding geometric center and center of orthotropy. In this case,

$$\overline{\Phi} \equiv 0;\ \xi = \bar{r};\ \overline{\alpha} \equiv 1. \qquad (1.32)$$

The plate equations that are derived from the simplified system (1.23), (1.24) are of the form

$$\Phi'' + \frac{\Phi'}{\rho} - \frac{K^2}{\rho^2}\cos\Phi\sin\Phi + \frac{1}{\rho}\left[m_{\theta T}\cos\Phi - \left(\rho m_{\xi T}\right)'\right]$$

$$=\frac{1}{\rho}\left[F\sin\Phi-(\rho v)\cos\Phi\right];\quad (0\le\rho\le 1). \tag{1.33}$$

$$F''+\frac{F'}{\rho}-\frac{FK^2}{\rho}\left(\frac{\cos^2\Phi}{\rho}-\nu_\xi\Phi'\sin\Phi\right)-vK^2\left(\frac{\cos\Phi\sin\Phi}{\rho}+\nu_\xi\Phi'\cos\Phi\right)$$

$$-\frac{K^2\nu_\xi}{\rho}(\rho v)'\sin\Phi+K^2 m_\xi\beta_{\xi T}\left[\frac{\lambda^2}{\rho}+\lambda^2\frac{\beta^1_{\xi T}}{\beta_{\xi T}}-\frac{\cos\Phi}{\rho}\right]$$

$$+q_H\left(1+K^2\nu_\xi\cos\Phi\right)+(\rho q_H)'=K^2 m_\xi(\cos\Phi-1);\quad (0\le\rho\le 1). \tag{1.34}$$

Equations (1.33) and (1.34) are valid for a homogeneous circular plate without a hole with constant thickness h, where the temperature field $T(\rho,\ \zeta)$ can vary along the radius and through the thickness. The dimensionless parameters that appear here are defined by :

$$\rho=\bar{r}/a;\ K^2=E_\theta/E_\xi=\nu_\theta/\nu_\xi;\ \lambda^2=\alpha_\theta/\alpha_\xi;$$

$$m_{\xi T}=-\frac{E_\xi a h^2}{(1-\nu_\xi\nu_\theta)D_\xi}(\alpha_\xi+\alpha_\theta\nu_\theta)\int_{-t/2}^{t/2}\zeta T d\zeta;\quad m_\xi=E_\xi a^2 h/D_\xi;$$

$$n_{\xi T}=(\alpha_\xi+\alpha_\theta\nu_\theta)\int_{-t/2}^{t/2}T d\zeta;\quad \beta_{\xi T}=\alpha_\xi\int_{-t/2}^{t/2}T d\zeta;\quad (\xi\Leftrightarrow\theta). \tag{1.35}$$

$$F=\Psi a/D_\xi;\quad v=Va^2/D_\xi;\quad q_H=p_H a^3/D_\xi. \tag{1.36}$$

$$D_\xi=E_\xi h^3/12(1-\nu_\xi\nu_\theta);\quad (\xi\Leftrightarrow\theta). \tag{1.37}$$

For constant thickness h, the simplified equations for a homogeneous spherical cap of radius $\overline{R}$ that follow from (1.23) and (1.24) take the following form:

$$\xi=\overline{\Phi};\ \overline{\alpha}=\overline{R};\ \bar{r}=\overline{R}\sin\xi. \tag{1.38}$$

$$\Phi''+\frac{1}{\sin\xi}\Big[(\Phi'-1)(\cos\xi-K^2\nu_\xi\cos\Phi)-(m_{\xi T}\sin\xi)'+K^2\nu_\xi(\Phi'\cos\Phi-\cos\xi)$$

$$-\frac{K^2\cos\Phi}{\sin\xi}(\sin\Phi-\sin\xi)+m_{\theta T}\cos\Phi-F\sin\Phi$$

$$+v\sin\xi\cos\Phi\Big]=0;\ (0\le\xi\le\overline{\Phi}_1) \tag{1.39}$$

$$F'' + \frac{1}{\sin\xi}\left\{F'\cos\xi + K^2 F\left(\frac{\cos^2\Phi}{\sin\xi} - \nu_\xi \Phi' \sin\Phi\right) - K^2(\sin\xi v)\right.$$

$$+\left(\frac{\cos\Phi\sin\Phi}{\sin\xi} + \nu_\xi\Phi'\cos\Phi\right) - K^2\nu_\xi\sin\Phi(\sin\xi v)' + (\sin\xi q_H)\left(\cos\xi + K^2\nu_\xi\cos\Phi\right)$$

$$+(\sin\xi q_H)'\sin\xi + K^2 m_\xi \beta_{\xi T}\left[\lambda^2\cos\xi + \lambda^2\frac{\beta^1_{\xi T}}{\beta_{\xi T}}\sin\xi - \cos\Phi\right]$$

$$\left.-K^2 m_\xi(\cos\Phi - \cos\xi)\right\} = 0; \quad \left(0 \le \xi \le \overline{\Phi}_1\right). \tag{1.40}$$

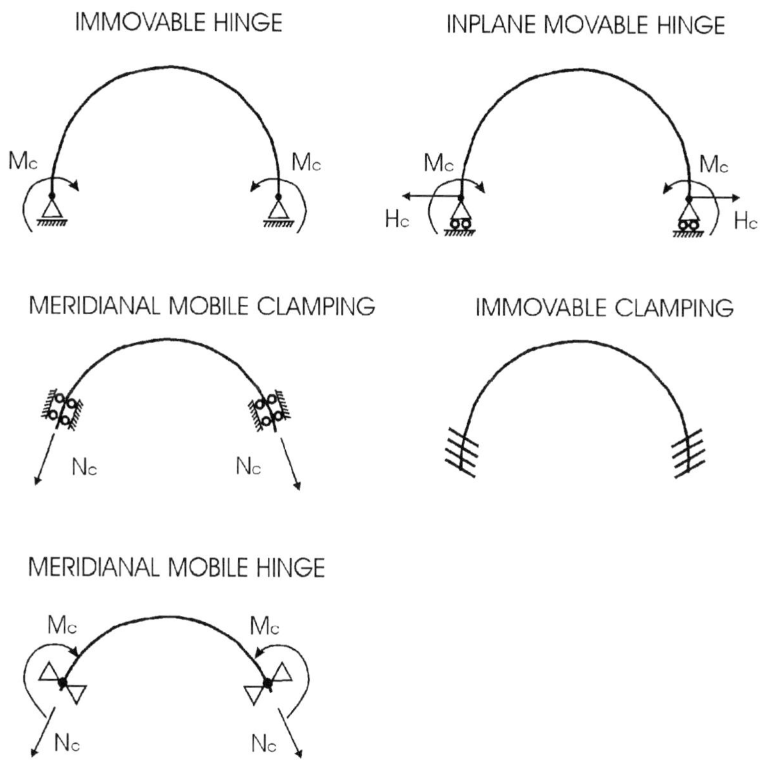

Figure 1.3 Classification of some edge supports.

$\overline{\Phi}_1$ is the value of the initial slope $\overline{\Phi}$ of the normal to the initial shape at the edge contour. The independent variable is ξ.

These equations are obtained for the same conditions as those of plate (i.e., no central hole, and $T(\xi, \zeta)$ can vary). With the exception of the first expression (1.35), the dimensionless parameters here are the same as those given by formulas (1.35), (1.36), and (1.37), where the radius a is now replaced by $\overline{R}$.

The equations for homogeneous circular plates and spherical shells of uniform rigidity that follow from the exact equations (1.25) and (1.26) are:

$$\Phi'' - \frac{(1+y')^2}{(\rho+y)\cos\Phi}\left\{-\frac{\Phi'\cos\Phi}{1+y'} + \frac{K^2\sin\Phi}{\rho+y} + m_{\xi T} - m_{\theta T} + \frac{\rho+y}{1+y'}\right.$$

$$\left.\times\left[m'_{\xi T} + \frac{\Phi'}{(1+y')^2}\left(y'' + \Phi'(1+y')\sin\Phi\right)\right]\right\} - \frac{12}{\beta^2}\frac{(1+y')^2\sin\Phi}{\cos^3\Phi}$$

$$\times\left[\frac{1+y'}{\cos\Phi} - 1 + \frac{K^2\nu_\xi y}{\rho} - n_{\xi T}\right] - \frac{(1+y')^2}{(\rho+y)\cos^3\Phi}\int\frac{(\rho+y)(1+y')}{\cos\Phi}q_V d\rho = 0; \quad (0 \le \rho \le 1). \quad (1.41)$$

$$y'' + (1+y')\Phi' tg\Phi + \frac{K^2\nu_\xi\cos\Phi}{\rho}\left(y' - \frac{y}{\rho}\right) - n'_{\xi T}\cos\Phi$$

$$+\left[\frac{1+y'}{\cos\Phi} - 1 - \frac{K^2\nu_\xi y}{\rho} + n_{\xi T}\right]\left(\frac{1+y'}{\rho+y}\cos\Phi + \Phi'\sin\Phi\right)\frac{1+y'}{\rho+y}K^2\left[\frac{y\cos\Phi}{\rho}\right.$$

$$\left.+\nu_\xi(1+y') - \nu_\xi\cos\Phi - n_{\theta T}\cos\Phi\right] + \frac{\beta^2\Phi'}{12(\rho+y)}\int\frac{(\rho+y)(1+y')}{\cos\Phi}q_V d\rho$$

$$+(1+y')\frac{\beta^2}{12}\left(q_H\cos\Phi + q_V\sin\Phi\right) = 0; \quad (0 \le \rho \le 1). \quad (1.42)$$

Besides the dimensionless parameters given in (1.35), (1.36), and (1.37), there are the additional parameters

$$\beta = h/a; \quad q_V = p_V a^3 / D_\xi; \quad y = u/a \quad (1.43)$$

For spherical shells, equations (1.25) and (1.26) give, respectively,

$$\Phi'' - \frac{(y'+\cos\xi)^2}{(y+\sin\xi)\cos\Phi}\left\{-\left(1-K^2\nu_\xi\right)\frac{\Phi'\cos\Phi}{y'+\cos\xi} + K^2\left(1-\nu_\xi\right)\frac{\sin\Phi}{y+\sin\xi}\right.$$

$$+m_{\xi T} - m_{\theta T} + 1 - K^2 + \frac{y + \sin\xi}{y' + \cos\xi}\left[m'_{\xi T} + \frac{\Phi'}{(y' + \cos\xi)^2} \right.$$

$$m_{\xi T} + \frac{\Phi'}{(y' + \cos\xi)^2}(-\sin\xi\cos\Phi + y''\cos\Phi + \cos\xi\Phi'\sin\Phi + y'\Phi'\sin\Phi)$$

$$\left. - \frac{K^2\nu_\xi}{y + \sin\xi}(\Phi'\cos\Phi - y' + \cos\xi\sin\Phi) \quad \right]\Bigg\}$$

$$- \frac{(y' + \cos\xi)^2}{\cos^3\Phi}\left\{ \frac{12}{\beta^2}\sin\Phi\left[\frac{y' + \cos\xi}{\cos\Phi} - 1 + \frac{K^2\nu_\xi y}{\sin\xi} - n_{\xi T} \right] + \frac{1}{y + \sin\xi} \right.$$

$$\left. \times \int \frac{(y + \sin\xi)(y' + \cos\xi)}{\cos\Phi} q_V d\xi \right\} = 0; \quad \left(0 \le \xi \le \overline{\Phi}_1\right). \tag{1.44}$$

$$(y + \sin\xi)\left[y'' - \sin\xi + (y' + \cos\xi)\Phi' tg\Phi + K^2\frac{\nu_\xi\cos\Phi}{\sin\xi}(y' - y\cot\xi) - n'_{\xi T}\cos\Phi \right]$$

$$+\left[\frac{y' + \cos\xi}{\cos\Phi} - 1 + \frac{K^2\nu_\xi y}{\sin\xi} - n_{\xi T} \right]\left[(y' + \cos\xi)\cos\Phi + (y + \sin\xi)\Phi'\sin\Phi \right]$$

$$-K^2(y' + \cos\xi)\left[\frac{y\cos\Phi}{\sin\Phi} + \nu_\xi(y' + \cos\xi - \cos\Phi) - n_{\theta T}\cos\Phi \right]\frac{\beta^2}{12}$$

$$\times\left[\Phi' \int \frac{(y + \sin\xi)(y' + \cos\xi)}{\cos\Phi} q_V d\xi + (y + \sin\xi)(y' + \cos\xi)(q_H\cos\Phi + q_V\sin\Phi) \right] = 0;$$

$$\left(0 \le \xi \le \overline{\Phi}_1\right). \tag{1.45}$$

The same dimensionless parameters are used here as in the previous case, except that a is replaced by $\overline{R}$.

An intermediate approximate system of governing equations may be derived from the exact equations (1.41)-(1.45) by a simplification of the type (1.21). Such an operation is legitimate, since it is done after all the differential operations have been carried out. The intermediate equations for the plate take the following form:

$$\Phi'' + \frac{1}{\rho}\cos\Phi\left\{-\Phi' + \frac{1}{\rho}K^2\sin\Phi + m_{\xi T} - m_{\theta T} + \frac{\rho}{\cos\Phi}\left[m'_{\xi T} + \frac{\Phi'}{\cos^2\Phi}\left(y'' + \Phi'\cos\Phi\sin\Phi\right)\right]\right\}$$

$$-\frac{12tg\Phi}{\beta^2}\left[\frac{1+y'}{\cos\Phi} - 1 + \frac{K^2\nu_\xi y}{\rho} - n_{\xi T}\right] - \frac{1}{\cos\Phi}\int\rho q_V d\rho = 0; \quad (0 \le \rho \le 1). \tag{1.46}$$

$$y'' + \Phi'\sin\Phi + \frac{K^2\nu_\xi}{\rho}\cos\Phi\left(y' - \frac{y}{\rho}\right) - n'_{\xi T}\cos\Phi + \left[\frac{1+y'}{\cos\Phi} - 1 + \frac{K^2\nu_\xi y}{\rho} - n_{\xi T}\right]$$

$$\times\left(\frac{1}{\rho}\cos^2\Phi + \Phi'\sin\Phi\right) - \frac{K^2\cos\Phi}{\rho}\left[\frac{y\cos\Phi}{\rho} + \nu_\xi\left(1 + y' - \cos\Phi\right) - n_{\theta T}\cos\Phi\right]$$

$$+\frac{\beta^2}{12}\left[\frac{\Phi'}{\rho}\int\rho q_V d\rho + \cos\Phi\left(q_H\cos\Phi + q_V\sin\Phi\right)\right] = 0; \quad (0 \le \rho \le 1) \tag{1.47}$$

In conformity with (1.21), the simplifications (1.48) were used here, with the exception of the reduction of expressions of type $(1 + y') - \cos\Phi$, which will be simplified using the approximations:

$$1 + y' = \alpha\cos\Phi \approx \overline{\alpha}\cos\Phi = \cos\Phi; \quad \rho + y \approx \rho. \tag{1.48}$$

For the spherical shell of radius R, expressions of type (1.48) are of the following form.

$$R\left(\cos\xi + y'\right) = \alpha\cos\Phi \approx \overline{\alpha}\cos\Phi = R\cos\Phi; \quad R\left(\sin\xi + y\right) \approx R\sin\xi. \tag{1.49}$$

Thus, the intermediate equations for spherical shells can be written as follows:

$$\Phi'' - \frac{\cos\Phi}{\sin\xi}\left\{-\left(1 - K^2\nu_\xi\right)\Phi' + K^2\left(1 - \nu_\xi\right)\frac{\sin\Phi}{\sin\xi} + m_{\xi T} - m_{\theta T} + 1 - K^2 + \frac{\sin\xi}{\cos\Phi}\right.$$

$$\left.\times\left[m'_{\xi T} + \frac{\Phi'}{\cos\Phi}\left(-\sin\xi + y'' + \Phi'\sin\Phi\right) - \frac{K^2\nu_\xi\cos\Phi}{\sin\xi}\left(\Phi' - \frac{\sin\Phi}{\sin\xi}\right)\right]\right\}$$

$$-\frac{1}{\cos\Phi}\left\{\frac{12\sin\Phi}{\beta^2}\left[\frac{y' + \cos\xi}{\cos\Phi} - 1 + \frac{K^2\nu_\xi y}{\sin\xi} - n_{\xi T} + \frac{1}{\sin\xi}\int\sin\xi q_V d\xi\right\} = 0;\right.$$

$$\left(0 \le \xi \le \overline{\Phi}_1\right). \tag{1.50}$$

$$\sin\xi\left[y'' - \sin\xi + \Phi'\sin\Phi + \frac{K^2\nu_\xi\cos\Phi}{\sin\xi}(y' - y\cot\xi) - n'_{\xi T}\cos\Phi\right]$$

$$+\sin\xi\left[y'' - \sin\xi + \Phi'\sin\Phi + \frac{K^2\nu_\xi\cos\Phi}{\sin\xi}(y' - y\cot\xi) - n'_{\xi T}\cos\Phi\right]$$

$$+\left[\frac{y' + \cos\Phi}{\cos\Phi} - 1 + \frac{K^2\nu_\xi y}{\sin\xi} - n_{\xi T}\cos\Phi\right]\left(\cos^2\Phi + \sin\xi\Phi'\sin\Phi\right) - K^2\cos\Phi$$

$$\times\left[\frac{y\cos\Phi}{\sin\xi} + \nu_\xi(y' + \cos\xi - \cos\Phi) - n_{\theta T}\cos\Phi\right]\frac{\beta^2}{12}$$

$$\times\left[\Phi'\int\sin\xi q_V d\xi + \sin\xi\cos\Phi(q_H\cos\Phi + q_V\sin\Phi)\right] = 0; \quad \left(0 \le \xi \le \overline{\Phi}_1\right). \tag{1.51}$$

Accordingly, it is also necessary to transform the boundary conditions, the general form of which is given by (1.28)-(1.31) and Table 1.1. Please note that this is the first time that equations with this specific (intermediate) range of validity have been formulated.

1.3 Axisymmetric thermo-elastic deformation of shallow shells of revolution.

The equations for shallow shells are often referred to as the von Kármàn equations for the axisymmetric case derived by Marguerre [1938]. The basic assumptions of these equations correspond to the Kirchhoff-Love hypotheses mentioned in Section 1.1, plus an assumption that follows from the shallowness of the initial curved shape of the shell. The assumption of shallowness implies that the metrics of the shell, as compared to that of a flat plate, differ only in the higher order terms.

In conformity with this approach, the equations of equilibrium (1.1)-(1.3) are simplified in the following way: the radial coordinate r of points of the deformed midsurface is approximated by the radial coordinate of the undeformed midsurface $\bar{r}$ (see Fig. 1.2), and $\xi = \bar{r}$; the force H is replaced by N_ξ; $ds \approx d\bar{s} = \bar{\alpha}d\xi$ and $\alpha = \bar{\alpha} = 1$; Φ, being small, leads to the simplifications $\sin\Phi \approx \Phi$ and $\cos\Phi \approx 1$. Replacing H by N_ξ and neglecting Φ^2 with respect to unity in equations (1.4) and (1.5), we find that

$$Q_2 = V - N_\xi\Phi. \tag{1.52}$$

The basic equations of equilibrium (1.1)-(1.3) then take the simplified form:

$$(\bar{r}v)' + \bar{r}p_V = 0; \tag{1.53}$$

$$\left(\bar{r}N_\xi\right)' - N_\theta + \bar{r}p_H = 0; \tag{1.54}$$

$$\left(\bar{r}M_\xi\right)' - M_\theta - \bar{r}\left(V + N_\xi\Phi\right) = 0; \tag{1.55}$$

where (′) means the derivative to $\xi = \bar{r}$. These equations must be added to the simplified form of the compatibility equation (1.13),

$$\left(\bar{r}\varepsilon_{\theta 0}\right)' - \bar{r}'\varepsilon_{\xi 0} = \frac{1}{2}\left(\Phi^2 - \overline{\Phi}^2\right), \tag{1.56}$$

which is based on the approximation

$$\cos\Phi - \cos\overline{\Phi} \approx \frac{1}{2}\left(\Phi^2 - \overline{\Phi}^2\right). \tag{1.57}$$

On the basis of the system (1.53)-(1.56), two governing equations can be derived. The type of these equations depends on the choice of the two basic unknowns. In dimensionless parameters, one such version, which is convenient for applications, is given below.

$$\mathrm{L}_K(\omega) + \frac{K^2(e_r t)'}{e_r t}\left(-\omega' + \frac{\nu_r \omega}{\rho}\right) = -q_t\left[2 + K^2\nu_r + \rho\frac{q_t'}{q_t} - \frac{\rho(e_r t)'}{e_r t}\right]$$

$$-K^2 e_r t\frac{\theta^2 + 2\theta\overline{\theta}}{2\rho} + K^2 e_r\left[\frac{1-\lambda^2}{\rho}N_T + \lambda^2\left(\frac{t'}{t}N_T - N_T'\right)\right]; \quad (0 \le \rho \le 1). \tag{1.58}$$

$$\mathrm{L}_K(\theta) + \frac{d_r'}{d_r}\left(\theta' + \frac{K^2\nu_r\theta}{\rho}\right) = \frac{m_a}{\rho d_r}\left[-\int_0^\rho \rho q \mathrm{d}\rho + \int_0^\rho \rho q_t\left(\theta + \overline{\theta}\right)\mathrm{d}\rho - P + \omega\left(\theta + \overline{\theta}\right)\right]$$

$$+\frac{1}{t^2}\left\{\frac{1 - K^2\nu_r - K^2\lambda^2(1-\nu_r)}{\rho}M_T + \left(1 + \nu_r K^2\lambda^2\right)\left[\frac{e_r'}{e_r}M_T + M_T'\right]\right\}; \quad (0 \le \rho \le 1). \tag{1.59}$$

The operator L_K is:

$$\mathrm{L}_K = \frac{\mathrm{d}^2}{\mathrm{d}\rho^2} + \frac{1}{\rho}\frac{\mathrm{d}}{\mathrm{d}\rho} - \frac{K^2}{\rho^2}. \tag{1.60}$$

These equations are valid for caps with no hole at the apex. The shell is circular orthotropic when the center of orthotropy is taken in the center of the base of the cap. $\rho = \bar{r}/a$ is the dimensionless radial coordinate, where a is the radius of the base of the cap, and (′) denotes the derivative with respect to ρ. The other dimensionless parameters are:

$$K^2 = E_\varphi / E_r = const. > 0; \quad \nu_\varphi = K^2 \nu_r; \quad \lambda^2 = \alpha_\varphi / \alpha_r = const. > 0;$$

$$m_a = 12\left(1 - \nu_r \nu_\varphi\right) = 12\left(1 - \nu_r^2 K^2\right) = const. > 0; \quad N_t = \frac{a^2}{\left(h^0\right)^2} \alpha_r \int_{-t/2}^{t/2} T d\zeta;$$

$$M_T = \frac{12a^2}{\left(h^0\right)^2} \alpha_r \int_{-t/2}^{t/2} T\zeta d\zeta; \quad \zeta = z/h^0; \quad \beta = h^0 / a; \quad t(\rho) = h(\rho)/h^0;$$

$$e_r(\rho) = E_r(\rho)/E_r^0; \quad D_r(\rho) = d_r(\rho)/D_r^0; \quad d_r = e_r t; \quad D_r^0 = E_r^0 \left(h^0\right)^3 / m_a;$$

$$h^0 = const.; \quad E_r^0 = const. \tag{1.61}$$

The thickness h of the cap and Young's modulus E_r can depend on ρ. If they are constant, then $t(\rho) \cong e(\rho) \cong 1$. The basic unknowns are the dimensionless membrane stress function $\omega(\rho)$ and the rotation of the normal to the midsurface of the shell $\theta(\rho) = \overline{\Phi} - \Phi$. The variable $\theta = -w'$ (see (1.63)) can be considered as the stress function for the bending stresses. All the parameters of the problem can thus be expressed in terms of these basic unknowns. Indeed, $\omega(\rho)$ is the basic unknown by which the normal membrane forces N_r, N_φ are generated. This means that $\omega(\rho)$ also represents the membrane stresses σ_r, σ_φ, and the elongations ε_r, ε_φ in the midsurface. (Here and henceforth, the additional indices r and φ refer to parameters connected with the radial and circumferential directions, respectively.) All the parameters connected with bending, such as the curvature changes χ_r, χ_φ and the resultants M_r, M_φ, can be expressed in terms of θ. All the parameters generated by the temperature field $T(r, \zeta)$ are denoted with the index T. In the definitions of the dimensionless relations given below, the true (physical) quantities are denoted by an asterisk.

$$N_r = N_r^* a^2 / E_r^0 \left(h^0\right)^3 = \omega / \rho; \quad N_\varphi = N_\varphi^* a^2 / E_r^0 \left(h^0\right)^3 = \omega' + \rho q_t;$$

$$\sigma_r = \sigma_r^* a^2 / E_r^0 \left(h^0\right)^2 = N_r / t + \frac{e_r\left(1 + K^2 \lambda^2 \nu_r\right)}{1 - K^2 \nu_r^2} \left[N_T / t - \alpha_r T(\rho, 0)/\beta^2\right];$$

$$\sigma_\varphi = \sigma_\varphi^* a^2 / E_r^0 \left(h^0\right)^2 = N_\varphi / t + \frac{K^2 e_r\left(\lambda^2 + \nu_r\right)}{1 - K^2 \nu_r^2}\left[N_T / t - \alpha_r T(\rho, 0) / \beta^2\right];$$

$$\varepsilon_r = \frac{\beta^2}{t}\left[\frac{1}{e_r}\left(\frac{\omega}{\rho} - \nu_r \omega' - \nu_r \rho q_t\right) + N_T\right];\ \varepsilon_\varphi = \frac{\beta^2}{t}\left[\frac{1}{K^2 e_r}\left(\omega' + \rho q_t - K^2 \nu_r \frac{\omega}{\rho}\right) + \lambda^2 N_T\right];$$

$$\chi_r = \chi_r^* a^2 / h^0 = \theta';\ \chi_\varphi = \chi_\varphi^* a^2 / h^0 = \theta / \rho;$$

$$M_r = M_r^* a^2 / h^0 D_r^0 = d_r\left(\theta' + \nu_r K^2 \theta / \rho\right) - e_r\left(1 + K^2 \lambda^2 \nu_r\right) M_T;$$

$$M_\varphi = M_\varphi^* a^2 / h^0 D_r^0 = K^2 d_r\left(\theta / \rho + \nu_r \theta'\right) - e_r K^2\left(\lambda^2 + \nu_r\right) M_T. \quad (1.62)$$

The bending moments are positive if they increase the curvature of the midsurface. The meridional curvature is positive when the convex side of the shell is directed downward (in the positive direction of the z-axis; see Fig. 1.4, where a cap is shown with a negative initial curvature).

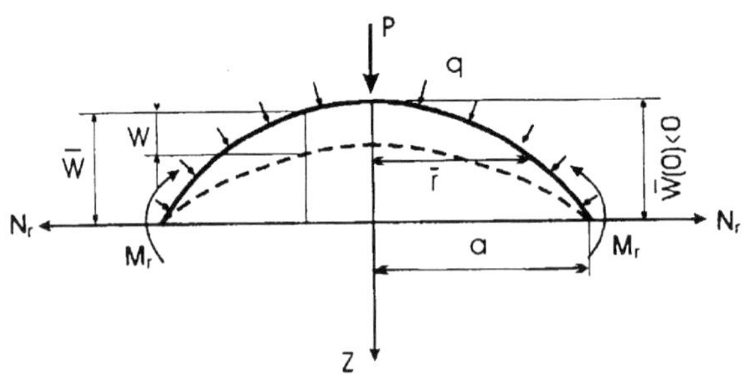

Figure 1.4 Basic conventions, general cap parameters.

The horizontal and vertical components u and w of the displacement of a point of the midsurface are:

$$u = u^* / h^0 = \rho \varepsilon_\varphi a / h^0 = \frac{\rho\beta}{K^2 e_r t}\left(\omega' + \rho q_t - K^2 \nu_r \omega / \rho\right) + \frac{\rho \lambda^2 \beta}{t} N_T$$

$$w = w^* / h^0 = -\int_0^\rho \theta d\rho + \xi;\ \xi = w(0)\quad \bar{\xi} = \bar{w}^*(o) / h^0. \quad (1.63)$$

If the vertical displacements of the boundary are restricted to ($w(1) = 0$), then

$$w(\rho) = \int_{\rho}^{1} \theta d\rho, \text{ and } \xi = \int_{0}^{1} \theta d\rho. \tag{1.64}$$

u is positive away from the apex, while the w is positive when directed downward and relative to the undeformed (original) meridian (see Fig. 1.4). The initial (undeformed) shape of the meridian of the cap is represented by the given function $\overline{w}(\rho)$, which is the initial displacement measured from the base plane of the shell. $\overline{w}(\rho)$, $\overline{\theta}(\rho)$, and $\overline{\xi}(\rho)$ are positive when the convex side of the undeformed shell is directed downward. The dimensionless, distributed load, with normal and tangential components (q, q_t), and the concentrated force P that acts at the apex (the sign convention is shown in Fig. 1.4), are given as follows:

$$q = q^* a^4 / E_r^0 \left(h^0\right)^4; \quad q_t = q_t^* a^3 / E_r^0 \left(h^0\right)^3; \quad P = P^* a^2 / 2\pi E_r^0 \left(h^0\right)^4. \tag{1.65}$$

The boundary conditions most frequently encountered can be written in the linear form

$$\omega(0) = \theta(0) = 0; \quad \alpha_1 \omega'(1) + \beta_1 \omega(1) = \gamma_1; \quad \alpha_2 \theta'(1) + \beta_2 \theta(1) = \gamma_2. \tag{1.66}$$

Conditions (1.66) at $\rho = 0$ serve to ensure the continuity and boundedness of ω and θ in the neighborhood of the origin. (Note that the operator L_K (see (1.60)) is singular at $\rho = 0$.) Conditions (1.66) at $\rho = 1$ reflect the way in which the boundary is supported. The condition for ω characterizes the degree of mobility of the boundary supports in the base plane of the shell; the second relationship describes the bending conditions at the boundary $\rho = 1$. The constants α_i, β_i, and γ_i ($i = 1,2$) cannot be equal to zero simultaneously. The constants for the nine basic boundary conditions at $\rho = 1$ are given in Table 1.2. The last two relationships of (1.66) are often independent of one another, but coupling can also occur.

Table 1.2 is used in the following way. At each edge point, it is necessary to have two conditions: one is connected with ω (these are in the upper half of the table), and the other with θ (lower part of the table). For example, for a mobile clamped edge where the edge is free to move in the radial direction, the corresponding conditions are given in Cells 2 and 4 of the table, while for the immovable hinge support without boundary moment, one can take data from Cells 3 and 5, and so on

The system of equations presented here is the primary version for axisymmetric thermoelastic deformations of shallow, geometrically nonlinear, orthotropic, non-homogeneous shells of revolution of variable thickness. The notion of shallowness cannot be defined exactly. It takes into consideration that both initial and deformed shapes must be shallow. Usually a shell is defined as shallow when the ratios of certain geometric parameters defining its initial shape do not exceed given values. For example, in conformity with Reissner [1958b], a shell of revolution is shallow when the maximum of the slope of the tangent for the initial shape does

Table 1.2

Nine basic combinations of the boundary conditions at $\rho = 1$.

	Movable Support without normal boundary forces	Movable Support with active normal boundary force N	Immovable Support
	1	2	3
Conditions for ω	$\alpha_1=0;\ \beta_1=1;\ \gamma_1=0.$	$\alpha_1=0;\ \beta_1=1;\ \gamma_1=N$	$\alpha_1=1;\ \beta_1=-\nu_r K^2;$ $\gamma_1=-q_t(1)-\lambda^2 K^2 e_r N_T(1)$
	4	5	6
Conditions for θ	$\alpha_2=0;\ \beta_2=1;\ \gamma_2=0$	$\alpha_2=1;\ \beta_2=\nu_r K^2;$ $\gamma_2=\dfrac{e_r}{d_r}\left(1+K^2\lambda^2\nu_r\right)M_T(1)$	$\alpha_2=1;\ \beta_2=\nu_r K^2;$ $\gamma_2=M/d_r+$ $e_r(1+K^2\lambda^2\nu_r)M_T(1)/d_r$
	Clamped	Hinged support without active boundary bending moment.	Hinged support with active boundary moment M

not exceed 1/3. For a spherical cap, this leads to $H/R < 1/6$ (where H is the height of the cap and R is its radius). In Vlasov [1958], the inequality $H/2a \le 1/5$ is recommended as a criterion of shallowness of a spherical cap, where a is the radius of its base. In Reissner [1946], it is required that this ratio be less than 1/8. Such formal approaches, however, cannot ensure the shallowness of the deformed shape of the shell; for this, other measures must be included, such as the magnitudes of the acting external forces. To the best of our knowledge, such criteria are not yet available.

For shallow spherical caps, the functions $\overline{w}(\rho)$, $\overline{\theta}(\rho)$, and $\overline{\xi}$ which describe the initial shape of the shell, are of the following form:

$$\overline{\theta}(\rho) = 2\overline{\xi}\rho;\quad \overline{w}(\rho) = -\int_0^\rho \overline{\theta}d\rho = \overline{\xi}\left(1-\rho^2\right);\quad \overline{\xi} = \overline{w}(0) = \overline{w}^*(0)/h^0. \tag{1.67}$$

$\overline{\xi}$ is negative when the convexity of the cap is directed upward (see Fig. 1.4).

1.4 Nonsymmetric shallow homogeneous isotropic caps of revolution of constant thickness.

An element of the cap and the forces that are acting on it are shown in Figure 1.5.

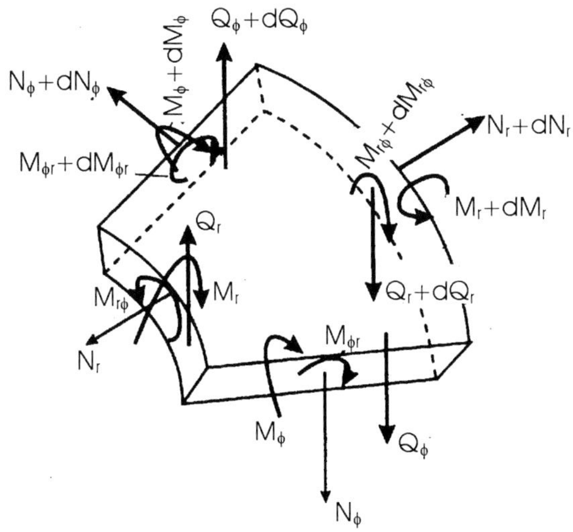

Figure 1.5 Element of a shallow cap with the resultants of forces and moments.

For a shallow cap, the equilibrium equations are:

$$\left(rN_{r\varphi}\right)' + \dot{N}_{\varphi} + N_{r\varphi} = 0; \quad \left(rN_r\right)' + \dot{N}_{r\varphi} - N_{\varphi} = 0. \tag{1.68}$$

$$\left(rM_r\right)' + \dot{M}_{r\varphi} - M_{\varphi} - rQ_r = 0; \quad \left(rM_{r\varphi}\right)' + \dot{M}_{\varphi} + M_{r\varphi} - rQ_{\varphi} = 0. \tag{1.69}$$

The equation of equilibrium for stress resultants in the z direction is:

$$rq + \left(rQ_r\right)' + \left[\left(rN_r\right)\left(w + \overline{w}\right)' + N_{r\varphi}\left(\dot{w} + \dot{\overline{w}}\right)\right]'$$

$$+\left[\frac{1}{r}N_{\varphi}\left(\dot{w} + \dot{\overline{w}}\right) + N_{r\varphi}\left(w + \overline{w}\right)' + Q_{\varphi}\right]^{\cdot} = 0. \tag{1.70}$$

($'$) and ($^\cdot$) denote partial derivatives with respect to r and φ, respectively. N_r, N_φ, and $N_{r\varphi}$ are the membrane stress resultants, M_r, M_φ, and $M_{r\varphi}$ the resultant bending and twisting moments, and Q_r, Q_φ are the shear force resultants (the sign convention is shown in Fig. 1.5). The function $\overline{w}$ characterizes the initial (undeformed) shape of the cap. For spherical caps, $\overline{w}$ is given by equation (1.67). q denotes the distributed normal load, which is defined positive when it is pointing downward.

The equations that express the components ε_r, ε_φ, and $\gamma = \varepsilon_{r\varphi}$ of the deformation of the midsurface and of the changes of curvature χ_r, χ_φ, and $\chi_{r\varphi}$ as functions of the displacement components u, v, and w of the midsurface (see Fig. 1.5) are written as

$$\varepsilon_r = u' + \overline{w}'w' + (w')^2/2; \quad \varepsilon_\varphi = u/r + \dot{v}/r + \dot{\overline{w}}\dot{w}/r^2 + (\dot{w}/r)^2/2;$$

$$\gamma = \dot{u}/r - v/r + v' + \dot{\overline{w}}w'/r + \overline{w}'\dot{w}/r + \dot{w}w'/r;$$

$$\chi_r = -w''; \quad \chi_\varphi = \ddot{w}/r^2 - w'/r; \quad \chi_{r\varphi} = -(\dot{w}/r)'. \tag{1.71}$$

The strain-stress relations for homogeneous, elastic, isotropic shells are:

$$Eh\varepsilon_r = N_r - \nu N_\varphi; \quad (r \Leftrightarrow \varphi); \quad Eh\gamma = 2(1+\nu)N_{r\varphi}; \quad M_r = D(\chi_r + \nu\chi_\varphi);$$

$$(r \Leftrightarrow \varphi); \quad M_{r\varphi} = (1-\nu)D\chi_{r\varphi}; \quad D = Eh^3/12(1-\nu^2). \tag{1.72}$$

As usual, we express the membrane stress resultants in terms of the stress function $F(r, \varphi)$ so that equations (1.68) are automatically satisfied.

$$N_r = F'/r + \ddot{F}/r^2; \quad N_\varphi = F''; \quad N_{r\varphi} = -(\dot{F}/r)'. \tag{1.73}$$

Eliminating the shear forces Q_r, Q_φ from equations (1.69) and employing the strain-stress relations (1.72) and strain displacement relations (1.71), we obtain the first governing equation.

$$D\Delta\Delta w = \left(\frac{F'}{r} + \frac{\ddot{F}}{r^2}\right)(w + \overline{w})'' + \frac{1}{r^2}F''(\ddot{w} + \ddot{\overline{w}}) + \frac{1}{r}F''(w + \overline{w})'$$

$$+2\left(\frac{\dot{F}'}{r} - \frac{\dot{F}}{r^2}\right)\left(\frac{\dot{w} + \dot{\overline{w}}}{r^2} - \frac{\dot{w}' + \dot{\overline{w}}'}{r}\right) + q; \quad \Delta = \frac{\partial^2}{\partial r^2} + \frac{1}{r}\frac{\partial}{\partial r} + \frac{1}{r^2}\frac{\partial^2}{\partial\varphi^2}. \tag{1.74}$$

The second governing equation results from the equation of compatibility, which can be obtained by eliminating the displacements u and v from the first three strain displacement equations (1.71):

$$(r\varepsilon_\varphi)'' - \varepsilon_r' - \frac{1}{r}(r\dot{\gamma})' + \frac{1}{r}\ddot{\varepsilon}_r = \frac{2}{r}\left(\dot{\overline{w}}' - \frac{1}{r}\dot{\overline{w}}\right)\left(\dot{w}' - \frac{1}{r}\dot{w}\right)$$

$$-\overline{w}''\left(w' + \frac{1}{r}\ddot{w}\right) - w''\left(\overline{w}' + \frac{1}{r}\ddot{\overline{w}}\right) + \frac{1}{r}\left(\dot{w}' - \frac{1}{r}\dot{w}\right)^2 - w''\left(w' + \frac{1}{r}\ddot{w}\right). \tag{1.75}$$

Expressing the left-hand side in terms of the membrane forces and then in terms of the stress function F (see (1.73)), we obtain the following form for the second equilibrium equation:

$$\frac{r}{Eh}\Delta\Delta F = \frac{1}{r}\left(\dot{w}' - \frac{\dot{w}}{r}\right)^2 - \left(w'' + \overline{w}''\right)\left(w' + \frac{1}{r}\ddot{w}\right)$$

$$+\frac{2}{r}\left(\dot{\overline{w}}' - \frac{1}{r}\dot{\overline{w}}\right)\left(\dot{w}' - \frac{1}{r}\dot{w}\right) - w''\left(\overline{w}' + \frac{1}{r}\ddot{\overline{w}}\right). \tag{1.76}$$

Equations (1.74) and (1.76) are the governing equations for the nonsymmetric deformation of shallow elastic isotropic caps.

In general, eight boundary conditions must be added, because equations (1.74) and (1.76) are both of the fourth order. At each point of the cap's edges, four conditions are usually given, which depend on the nature of the restrictions put on the displacements at the boundary point and on the type of external forces applied.

Let C denote a point situated on the circular boundary of the shell of radius a. For a completely clamped edge we have:

$$u|_C = v|_C = w|_C \equiv 0; \quad \frac{\partial w}{\partial r}\Big|_C \equiv 0. \tag{1.77}$$

When the point C is fixed, but the edge is hinged, the boundary conditions have the following form:

$$u|_C = v|_C = w|_C \equiv 0; \quad M_r|_C = M. \tag{1.78}$$

Here, M is the applied external moment.

If there are no restrictions on the displacements u and v at C, then the boundary conditions that replace the first two constraints in (1.77) and (1.78) are:

$$N_r|_C = N; \quad N_{r\varphi}|_C = \overline{N}, \tag{1.79}$$

where N and $\overline{N}$ are the corresponding external membrane forces along the edge. All the boundary conditions must be expressed through the basic unknowns w and F. In order to realize this, it is convenient to use the following expressions based on (1.71):

$$r\varepsilon'_\varphi - \varepsilon_r - \dot{\gamma}|_C = \left[-\frac{1}{r}(u + \ddot{u}) - \frac{\dot{w}}{r^2}\left(\dot{w} + 2\dot{\overline{w}}\right) - w'\left(\overline{w}' + \frac{1}{2}w' + \frac{\ddot{w} + \ddot{\overline{w}}}{r}\right) - \frac{\ddot{w}\overline{w}'}{r}\right]_C. \tag{1.80}$$

$$\left(r\varepsilon_\varphi\right)' - \varepsilon_r|_C = \left[\dot{v}' - \frac{1}{r}\dot{w}\left(\frac{1}{r}\dot{\overline{w}} - \dot{w}'\right) + \frac{1}{r}\left(\dot{\overline{w}}\dot{w}\right)' - \overline{w}'w' - \frac{1}{2}w'^2 - \frac{1}{2r^2}\dot{w}^2\right]_C. \tag{1.81}$$

$$\varepsilon_\varphi\big|_c = \left[\frac{1}{r}(u + \dot{v}) + \frac{\dot{w}}{2r^2}\left(\dot{w} + 2\dot{\bar{w}}\right)\right]_c . \tag{1.82}$$

In the last three relations, it is necessary to express the components of deformations ε_r, ε_φ, γ via the stress function F by means of equations (1.72) and (1.73). In such a form, equations (1.80)-(1.82) can serve as boundary conditions. For example, instead of the first condition (1.77) or (1.78), equation (1.80) can be used, where, in its right side, $u\big|_c = \ddot{u}\big|_c = 0$. Analogously, the condition $v\big|_c = 0$ can be replaced by (1.81), where the term with v is discarded. The fourth condition from (1.78) and (1.79) can be easily expressed by w and F using corresponding formulae (1.71) and (1.73).

A special investigation is required to formulate the boundary conditions at the pole of the cap. This can be done by assuming that all second derivatives of $\overline{w}(r,\varphi)$ exist in the neighborhood of $r = 0$. Let the stress function F in the neighborhood of $r = 0$ be presented by the Taylor series (see (1.83)).

$$F(r,\varphi) = \sum_{j=0}^{\infty} F^{(j)}(0,\varphi) r^j . \tag{1.83}$$

The boundedness of $N_{r\varphi}$ in this neighborhood (see (1.84)) yields the boundary condition (1.85).

$$N_{r\varphi} = \left(\frac{1}{r}\dot{F}\right)' = \left[\sum_{j=0}^{\infty} \dot{F}^{(j)}(0,\varphi) r^{j-1}\right]' . \tag{1.84}$$

$$\dot{F}(0,\varphi) \equiv 0 . \tag{1.85}$$

Similarly, because N_r must be bounded when $r \to 0$, equations (1.73), (1.83) and (1.85) lead to

$$F'(0,\varphi) \equiv 0 . \tag{1.86}$$

The demand that the changes of curvatures $\chi_{r\varphi}$ and χ_r remain bounded at $r = 0$ leads to

$$\dot{w}(0,\varphi) \equiv 0; \quad w'(0,\varphi) \equiv 0 . \tag{1.87}$$

Thus, there are four boundary conditions ((1.85), (1.86) and (1.87)) at the pole.

When the deformation of the cap is axisymmetric in the region of the pole, F and w turn out to be odd and even functions of r, respectively. There are two particular situations where this occurs. When the deformation of the cap is axisymmetric altogether and when the axisymmetric pre-buckling state bifurcates into a nonsymmetric post-buckling state whereby periodic waves are formed in a circular, belt-like region near the edge or in a region between the apex and the edge. In the latter case, the nonsymmetric buckling pattern leaves the axisymmetrically deformed area around the pole unaffected (and also the edge itself!)[Shilkrut, 1980].

To conclude this section, we present here another, rather elegant form of the basic equation (1.76):

$$\frac{1}{Eh}\Delta\Delta F = -\left(\Gamma_\Delta - \overline{\Gamma}\right), \tag{1.88}$$

where Γ_Δ and $\overline{\Gamma}$ are the Gaussian curvatures of the deformed and initial shapes of the shell, respectively. The derivation of (1.88), which follows from the compatibility equation (1.75), can be realized by formal algebra, keeping in mind the following approximate formulae, which are valid for shallow surfaces:

$$\Gamma_\Delta = \chi_{r\Delta}\chi_{\varphi\Delta} - \chi_{r\varphi\Delta}^2; \quad \overline{\Gamma} = \overline{\chi}_r\overline{\chi}_\varphi \tag{1.89}$$

The curvature expressions here are given by the last formulae of (1.71) with the variable expression $w_\Delta = w + \overline{w}$ appearing in Γ_Δ and $\overline{w}$ in $\overline{\Gamma}$. In Cartesian coordinates, equation (1.88) is given in the book by Volmir [1956].

Chapter 2
Qualitative Investigations of Geometrically Nonlinear Shells

This chapter deals with a qualitative investigation of the general properties of the solutions of shell equations. It is aimed at offering a better understanding of the behavior of nonlinear shells — a behavior that is often dominated by very complicated, essentially nonlinear effects that sometimes lead to unexpected phenomena. The qualitative investigation reported here is mathematical in character and is based on the properties of the governing equations, rather than on their actual solution.

2.1 Some important properties of shallow spherical caps and circular plates subjected to transverse loads.

To understand the behavior of shells in a qualitative way, a summary of the principal nonlinear effects that govern this behavior is in place at this point. While extensive discussion and analysis of these effects will be given in other chapters, attention in this section is focused on the principal characteristics of the solutions, rather than on the details of the methods that are used to obtain them.

We start with the simplest case: an axisymmetric, isotropic, homogeneous spherical cap of uniform rigidity subjected to a single parameter load; i.e., an axisymmetrically distributed set of transverse forces. Although this case belongs to the simplest of its kind, its solutions already show very complicated behavior.

The initial (undeformed) shape of the shallow spherical cap is specified by the single, nondimensional, geometric parameter $\bar{\xi}$ (the apex rise), given by equation (1.67). We shall begin the survey with several characteristic examples of axisymmetric deformations. We assume that the boundary of the cap is clamped and that the load consists of a uniformly distributed pressure, the intensity of which is prescribed by the dimensionless parameter q (see 1.64). (Please note: if the apex rise $\bar{\xi}$ is negative and q is positive, the pressure is directed toward the inside of the cap and we call this the case of external pressure. If $\bar{\xi}$ is negative and q is also negative, the situation is reversed and we call this the case of internal pressure). In Figure 2.1, the so-called *characteristic curves* of the cap are given for $\bar{\xi} = 0,-1,-2,-3,-4$. Each characteristic curve is the image of a load-displacement relation $q(\xi)$ which represents the equilibrium states of the cap for a range of values of q, where ξ denotes the nondimensional displacement of the apex (see (1.64)). Alternative terms for *characteristic curves* are: equilibrium curves, equilibrium branches, equilibrium paths, or deformation paths. We will also use the term "response curve" or simply "response."

When $\bar{\xi} = 0$ (circular plate), there is a unique state of equilibrium for each value of q. As $|\bar{\xi}|$ increases, the shapes of the characteristic curves change qualitatively, as can be seen in Figure 2.1.

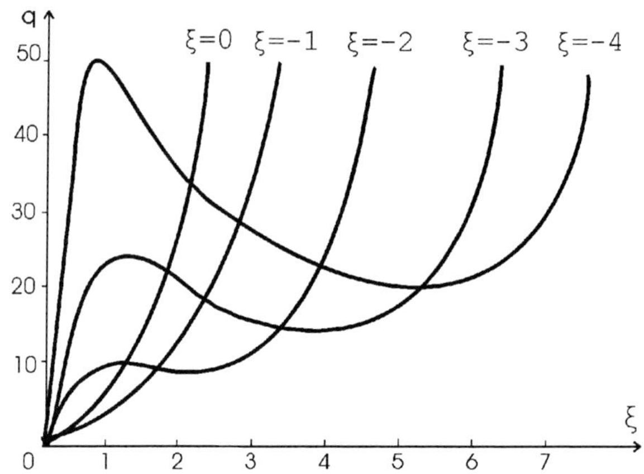

Figure 2.1 The axisymmetric response of a (clamped) spherical cap under uniform pressure q (the cases $\bar{\xi} = 0 \div \bar{\xi} = 4$).

Starting from the value $|\bar{\xi}| \approx 2$, the equilibrium curves cease to be monotonic and there are maxima and minima. In these cases, there is the possibility of loss of stability in terms of a snap-through process from an equilibrium state A to another state B, as indicated in Figure 2.2.

This snap-through type of instability, already identified by Theodore von Kármàn [see von Kármàn & Tsien, 1939; Biezeno, 1935], is a loss of stability *in the large*. It differs essentially from the loss of stability in the classical sense related to bifurcation (Euler), which is a loss of stability *in the small* of the initial, pre-critical state. In the case of snap-through, the pre-critical state A and post-critical state B are separated from each other by a finite dis-tance; in the case of Euler loss of stability, the immediate post-critical state is infinitely close to the pre-critical one. The transition from A to B is a snapping process under an increasing load, whereas the transfer from position C to D (see Fig. 2.2) is a snapping process which occurs as a result of a decreasing load. The latter form of snap-through is usually called snap-back.

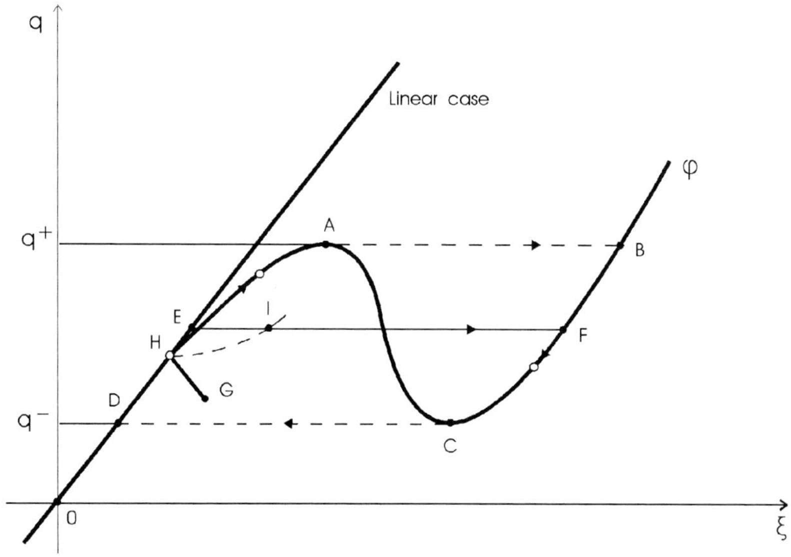

Figure 2.2 Various types of nonlinear behavior.

The values of the loading parameter q which correspond to points A and C are the *upper* and *lower* critical load, respectively. In this case they are limit loads. We emphasize that we cannot strictly justify the occurrence of these snapping phenomena on the basis of the static equilibrium curve $q(\xi)$ only, because the snapping process is a transient process and it can only be properly demonstrated by solving the corresponding nonlinear equations of motion. We may conclude, however, that one of the main characteristics of the nonlinear behavior of shells is the possibility of the existence of more than one equilibrium state for one value of the external loading parameter. There is no longer a unique solution to the corresponding boundary value problem. As is shown in the next chapter, the degree of non-uniqueness (that is, the number n of possible solutions of the boundary value problem for the given value of q) increases with $|\xi|$, and this in such a way that, when $|\xi|$ grows beyond bounds, this also occurs with n. This important fact was discovered by the author in the mid-1960s and published in Shilkrut [1967, 1970, 1974]. The following hypothetical (heuristic) asymptotic formula may be found in Vorovich and Minakova [1973] for the number n of possible equilibrium states of a closed shell at a fixed value of loading q.

$$n \sim A(q/E)^{1/2}(h/R)^{-3/2}$$

Here, A depends on the form of the shell and on Poisson's ratio ν of the material, but it does not depend on its dimensions. We believe that this relation can be a guideline for very deep caps, but for general use, this formula needs a more rigorous justification.

Deformation curves for large values of $|\bar{\xi}|$ are presented in Figures 2.3-2.6. As shown in these figures, increasing $|\bar{\xi}|$ complicates the topological features of the characteristic curves. At $|\bar{\xi}| \cong 12$, the main part of the characteristic curve is simple, but there are also separate closed loops. (See Fig. 2.6, where such loops, II and III, are shown; I denotes the main part of the characteristic curve for $|\bar{\xi}|=15.15$.) Similar loops were discovered by Max Born in his dissertation (1906) when he investigated the finite deflections of flexible beam that loses its stability when it is loaded in compression. Independently from our own investigations, such loops were also described by Keller and Wolfe [1967] and later in a number of other publications.

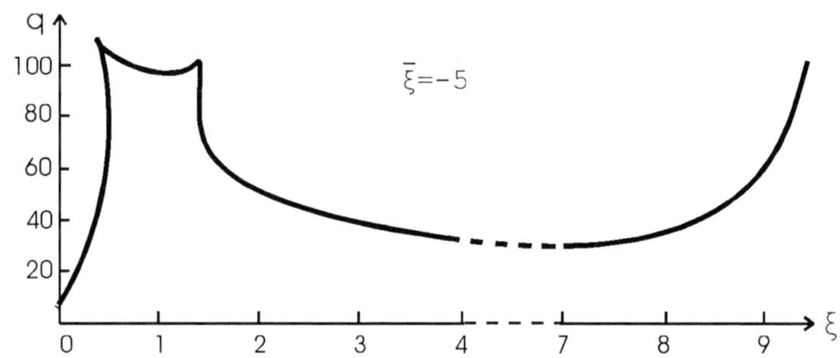

Figure 2.3 The axisymmetric response of a clamped spherical cap under uniform pressure q (the case $\bar{\xi} = -5$).

Separate loops or separate branches generally exist, and they are not a result of the selection of the plane onto which the (one dimensional) equilibrium curves are projected. In the context of a quasi-static treatment of our problems, an isolated branch means a closed solution curve that cannot be reached by continuously changing the loading parameter starting from the undeformed, unloaded state. For example, it is not possible to pass from branch I to branch II in Figure 2.6. This transition can only be realized dynamically by means of, for example, an impulse. It is necessary to emphasize here that, in the area between branches I and II in Figure 2.6, only a single equilibrium state can exist. This property is invariant and is independent of the selection of the displacement of a point along the meridian, which is given as the parameter ξ. See also Sections 3.4, 4.4 and 6.3.

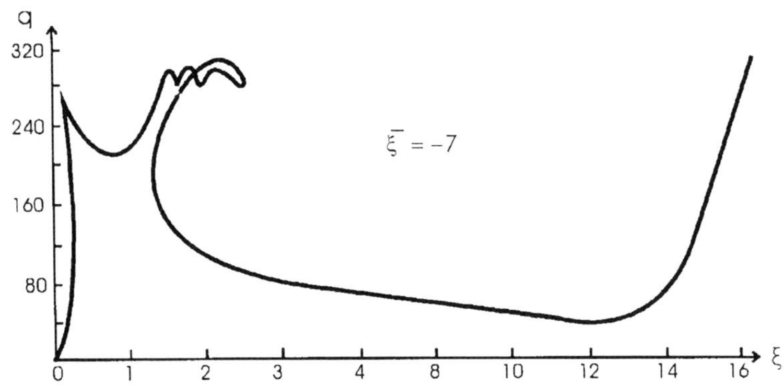

Figure 2.4 The axisymmetric response of a clamped spherical cap under uniform pressure q (the case $\bar{\xi} = -7$).

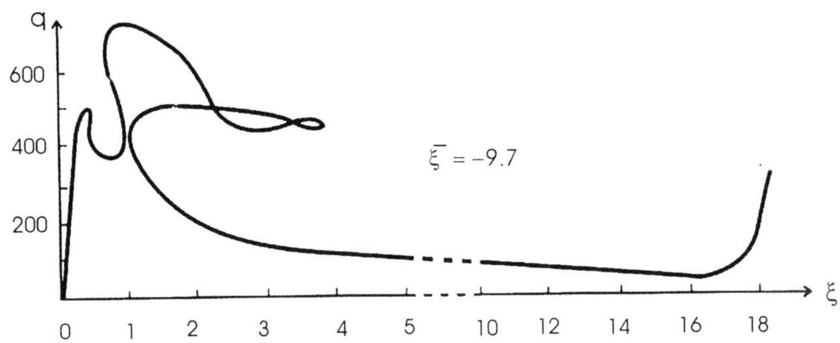

Figure 2.5 The axisymmetric response of a clamped spherical cap under uniform pressure q (the case $\bar{\xi} = -9.7$).

As is shown for a clamped cap, the lower critical load q^- is always positive. But the lower critical load can become negative, beginning from some value of $|\bar{\xi}|$, when the cap is hinged (see, for example, Figs. 2.7-2.9 and others given below, where the active load, the edge moment M, replaces the role of the external pressure q in the previous case). This means that the reverse snapping (from C to D; see Fig. 2.2) can take place under internal pressure. The negative value of the lower critical pressure q^- is at the same time an indication that non-trivial (i.e., deformed states) exist which correspond to external forces that are zero, so that the internal forces in the cap in these states comprise a self-equilibrated system. These particular (zero load) states correspond to points of intersection of the equilibrium curve with the ξ-axis.

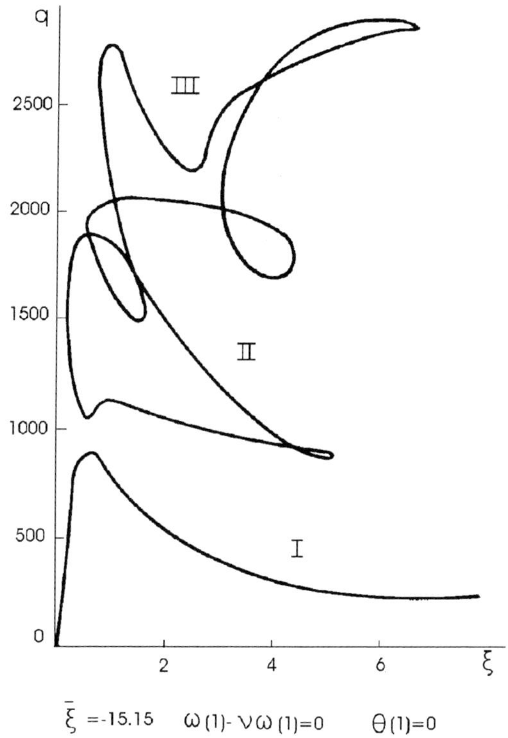

Figure 2.6 The axisymmetric response of a clamped spherical cap under uniform pressure q (the case $\bar{\xi}$=−15.15).

In Figure 2.1, only the curves for positive q and ξ are depicted. The part of the characteristic curve where q is negative (internal pressure) is a monotonic curve of the type shown in the third quadrant of Figure 2.2. It is noted here, that these parts of the characteristic curves are usually not presented, since they are not very interesting due to their monotonic character. But they are, for example, shown in Figure 2.7, in terms of the relation $M = M(\xi)$ and in Figure 2.10 in terms of $M = M(\xi, q)$ (for $q = 0$). In the first case (Figure 2.7), this part of the characteristic curve is located in the fourth quadrant of the plane (ξ, M).

The increase of complexity of the shape of the characteristic curves with the increase of $|\bar{\xi}|$ may also be observed for other cases with other types of support and loading. For example, Figures 2.7 and 2.8 give the characteristic curves for *pure* bending of shells with fixed hinged supports. The loading here is an external, distributed bending moment M^* applied along the edge, the intensity of which is defined by the dimensionless parameter $M = 12\left(1-\nu^2\right)M^* a^2 / Eh^4$. When the value of $\left|\bar{\xi}\right|$ increases, the multiplicity of equilibrium states at the same value of the external loading parameters also increases. Again, when $\left|\bar{\xi}\right| \to \infty$ (that is, when $h \to 0$, see (1.64)), the degree of multiplicity of equilibrium states also tends to infinity.

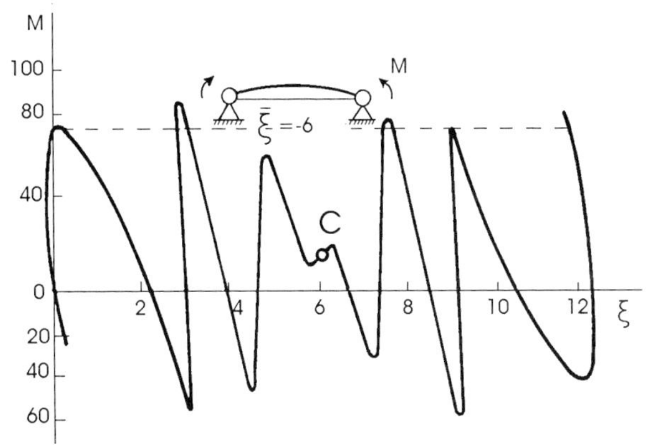

Figure 2.7 Axisymmetric solutions for a cap loaded by an edge moment M; $\bar{\xi} = -6$.

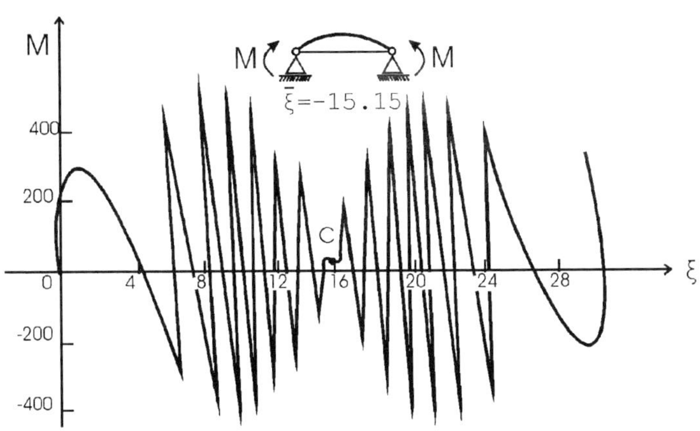

Figure 2.8 Axisymmetric solutions for a cap loaded by an edge moment M; $\bar{\xi} = -15.15$.

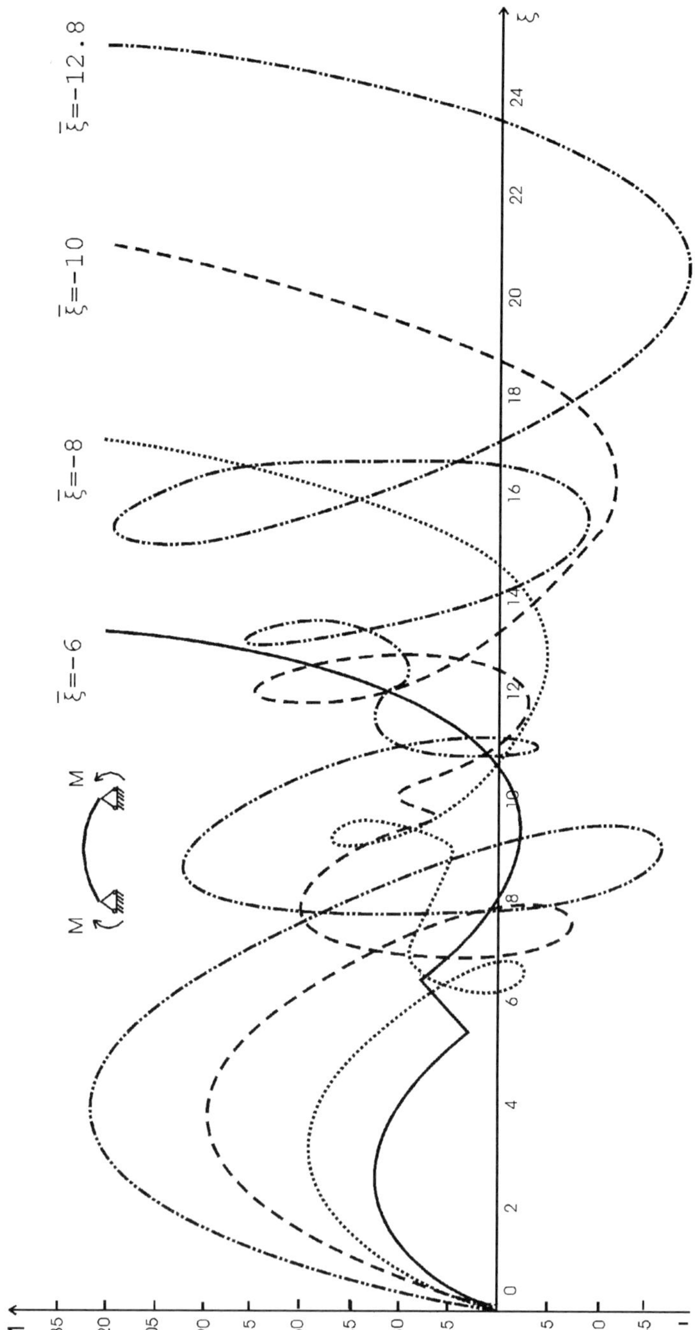

Figure 2.9 Additional results for the cap loaded by an edge moment (see also Figs. 2.7 and 2.8).

Some characteristic curves for pure bending with a movable hinged edge are given in Figure 2.9. We note that the case of pure bending corresponds at the same time to an important problem of thermo-elastic deformations of the shell (see Section 2.7).

The complexity of the topological structure of a characteristic curve (i.e., the occurrence of many limit points) raises the following important questions. How does one define the upper or lower critical load? Does the upper critical load correspond to the first limit point on the characteristic curve or to one of the other limit points? Is it possible that more than one upper (or lower) critical load exist? It turns out that the answers to these questions can only be obtained after an investigation of the stability of the entire set of equilibrium states. The problem of the stability of the solutions is the subject of the discussion presented below, but we shall first present a short survey of the behavior of shallow caps subjected to multi-parameter loading in order to reveal the contrasting difference of this generalization with the single-parameter loading cases that have been considered so far.

Three examples of the behavior of shells under a combined system of loading are illustrated in Figures 2.10-2.12. Figures 2.10 and 2.11 show cases when the shell is subjected to a boundary moment M and uniform pressure q.*

The edge is hinged by movable supports and the geometry is given by $|\bar{\xi}| = 4$. The graph $M(\xi)$ for $q = 0$ (see Fig. 2.10) is a skew-symmetric curve with the center C ($\xi = -\bar{\xi}$; $M = -2\bar{\xi}(1 + \nu)$), in agreement with the theory of reciprocal systems that we will discuss in Section 2.4. Each curve for $q \neq 0$ does not have any symmetry properties, but two curves defined by $q = \pm K$ are mutually skew-symmetric; i.e., they are reciprocal with respect to the center C. The same characteristics are recreated in the plane (ξ, q) (see Fig. 2.11). In this plane, there also exists a center of symmetry with coordinates $\xi = -\bar{\xi}$; $q = 0$; i.e., for each equilibrium point on this plane corresponding to $q = q_1$ and $M = M_1$, there exists a twin (symmetric relative to the above-mentioned center) for which $q_2 = -q_1$ and $M_2 = -4\bar{\xi}(1 + \nu) - M_1$. Figure 2.12 illustrates the case where, in addition to the loading q and M, a uniformly distributed compressive force $N = N^*a^2/Eh = -5$ is applied at the edge.

---

* The cap itself with its loads is drawn in Figure 1.4.

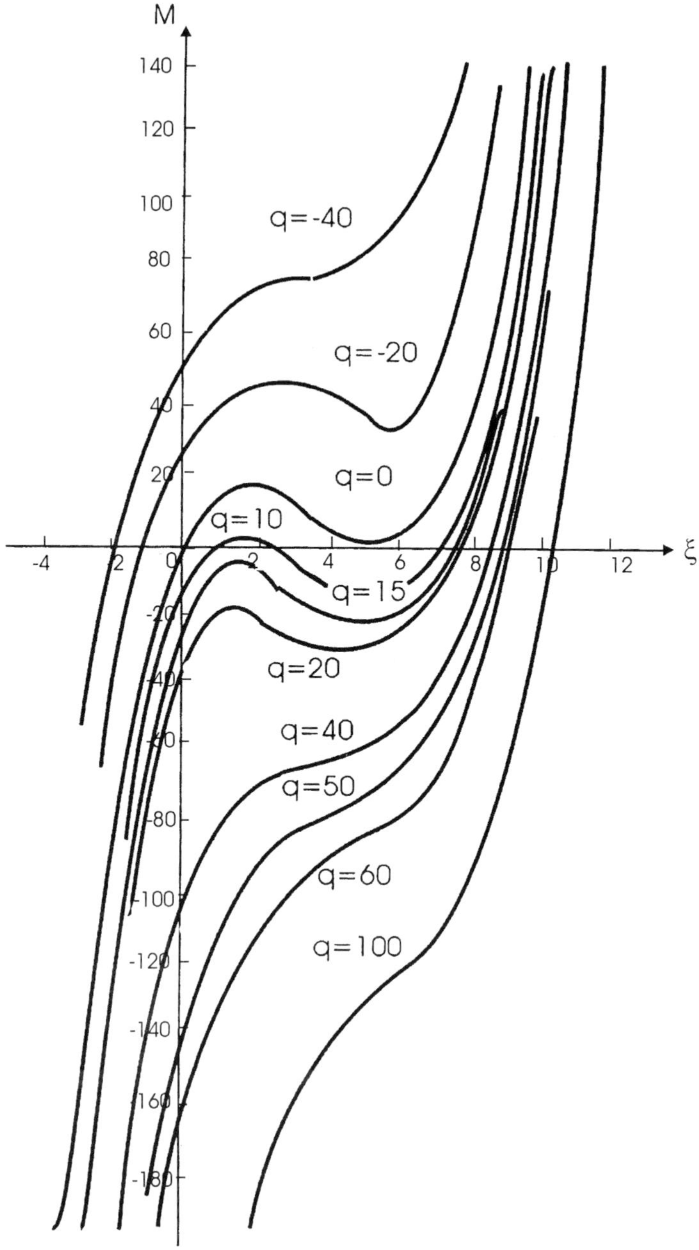

Figure 2.10 Results for a cap loaded by uniform pressure q and an edge moment M.

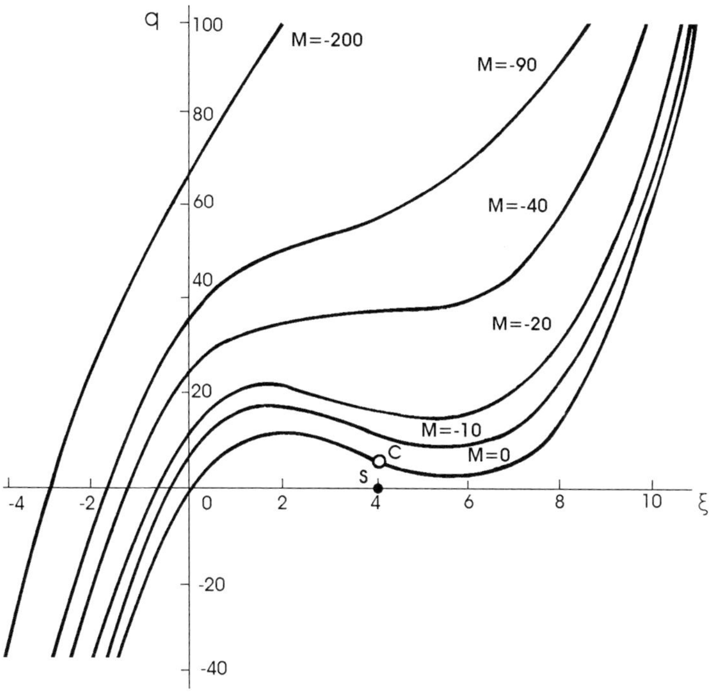

Figure 2.11 Results of Figure 2.10 in the (q, ξ) plane.

It can be deduced from these examples, that an additional load can stabilize the structure so that the difference between upper and lower critical loads becomes smaller and smaller the higher this load becomes, so that, eventually, the danger of snap-through is eliminated altogether (in the last example, Fig. 2.12, the additional load N should of course be positive to achieve this effect). This observation implies that, beyond a certain intensity of the additional load, the deformation curves become monotonic. Thus, this is one of the possibilities that is available to a designer to eliminate the danger of (axisymmetric) snap-through of the shell.

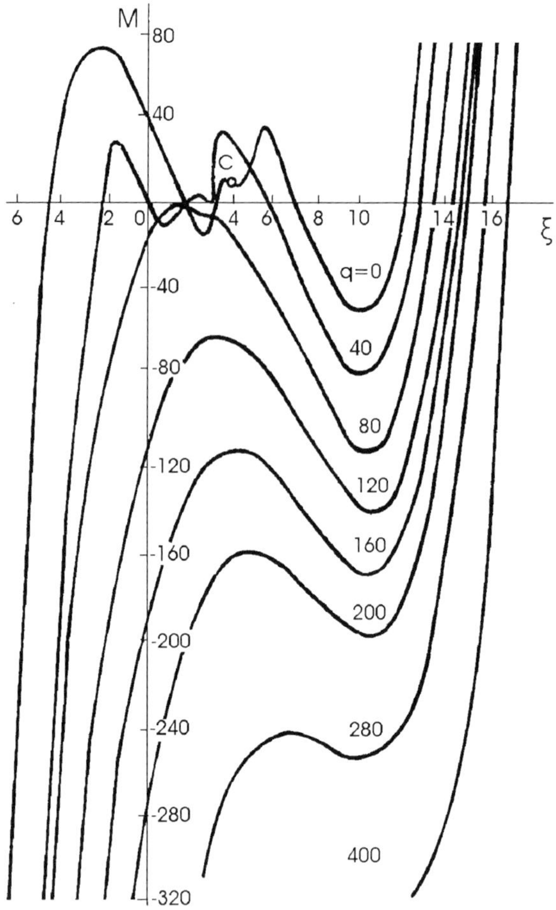

Figure 2.12 The effect of an additional edge load $N = -5$.

The properties of the solutions discussed above concern the behavior of shallow shells. It turns out that these properties are also found for non-shallow caps [Shilkrut, 1969b, 1974; Shilkrut, Morar & Gerlaku, 1969]. By non-shallow caps, we mean shells with their initial shape in the form of a segment of a spherical surface with large opening angles not exceeding π. (The central angle is equal to π for a hemispherical shell). At the same time, it is necessary to emphasize here that the behavior of very deep spherical shells (when the opening angle is more than π) can differ from what is described above. The deformation of such deep shells must be investigated separately, but this task is not considered in this book.

a) Stability of axisymmetrically loaded shells (strictly restricted to axisymmetric deformations only).

Equilibrium states that are unstable *in the small* are also unstable *in the large*, but the converse is not true; a system can be stable in the small, but not be stable in the large. We shall illustrate the relation between these two types of stability losses by considering the stability of the axisymmetric equilibrium states of shallow and non-shallow caps. An illustration of the possible solutions in this case is given in Figure 2.2.

Let us consider an axisymmetric (prebuckling) equilibrium state marked by some point E on the rising branch O-A of the deformation curve presented in Figure 2.2. Let us assume, for the time being, that all other possible deformations of the shell are also axi-symmetric. In this situation, we shall now apply a small impulsive perturbation to the equili-brium state represented by *E*, keeping the value of the loading which generated state *E* unchanged. A small oscillation will then take place around this state (i.e., around point E). In the terminology used here, the state *E* will be declared *stable in the small* if the oscillation does not result in a motion away from E (for sufficiently small initial perturbations). But if the magnitude of the perturbation is increased and rises above a certain limit, a transition (a jump) will take place from the tested equilibrium state *E* to a new position F, which is far removed from position E (provided that there is damping present, which we tacitly assume). This means that, for sufficiently large perturbations, there will be a loss of stability *in the large* of state *E*. This observation implies that we can consider the tested state *E* stable in the small, but unstable in the large. The same property is also characteristic for the new equilibrium state *F*.

The foregoing concept was established for perturbations and responses that are purely axisymmetric. But if the class of perturbations is enlarged by also taking into account non-symmetric perturbations, the stability properties of the equilibrium states considered above can change radically. For example, it can then happen that there is a bifurcation point H (Fig. 2.2) before E, where a transition from the axisymmetric state to a nonsymmetric state takes place. In Figure 2.2, the initial part of such a bifurcation that starts at H is denoted by a dashed line, and we will assume that this branch is stable in the small. In this situation, state *E*, which was stable in the subspace of axisymmetric disturbances and responses, has become unstable (in the small) in the larger space of nonsymmetric disturbances and responses. Thus, when the restriction to axisymmetric deformations is removed, the shell will transfer its state from *E* to a nonsymmetric equilibrium state *I* on the postbifurcation branch even for small disturbances that vanish. On the other hand, if the postbifurcation path starting from H is not rising, but is directed downward (HG, for example), the transition from state E will happen by a snap-through process to another equilibrium state, in this case the state F. In fact, if the bifurcation branch is descending, snap-through (or snap-buckling) will already occur at the bifurcation point H. This follows, because a simple bifurcation point (H in our example) that marks the boundary between the stable and unstable parts of the primary branch O-A already belongs to the unstable part of the primary branch when there is at least one descending secondary branch [see Koiter, 1945, 1963]. (By simple bifurcation point, it is meant that only two equilibrium curves cross at H).

The establishment of the stability of the complete set of axisymmetric equilibrium states has only recently been given proper attention. Traditionally, the equilibrium states near bifurcation points are determined by means of Koiter's asymptotic method [Koiter, 1945, 1963b] or by the Liapunov-Schmidt reduction approach [see Golubitsky & Schaeffer, 1985]. The stability of the complete set of axisymmetric states has been examined from the standpoint of Poincaré's theory, which was proved rigorously only for systems with finite numbers of degrees of freedom. But a direct generalization of this theory to continuous systems may lead to erroneous results, as was mentioned (without any further comment) by Koiter [1963b], and was demonstrated in our own work in the context of applications to nonlinear shells [Shilkrut, 1980; Shilkrut & Vyrlan, 1975a,b, 1976a,b, 1977; Shilkrut, Barladeanu, Vyrlan, Gerlaku, Fishberg & Tcherkezov, 1977; Vyrlan & Shilkrut, 1978]. These matters will be considered extensively in Chapter 6.

In our own investigations of the stability (in the small) of the solutions, we used the method of small oscillations around the equilibrium state or the principle of the minimum of the potential energy. But we also determined the stability by solving the nonlinear equations of motion of a shell when the load is slowly increasing in a quasi-static fashion. With all these methods, results were produced that were in agreement with each other. For single parameter load systems these results can be summarized as follows:

1) All equilibrium states along the characteristic curve that precede its first maximum and which follow from its last minimum are stable in the small in the class of axisymmetric deformations. All other intermediary equilibrium states, regardless of their number, including the states that belong to separate loops, are unstable. As a consequence, no more than two stable (in the small) equilibrium forms can exist at the same value of the load parameter, regardless of the topological structure of the characteristic curve. This property is probably a universal feature of geometrically nonlinear shells, at least for one-parameter loading. (See more complicated cases in Section 6.8.) For example, the two stable branches of the curve in Figure 2.7 can be described as follows. The first stable part is the left-most branch, which passes through point 0 and ends at the first extremum A, while the second part is the right-most branch, which

originates at the last minimum B and rises upward. All the intermediate points of the characteristic curve between points A and B are unstable states.

2) It may be possible to define critical loads for axisymmetric deformations. The upper critical load corresponds to the first maximum on the characteristic curve, and the lower critical load to the last of its minima. There is only one upper and one lower critical load. Successive snaps are impossible, in spite of the fact that the second and the next maximum points of the deformation curve can correspond to loads that are higher than the first (see, for example, Figs. 2.5-2.8). This is proved and discussed in detail in Chapter 6, which is devoted to investigations of the stability of the equilibrium states of the shell.

3) The algebraic sign of the derivative $\mathrm{d}q/\mathrm{d}V$ (Poincaré's criterion for a system with one degree of freedom) is not a suitable criterion for multi-dimensional systems, nonlinear shells, and other nonlinear continua. q is here the intensity of the load and V the corresponding generalized deflection in the sense that the inner product (qV) yields the work that is done by the external forces. The derivative $\mathrm{d}q/\mathrm{d}V < 0$ is both the necessary and sufficient condition for instability; $\mathrm{d}q/\mathrm{d}V > 0$ is a necessary condition for stability, but not a sufficient one. When $\mathrm{d}q/\mathrm{d}V$ is positive, the corresponding equilibrium state may be either stable or unstable. This circumstance can be easily seen from the curves presented in Figures 2.4-2.9, where all the intermediate branches between the two furthest separated (stable) branches are unstable, independent of the signs of the derivatives (directions of tangents) $\mathrm{d}q/\mathrm{d}\xi$ or $\mathrm{d}M/\mathrm{d}\xi$, derivative terms which are representative of the previously mentioned $\mathrm{d}q/\mathrm{d}V$.

b) Bifurcation phenomena in axisymmetrically loaded shells (with no restrictions imposed on the states of deformation).

As is well-known, under certain conditions, axisymmetric, transversely-loaded spherical shells can exhibit nonsymmetric states in addition to axisymmetric ones; i.e., bifurcation phenomena can take place (see, e.g., point H in Fig. 2.2). Specific cases of bifurcation points that are numerically determined are presented in Figures 2.13-2.15 and 2.17. The nonsymmetric modes have periodic waves in the circumferential direction that manifest themselves in a belt-like region lying between the pole and the boundary of the shell.

The first approximate solutions of this problem for a plate were given by Biezeno and Grammel [1956; see also Panov & Feodosiev, 1948; Bodner, 1955] and, for shells, by Krivosheev [1959] and Grigolyuk [1960]. The existence of these phenomena in plates was clearly validated mathematically by Morozov [1961]. A qualitative discussion of the possibility of this kind of bifurcation for shells is given by Shilkrut [1967], while a series of accurate numerical solutions of this problem, first presented by Huang [1964], may also be found in Weinitschke [1962, 1965], Thurston, [1964], Archer and Famili [1965], Bushnell [1967a], Shilkrut, Morar and Gerlaku [1969], Shilkrut and Gerlaku [1969], Gerlaku and Shilkrut [1970], Fitch and Budiansky [1970], and Valishvili [1976]. In all these works (apart from Bushnell [1967a], Shilkrut and Gerlaku [1969], and Gerlaku and Shilkrut [1970]) the bifurcation problem is considered for shallow spherical shells loaded by external forces, normal to the shell surface, which point toward the inside of the shell. In this case, bifurcation occurs only if the rise parameter $|\bar{\xi}|$ is greater than some particular value $|\xi_0| > 0$ that depends on the composition of the external loading and the particular type of edge supports [Huang, 1964; Shilkrut, 1974, 1980; Shilkrut & Gerlaku, 1969; Gerlaku & Shilkrut, 1970; Valishvili, 1976]. This feature is very clearly illustrated by the data in Figure 2.14, where (as in Figs. 2.13, 2.15, and 2.17) the bifurcation points are marked by open dots.

The characteristic curves in Figure 2.14 are derived for the case of a simply supported spherical cap subjected to a uniformly distributed edge moment $M > 0$, in the way it is depicted in the Figure. There are no bifurcation points when $|\bar{\xi}| = 3$, while for $|\bar{\xi}| \geq 6$, bifurcation points have appeared. Characteristic of this type of bifurcation is that it occurs when the meridional stresses (along the meridian) of the prebuckling state are negative along some part of the shell or negative in the whole domain. Thus, in general, bifurcation occurs because the sign of the meridional stresses is negative in some region. In contrast, there also exists a type of bifurcation that takes place when the meridian is in tension everywhere [Panov & Feodosiev, 1948; Shilkrut, 1980; Shilkrut & Gerlaku, 1969; Gerlaku & Shilkrut, 1970]. This happens, for example in plates subjected to transverse forces of arbitrary type [Shilkrut, 1967, 1974] (see Sections 3.2 and 3.7), but it also happens with caps that are loaded by transverse forces that are directed toward the outside of the cap, such as uniform internal pressure.

Thus, there are two different, distinct types of bifurcation phenomena that can take place in axisymmetrically loaded circular plates and shells of revolution. The first type we will call *bifurcation in compression* (see the discussion above) and the second we will call *bifurcation in tension* (the opposite type) [Shilkrut, 1974; Shilkrut & Gerlaku, 1969; Gerlaku & Shilkrut, 1970]. Correspondingly, there exist two different spectra (series) of bifurcation loads that correspond to two different sets of bifurcation points. Incidentally, these bifurcation loads are determined by the Euler approach; that is, they are eigenvalues of a homogeneous boundary value problem for a linearized set of differential equations that has an infinite discrete spectrum of eigenvalues and eigenmodes (see, e.g., [Shilkrut, 1974, 1980;

Gerlaku & Shilkrut, 1970; Fitch, 1968].

Apart from the question of the sign of the meridional stresses in the prebuckling state, there is also another distinction between the two types of bifurcations that we described above that is essential here. Bifurcation in "compression" can occur for any conditions of edge support, but it is restricted to values of $|\bar{\xi}|$ greater than a certain threshold value $|\xi_0|$ mentioned above. For bifurcation in "tension," the situation is the opposite: it can take place for any value of $\bar{\xi}$, including $\bar{\xi} = 0$ for the plate, but not for all types of edge support. For example, for immovable supports in the direction of the base of the cap, bifurcation in tension is impossible. These particular properties of the various phenomena that can be encountered will be discussed in detail in Sections 3.2 and 3.7.

We shall now illustrate the foregoing observations with the use of an example of a shallow spherical cap supported by movable hinges that is loaded by a uniformly distributed edge-bending moment (see also, Shilkrut & Gerlaku, 1969; Gerlaku & Shilkrut, 1970]. Figure 2.13 shows the characteristic curves for the axisymmetric prebuckling states and the first bifurcation points that are encountered along these states (corresponding to the lowest loads) for a plate $\bar{\xi} = 0$ and a shell with $|\bar{\xi}| = 1$.

That these bifurcation points belong to a prebuckling state that is in "tension" can already be deduced from the fact that the prebuckling states are monotonically rising and therefore have no limit points. Limit points start to appear along the axisymmetric characteristic curve when $|\bar{\xi}| \geq 1.92$ [Shilkrut, 1967; Shilkrut & Gerlaku, 1969]. This means that, in order to get bifurcation in compression, $|\xi|$ must in any case be greater than 1.92. This can be seen (as was noted above) from the graph shown in Figure 2.14, where the bifurcation points in compression (marked here by small dots) occur only after $|\xi|$ has exceeded $|\bar{\xi}| = 3$, even though limit points appear when $|\xi| < 3$.

Consequently, the picture is the following. When $|\bar{\xi}|$ is small and the external axisymmetric transverse forces are pointed to the inside of the cap, the deformation curve is monotonic and loss of stability does not occur in general (see, e.g., Fig. 2.1). Starting from some value of $|\bar{\xi}|$, which we denote by $|\bar{\xi}_{00}|$, limit points appear on the characteristic curve and loss of stability takes place in the form of snap-through toward a new axisymmetric state. Bifurcation in compression appears later, when $|\bar{\xi}|$ exceeds a threshold value $|\bar{\xi}| = |\bar{\xi}_0|$, where $|\bar{\xi}_0| > |\bar{\xi}_{00}|$. The change of stability can then occur at a bifurcation point whereby the original axisymmetric state goes over into a new nonsymmetric form. The magnitudes of $|\bar{\xi}_{00}|$ and $|\bar{\xi}_0|$ are dependent on the type of loading and edge support. The threshold values in the case given in Figure 2.14, are $|\bar{\xi}_{00}|=1.92$; $|\xi_0| \approx 4$. Please note, that the bifurcation points that are detected for the case $|\bar{\xi}| = 1$ can only belong the type that occurs in "tension" (see Fig. 2.13).

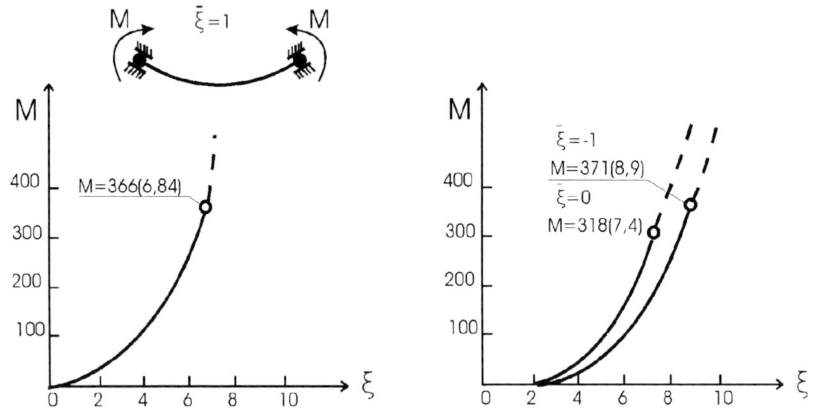

Figure 2.13 Examples of the occurrence of bifurcation in tension.

In order to illustrate, in a qualitative way, the process of the emergence of bifurcation points in compression, we shall use the specific class of shells called *reciprocal systems*, which are described and investigated below in Sections 2.2-2.4. To this class of shells belong hinged spherical caps subjected to edge-bending moments M (*pure* bending). One of the basic features of reciprocal systems is the specific form of the characteristic curves, which are skew-symmetric with a mirror point called center of symmetry. All curves presented in Figures 2.14, 2.15, and 2.17 are of this type, where their centers of symmetry are labeled by C.

From the beginning, when $|\bar{\xi}| = |\bar{\xi}_0|$, the bifurcation points occur in the center of symmetry of the deformation curve, which is situated along an unstable branch. (Please note that the branches of the characteristic curve counted from the zero value of the loading parameter up to the upper critical load, and from the lower critical value up to infinity, are called the *stable* branches, while the intermediate branches are called the *unstable* branches). As $|\bar{\xi}|$ increases beyond $|\bar{\xi}_0|$, the bifurcation points move away from the center of symmetry and approach the stable branches. These points are distributed on the deformation curve as symmetric pairs (see Fig. 2.14), in accordance with the theory of reciprocal systems (see Sections 2.2-2.4). Explicit numerical data are given in Table 2.1, which shows the first two bifurcations points in compression for different $\bar{\xi}$. The parameter n is the number of circumferential waves of the corresponding bifurcation modes.

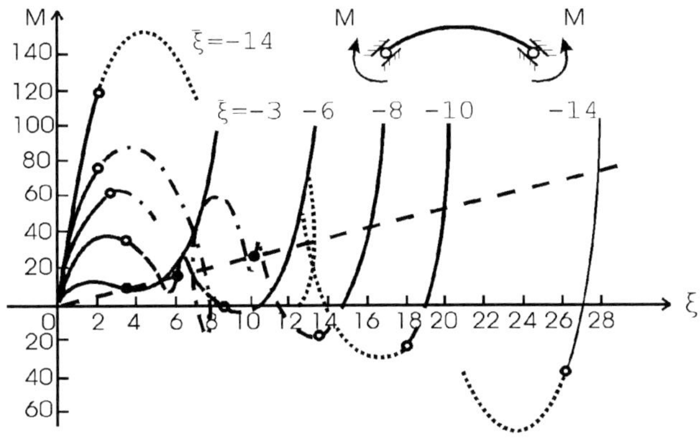

Figure 2.14 An example of bifurcations in compression.

Table 2.1

ν = 0.3

$\bar{\xi}$	Symmetric Solution Upper Critical Points		Bifurcation Points			Footnote
	ξ^+	M^+	n	ξ_b	M_b	
−3	1.81	10.88	-	-	-	No bifurcations
−5	2.24	27.73	2	3.65	22.32	Bifurcation points on
			3	3.76	21.55	unstable branches
−6	2.51	37.64	2	3.36	35.02	" "
			3	3.62	33.33	
−8	2.98	60.39	3	2.70	60.10	Bifurcation points on
			2	2.74	60.19	stable branches
−10	3.18	86.54	4	2.11	77.09	" "
			6	2.12	77.31	
−14	4.37	149.11	6	2.06	116.29	" "
			5	2.07	116.73	

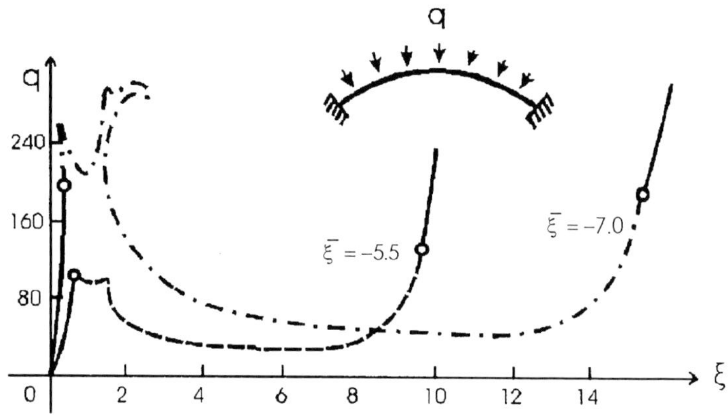

Figure 2.17 Bifurcation points, marking the beginning of the second stable branch.

Table 2.2

ν = 0.3

$\bar{\xi}$	First Stable Branch					Second Stable Branch				
	Symmetric Solutions Upper Critical Points		First Bifurcation Points			Symmetric Solutions Lower Critical Points		Second Bifurcation Points		
	ξ_{cr}^{+}	q_{cr}^{+}	n	ξ_b	q_b	ξ_{cr}^{-}	q_{cr}^{-}	n	ξ_b	q_b
−5.5	0.4	122	2	0.424	110.5	8.3	33	2	10.67	136.1
			3	0.427	114.7			3	10.48	126.6
−7	0.2	260	3	0.296	178.2	12.2	41	2	15.35	199.5
			2	0.296	180.4			4	15.18	168
−9.7	0.5	512	4	0.264	333.1	16.4	60	4	18.17	275.8
			6	0.264	354			6	18.15	270.1

On the other hand, the existence of bifurcation points does not necessarily lead to the loss of stability for nonsymmetric shapes. Everything depends on the stability of the symmetric and nonsymmetric equilibrium states at and around the bifurcation points. This question may be investigated by means of Koiter's method [see, e.g., Koiter, 1945, 1963b; Budiansky, 1974], which makes it possible to affirm the following. If the post-bifurcation branch of nonsymmetric states that emerges from the bifurcation point of the prebuckling state is falling (the corresponding dq/dV is negative), then at this point a loss of stability takes place in the form of a snapping process (see Fig. 2.2), where the new stable post-bifurcation state can be either axisymmetric or nonsymmetric.

In the description of the properties above, we used the load q versus displacement diagram ξ, where ξ is the normal displacement of the apex. But instead of ξ, we could also have used the displacement of any other point of the shell, because the particular properties of the load displacement curves that interest us here are invariant with respect to this choice.

With this last remark, we shall end this short survey of the most important qualitative features of the solutions in question. A more detailed description will be given below.

2.2 Reciprocal systems of shallow shells.

There exist a class of shells called *reciprocal* systems that possess important symmetry properties [Shilkrut, 1967, 1969a, 1974; Shilkrut & Vyrlan, 1977]. By investigating these properties, we can clarify several complicated aspects of the behavior of nonlinear shells in general. This classification, which is given in this section, deals with isotropic, homogeneous, shallow shells of uniform rigidity when thermal stresses are absent (described by equations (1.74) and (1.76)). This case has a direct and close link with spherical caps, as will be shown later.

The basic condition for this classification is the existence of a reciprocal symmetric pair of equilibrium states. Two different equilibrium states, $w_1(r, \phi)$ and $w_2(r, \phi)$, will be called reciprocal if they satisfy the relations

$$w_1 + \overline{w} = -(w_2 + \overline{w}); \quad F_1 \equiv F_2 \text{ in } \Omega, \tag{2.1}$$

where the (given) function $\overline{w}$ describes the initial geometry of the shell (see Figs. 1.4 and 2.8); F_1, F_2 are the membrane stress functions of the reciprocal states w_1, w_2, respectively; and Ω denotes the domain of the shell.

The first relation of (2.1) has a very clear geometrical meaning. It implies that the modes w_i are symmetric relative to the base plane of the cap; that is, one of these states is the mirror image of the other relative to the base plane of the shell (see Fig. 2.18). (Hence-forth, all the parameters related to one of the reciprocal states will be marked by the additional index "1" or "2.") A reciprocal system corresponds to a restricted set of geo-metries $\overline{w}$. To see this, we note that the boundary conditions (for w_i and F_i (i = 1, 2)) should of course be compatible with conditions (2.1), which, first of all, implies that the boundary

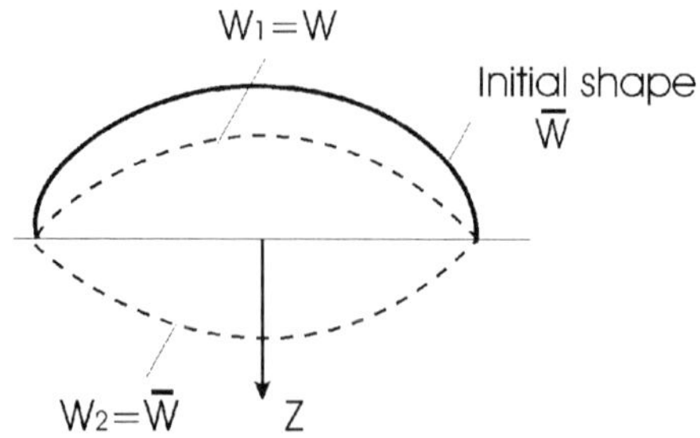

Figure 2.18 Definition of a reciprocal state.

conditions for w_i and F_i must be identical. However, the compatibility requirement also leads to a restriction on the class of admissible initial geometries. Let the given boundary conditions for w be expressed by

$$\mathrm{O}(w)\big|_S = 0, \tag{2.2}$$

where O is an operator that depends on the specific type of shell support on the edges S. If w_1 and $w_2 = -2\overline{w} - w_1$ are equilibrium states (see (2.1), they should of course satisfy

$$\mathrm{O}(w_1)\big|_S = 0; \quad \mathrm{O}(-2\overline{w} - w_1)\big|_S = 0. \tag{2.3}$$

Thus, relations (2.3) present a restriction on the initial geometry $\overline{w}$ if the shell is a reciprocal system. In the very important special case when the operator O is linear, the second condition (2.3) reduces to the form

$$\mathrm{O}(\overline{w})\big|_S = 0. \tag{2.4}$$

The theory of reciprocal systems is further based on a central theorem, which will be proved for the general case of arbitrary deformations of shallow shells (equations (1.74), (1.76)) in the case of a linear operator O for the kinematic edge condition (2.4).

Theorem of Reciprocal Systems. A system with initial geometry $\overline{w}$ is reciprocal if $\overline{w}$ satisfies the equation

$$D\Delta\Delta\overline{w} = -\frac{1}{2}(q_1 + q_2); \text{ in } \Omega, \tag{2.5}$$

together with the conditions (2.4). Conditions (2.4) and (2.5) are necessary and sufficient for a system to be reciprocal. If the geometry $\overline{w}$ satisfies (2.4) and (2.5), the complete set of solutions of equations (1.74) and (1.76) for the given boundary conditions and lateral forces consists of pairs of reciprocal equilibrium states defined by relations (2.1). q_1 and q_2 are the acting distributed transverse forces that generate the states w_1 and w_2, respectively.

That conditions (2.4) and (2.5) must be necessary can be shown as follows. Keeping in mind relations (2.1) and the circumstance that w_1, F_1 and w_2, F_2 are solutions of system (1.74), (1.76) for q_1 and q_2, respectively, relations (2.6) can be obtained by a simple addition and subtraction of the corresponding equations.

$$D\Delta\Delta(w_1 + w_2) = q_1 + q_2; \quad \Delta\Delta(F_1 - F_2) = 0. \tag{2.6}$$

Both equations (2.6) are linear and elliptical, and they have unique solutions for conditions (2.4) [see e.g., Courant & Hilbert, 1962] that are almost always satisfied for real shells. If we have two solutions (w_1, F_1) and (w_2, F_2), for which the boundary conditions on F_i are identical, then, from the second equation (2.6), we get $F_1 - F_2 \equiv 0$, which satisfies the second relation (2.1). With regard to the first equation of (2.6), one can find its solution $w_1 + w_2$, which satisfies condition (2.4), and which is denoted by $-2\overline{w}$.

The sufficiency of conditions (2.4) and (2.5) can be proved in the same manner. Let w_1, F_1 and w_2, F_2 be two solutions of basic equations (1.74) and (1.76) generated by transverse forces q_1 and q_2, which satisfy conditions of type (2.2) for w_i and the boundary conditions for F_i. Let $\overline{w}$ be a solution of (2.5), which satisfies condition (2.4). We formally complete the equations (2.7) using equations (1.74), (1.76), and (2.5).

$$D\Delta\Delta(w_1 + w_2 + 2\overline{w}) = A(w_1, \overline{w}, F_1, q_1) + A(w_2, \overline{w}, F_2, q_2) - \frac{1}{2}(q_1 + q_2);$$

$$r\Delta\Delta(F_1 - F_2) = B(w_1, \overline{w})Eh, \; - B(w_2, \overline{w})Eh, \tag{2.7}$$

where A and B are the right-hand expressions of (1.74) and (1.76), respectively. It is now easy to verify that the relations $w_1 + w_2 - 2\overline{w} = 0$ and $F_1 - F_2 = 0$ satisfy the system (2.7), if $\overline{w}$ satisfies equation (2.5) and corresponding conditions (2.4). This can be done by direct computation. It has thus been shown that conditions (2.4) and (2.5) are indeed necessary and sufficient for the existence of reciprocal symmetric states.

In order to obtain a pair of such states, the following method is recommended. Let the given function $\overline{w}$ satisfy conditions (2.4) and let w_1, F_1 be a solution of basic equations (1.74), and (1.76) generated by the forces q_1 and subject to conditions (2.2). Then the corresponding reciprocal solution, relative to the considered state w_2, F_2 is completely defined by relations (2.1). The corresponding loading q_2 that generates this solution can be obtained using equation (2.5) as follows:

$$q_2 = -2\Delta\Delta\overline{w} - q_1. \tag{2.8}$$

From (2.8), it can be seen that q_1 and q_2 are generally of different form. However, we are mainly interested here in the case when all the solutions for the given shells, boundary conditions and types of forces consist of reciprocal symmetric pairs; that is q_1 and q_2 differ only by a constant factor. In that case the load q is defined by $q(r, \varphi) = q_0\lambda(r, \varphi)$, where $\lambda(r, \varphi)$ is a given function, and only the value of the intensity parameter q_0 defines the corresponding loading. The basic equation (2.5) for $\overline{w}$ then contains the term $-(q_{01} + q_{02})\lambda(r, \varphi)/2$ in the right-hand side; consequently, the solution $\overline{w}$ of (2.5) will be of the form $\overline{w}(r, \varphi) = (q_{01} + q_{02})f(r, \varphi)$.

Now, let the rise H of the cap be defined by the coordinates $r = r_H$, $\varphi = \varphi_H$, so that $H = \overline{w}(r_H, \varphi_H)$. We then have

$$q_{01}+q_{02} = H/f(r_H, \varphi_H). \tag{2.9}$$

Thus, each pair of reciprocal states is generated by forces with their parameters prescribed by (2.9). It is of interest to note that equation (2.5) is the well-known Sophie Germain-Lagrange equation for the bending of linear plates subjected to transverse loading $-(q_1 + q_2)/2$.

We shall thus use the term *reciprocal system* for a shell, the equilibrium states of which consist of pairs of reciprocal states. According to this definition, a structure can be a reciprocal system under the established conditions, but ceases to be a reciprocal system for other transverse forces or edge supports. We shall now proceed with a discussion of some characteristic examples.

Assume, for example, that $q \equiv 0$ and that the initial shape of the shell is a surface of revolution with a hinged edge. The reciprocal condition (2.5) and the boundary conditions for $\overline{w}$ are then given by

$$\Delta\Delta\overline{w} = 0, \tag{2.10}$$

$$\overline{w}\big|_C = 0; \quad M_r\big|_C = M, \tag{2.11}$$

where index C denotes the boundary and M stands for the externally applied edge moment. At the pole ($r = 0$), the standard conditions (1.87) remain valid; namely,

$$\frac{\partial \overline{w}}{\partial r}(0,\phi) = 0 = \frac{\partial \overline{w}}{\partial \phi}(0,\phi).$$

Keeping only the periodic terms of ϕ, the general solution of equation (2.10) (Michell's solution of the plane problem of the theory of elasticity in polar coordinates) is given by (2.12) [see, e.g., Timoshenko & Goodier, 1970].

$$\overline{w}(r,\phi) = a_0 \ln r + b_0 r^2 + c_0 r^2 \ln r + d_0 + \left(a_1 r + b_1 r^3 + A_1 r^{-1} + B_1 r \ln r\right)\cos\phi$$

$$+\left(c_1 r + d_1 r^3 + C_1 r^{-1} + D_1 r \ln r\right)\sin\phi + \sum_{n=2}^{\infty}\left(a_n r^n + b_n r^{n+2} + A_n r^{-n} + B_n r^{-n+2}\right)$$

$$\cos n\phi + \sum_{n=2}^{\infty}\left(c_n r^n + d_n r^{n+2} + C_n r^{-n} + D_n r^{-n+2}\right)\sin n\phi. \qquad (2.12)$$

The solution can only be periodic in terms in ϕ because the shell is a shell of revolution. The solution given in Timoshenko and Goodier [1970] contains some non-periodic terms, while other terms were used in Biezeno and Grammel [1939] and Filonenko-Boroditch [1965]. The coefficients in (2.12) are still arbitrary, but they can be found by satisfying the boundary conditions. The conditions (1.87) on $\overline{w}$ at $r = 0$ demand

$$a_0 = a_1 = c_1 = A_1 = B_1 = C_1 = D_1 = A_n = B_n = C_n = D_n = 0; \quad (n = 2,3,...) \qquad (2.13)$$

and

$$c_0 = 0. \qquad (2.14)$$

Relation (2.13) is necessary to ensure the boundedness of the second derivatives of $\overline{w}$ at the pole of the cap. The first condition (2.11) at the boundary $r = a$, and equations (2.13) and (2.14) yield:

$$b_1 = d_1 = 0;\ b_0 a^2 + d_0 = 0;\ a_n + b_n a^2 = 0;\ c_n + d_n a^2 = 0; \quad (n = 2,3,...). \qquad (2.15)$$

When the load q is absent, the edge $r = a$ is hinged and immovable in the z-direction, and when the geometry $r = a$ is continuous up to at least the second order derivatives, there is an infinitely large family of reciprocal systems defined by

$$\overline{w}(r,\phi) = \left(r^2 - a^2\right)\left[b_0 + \sum_{n=2}^{\infty} r^n (b_n \cos n\phi + d_n \sin n\phi)\right] \tag{2.16}$$

The particular case in which the sum in the right-hand side of (2.16) disappears defines a shallow spherical cap without a central hole. The corresponding function $\overline{w}$ is given by (1.67) in dimensionless parameters. In real (geometrical) parameters, we have:

$$\overline{w}(r) = -b_0\left(a^2 - r^2\right) = -H\left(1 - \frac{r^2}{a^2}\right), \tag{2.17}$$

where $H > 0$ denotes the rise of the apex of the cap. It should be remembered that $\overline{w} < 0$ for the convex cap, according to the sign convention.

Let us now assume that q = uniform $\neq 0$, the edge $r = a$ is clamped, and the geometry $\overline{w}$ is axisymmetric. In this case, $\overline{w}$ is a solution of the non-homogeneous equation (2.5), which corresponds to the boundary conditions

$$\overline{w}\big|_C \equiv \overline{w}'\big|_C \equiv 0. \tag{2.18}$$

The solution is:

$$\overline{w}(r) = \frac{\alpha}{32D}\left(a^2 - r^2\right)^2, \tag{2.19}$$

where $\alpha = -(q_1 + q_2)/2$. Thus,

$$\overline{w}(0) = -H = \frac{\alpha a^4}{32D}, \tag{2.20}$$

and

$$q_1 + q_2 = \frac{64DH}{a^4}. \tag{2.21}$$

Formula (2.21) defines the forces that generate the reciprocal pair of equilibrium states. A similar relation which connects the loading parameters M_1 and M_2 is obtained in the next section. Thus when $q \equiv 0$, the geometry of the shell $\overline{w}(r)$ is axisymmetric, as in (2.17), or nonsymmetric, as expressed by (2.16). When $q \neq 0$, the corresponding shell can only be a shell of revolution (2.19).

We are now interested in the simplest version of a reciprocal system; i.e., when the shell has a spherical shape (see (2.17)).

2.3 Properties of reciprocal systems: Shallow shells.

As was shown in the preceding section, the shallow spherical cap is a reciprocal system when its edge is hinged, and it is loaded only by an edge moment M. In the sequel we will use the equations for axisymmetric and nonsymmetric deformations of shallow spherical shells given in Sections 1.2 and 1.3, respectively. The reciprocal systems are considered for spherical isotropic caps of uniform rigidity whereby we disregard the possibility of thermal loading. We will investigate the main properties of such systems below, but we will not consider shells of other shapes or shells under other conditions; these more general cases were studied by Shilkrut [1967, 1969a,b, 1974] and in other papers by the author.

At first sight, it seems that a reciprocal system based on the spherical shape corresponds to a very special case. However, we will show later that this system is a starting point for the investigation of a wider class of shells that is of considerable theoretical and practical importance.

Property 1. The complete set of possible equilibrium states of a reciprocal system consists of pairs of reciprocal states (see (2.1), (2.17), (1.78)):

$$w_1 + w_2 \equiv 2H\left(1-\frac{r^2}{a^2}\right); \quad (0 \le r \le a); \quad (0 \le \phi \le 2\pi); \tag{2.22}$$

$$F_1 \equiv F_2 \; . \tag{2.23}$$

Because reciprocal states share the same stress functions F, all the parameters determined by the membrane part of the deformation of the shell are identical for both reciprocal symmetric states; that is,

$$\varepsilon_{r_1} \equiv \varepsilon_{r_2}; \; \varepsilon_{\phi_1} \equiv \varepsilon_{\phi_2}; \; \gamma_1 \equiv \gamma_2; \; N_{r_1} \equiv N_{r_2}; \; N_{\phi_1} \equiv N_{\phi_2}; \; N_{r\phi_1} \equiv N_{r\phi_2};$$

$$u_1 \equiv u_2; \; v_1 \equiv v_2; \; U_{m_1} \equiv U_{m_2}, \tag{2.24}$$

where U_m is the membrane part of the potential strain energy, which can be computed by means of the formula (2.25) [see, e.g., Brush & Almroth, 1975].

$$U_m = \frac{Eh}{2\left(1-\nu^2\right)} \int_0^a \int_0^{2\pi} \left[\varepsilon_r^2 + \varepsilon_\phi^2 + 2\nu\varepsilon_r\varepsilon_\phi + \frac{1}{2}(1-\nu)\gamma^2\right] r d\phi dr . \tag{2.25}$$

This expression is valid for a thin shell, with linearly elastic material behavior, for which the Kirchhoff hypotheses apply.

Additional relations between the parameters of the two reciprocal states follow from equations (2.22), (1.69), (1.71), and (1.72).

$$\chi_{r_1} + \chi_{r_2} = \chi_{\phi_1} + \chi_{\phi_2} \equiv 4H/a^2; \quad \chi_{r\phi_1} + \chi_{r\phi_2} \equiv 0;$$

$$\ddot{w}_1 + \ddot{w}_2 \equiv 0; \quad M_{r_1} + M_{r_2} = M_{\phi_1} + M_{\phi_2} \equiv D(1+\nu)4H/a^2;$$

$$M_{r\phi_1} + M_{r\phi_2} \equiv 0; \quad Q_{r_1} + Q_{r_2} = Q_{\phi_1} + Q_{\phi_2} \equiv 0; \tag{2.26}$$

$$U_{b_1} = U_{b_2} + \frac{4(1+\nu)HD}{a^2}\left[4\pi H - \int_0^a \int_0^{2\pi} \left(\chi_{r_1} + \chi_{\phi_1}\right) r dr d\phi\right], \tag{2.27}$$

where the bending part of the strain energy is computed by (2.28) [see Brush & Almroth, 1975].

$$U_b = \frac{D}{2}\int_0^a \int_0^{2\pi} \left[\chi_r^2 + \chi_\phi^2 + 2\nu\chi_r\chi_\phi + 2(1-\nu)\chi_{r\phi}^2\right] r d\phi dr. \tag{2.28}$$

The boundary conditions expressed in terms of F must be identical for both members of a reciprocal pair of modes, as mentioned above. Relations on the edge C connected with the boundary bending moments follow from (2.26).

$$M_{r_1} + M_{r_2}\big|_C \equiv 4HD(1+\nu)/a^2. \tag{2.29}$$

Condition (2.29) is valid for general deformations of spherical caps; that is, for deformations caused by nonsymmetric load systems or for bifurcation into nonsymmetric deformations of a cap under symmetric loading conditions. In the specific case of symmetric deformations, the relations (2.22)-(2.29) can be reduced to:

$$w_1 + w_2 \equiv -2\overline{w} \equiv -2\overline{\xi}\left(1-\rho^2\right); \quad \theta_1 + \theta_2 \equiv -2\overline{\theta} \equiv -4\overline{\xi}\rho \text{ (see (1.67))};$$

$$\omega_1 \equiv \omega_2; \quad M_{r_1} + M_{r_2} \equiv M_{\phi_1} + M_{\phi_2} \equiv 4(1+\nu)\overline{\xi}; \text{ (see (1.62))};$$

$$\xi_1 + \xi_2 = -2\overline{\xi}; \text{ (see (1.64))}. \tag{2.30}$$

using the dimensionless quantities, which were defined in Section 1.2 for an isotropic shallow spherical cap with uniform rigidity in the absence of thermal loads. Note that q_t is the same for both states and that ξ denotes the dimensionless deflection of the cap's apex.

For the strain energy in dimensionless form, the corresponding relations are:

$$U_{m_1} = U_{m_2};\ U_{b_2} = U_{b_1} + \frac{2\pi\bar{\xi}}{3(1-\nu)}\left[\int_0^1 \left(\theta_1' + \frac{\theta_1}{\rho}\right)\rho d\rho + 2\bar{\xi}\right]$$

$$= U_{b_1} + \frac{2\pi\bar{\xi}}{3(1-\nu)}\left[\theta_1(1) + 2\bar{\xi}\right]. \tag{2.31}$$

where

$$U_m = \frac{U_m^* a^2}{Eh^5} = \pi\int_0^1 \left[\left(\frac{\omega}{\rho}\right)^2 + \omega'^2 - 2\nu\frac{\omega'\omega}{\rho}\right]\rho d\rho; \tag{2.32}$$

$$U_b = \frac{U_b^* a^2}{Eh^5} = \frac{\pi}{12\left(1-\nu^2\right)}\int_0^1 \left[\theta'^2 + \left(\frac{\theta}{\rho}\right)^2 + 2\nu\frac{\theta'\theta}{\rho}\right]\rho d\rho. \tag{2.33}$$

The corresponding boundary relation (2.34) for the loading M follows from (2.29) and (2.30).

$$M_{r_1} + M_{r_2}\big|_{\rho=1} \equiv -4(1+\nu)\bar{\xi}. \tag{2.34}$$

Expression (2.34), like (2.29), is also valid for general states of deformation of spherical caps. In order to ensure the symmetry of the deformation states, it is necessary (but not sufficient) that $M_r\big|_C$ be uniform along the edge. Of course, in general, $M\big|_C$ is variable along the edge, but here we shall only consider the case where M is uniformly distributed. Nonsymmetric deformation states will be presented as a special case.

For convenience, we recapitulate the basic equations (1.58) and (1.59) which describe the axisymmetric state of deformation for an isotropic cap of uniform rigidity in the absence of thermal loading

$$\mathrm{L}(\omega) = -\frac{1}{2\rho}\left(\theta^2 + 2\theta\bar{\theta}\right) - (2+\nu)q_t - \rho q_t' \ . \tag{2.35}$$

This follows immediately from (2.38). The deformation curves presented in Figures 2.7-2.9, 2.10 (only for $q = 0$), and 2.14, 2.15 are skew-symmetric. In the latter two, the corresponding straight lines (2.42) are also drawn.

Property 3. When $q \equiv q_t = P = 0$, reciprocal systems possess the following two, exact solutions:

1) $$\theta(\rho) \equiv -\bar{\theta}(\rho) = -2\bar{\xi}\rho; \quad \omega(\rho) = A\rho + \frac{1}{4}\bar{\xi}^2\rho^3; \quad M_r(\rho) \equiv M_r(1) = -2\bar{\xi}(1+\nu), \tag{2.43}$$

2) $$\theta(\rho) \equiv -2\bar{\theta}(\rho) = -4\bar{\xi}\rho; \ \omega(\rho) \equiv 0; \quad M_r(\rho) \equiv M_r(1) = -4\bar{\xi}(1+\nu). \tag{2.44}$$

where the constant A is determined by the boundary condition for ω. The boundary conditions for ω must be homogeneous of the type

$$\alpha\omega'(1) + \beta\omega(1) = 0. \tag{2.45}$$

Boundary condition (2.45) covers immovable, free, or flexible edge supports (in the radial direction) where radial, externally applied, edge forces are absent. The specific type of support is determined by values of the constants α and β. To verify this solution, one only has to substitute relations (2.43) and (2.44) into equations (2.35) and (2.36) setting $q \equiv q_t = P = 0$.

Solution (2.43) describes the case where the deformation turns the cap into a flat circular plate which fits into the base of the cap. Here, both reciprocal states coincide $\left(\theta_1 \equiv \theta_2 \equiv -\bar{\theta}\right)$, and they belong to the center of symmetry C of the deformation curve. Solution (2.44) describes the deformed state, which corresponds to the initial shape of the cap turned inside-out. This is a pure bending state (in the sense that the membrane stresses are zero), the image of which is given by point 0_2 in Figure 2.19. This configuration, together with the initial (undeformed) shape (point 0 in Fig. 2.19), constitutes a couple of reciprocal states. It is easy to prove that only a reciprocal system is able to acquire the shape of the type described by formulas (2.44).

Property 4. This is a property of an axisymmetrically deformed reciprocal system which exhibits snap-through type of behavior (see Section 2.1). When $\left|\bar{\xi}\right|$ is small enough, snapping cannot occur. In that case, the equilibrium branches are monotonically rising, and there is then only a single state of equilibrium for each value of the loading parameters. Beginning from some value of $|\bar{\xi}| = |\bar{\xi}_{00}|$, the equilibrium branches stop being monotonic and extremal points appear along them (see Section 2.1.b). $|\bar{\xi}_{00}|$ is the value of $|\bar{\xi}|$ at which the solution branch begins to contain points with a horizontal tangent. The first point with a horizontal slope at $|\bar{\xi}_{00}|$ is an inflexion point at the center of symmetry C.

C is a point of inflection for each other deformation curve that passes through it, where the angle ϕ of this point between the tangent to the curve and the horizontal axis depends on the parameter $\bar{\xi}$. When $\bar{\xi} = 0$ (plate), the angle ϕ coincides with the slope of the straight line $M = K\xi$, which represents the solution of the classical linear plate theory. The center C is then located at the origin in the (ξ, M) plane. When $|\bar{\xi}|$ increases, the center C slides along the line (2.42) and the angle ϕ decreases, so that, when $|\bar{\xi}|$ reaches the value $|\bar{\xi}_{00}|$, the angle ϕ becomes zero. When $|\bar{\xi}|$ is larger than $|\bar{\xi}_{00}|$ and continues to increase, the angle ϕ changes and the described phenomenon is repeated for other values of $|\bar{\xi}|$. (The angle ϕ is positive when it is measured from the horizontal axis ξ in a counter-clockwise fashion, as, for example, is done in Fig. 2.7.)

As will be shown, there is an infinite discrete spectrum $|\bar{\xi}_{0i}|$ ($i = 0,1,2...$) of values $|\bar{\xi}|$, where the angle ϕ becomes zero at the point of inflection C. The parameter $|\bar{\xi}_{00}|$ is the first (lowest) member of this spectrum. At $|\bar{\xi}| = |\bar{\xi}_{00}|$, this inflection point unfolds into maxima and minima. Thus, in the interval $|\bar{\xi}_{00}| < |\bar{\xi}| < |\bar{\xi}_{01}|$, the characteristic curves possess only one maximum and one minimum. An additional pair of extremal points exists when $|\bar{\xi}_{01}| < |\bar{\xi}| < |\bar{\xi}_{02}|$, and so on. The dynamics of the variation of ϕ when $|\bar{\xi}|$ varies is clearly illustrated by the plots in the figures presented above. Figure 2.1 shows that the deformation curves are monotonically rising for very small $|\bar{\xi}|$, where $\phi > 0$ at inflection points, while, at $|\bar{\xi}| \cong 1.5$, the angle reaches $\phi = 0$. Immediately after this (when $|\bar{\xi}|$ increases), there are three solutions in the region of the inflection point of the corresponding curve. Although Figure 2.1 describes the case of a clamped spherical cap subjected to a pressure load q that is not a reciprocal system, its behavior for small $|\bar{\xi}|$ is qualitatively the same as that of a reciprocal system [Shilkrut 1969b, 1980].

We proceed now to results presented in Figure 2.15, which represent the behavior of immovable hinged caps subjected to an uniformly distributed edge moment M (an example of a reciprocal system). These plots show that, at $|\bar{\xi}| = 1$, there are three solutions for $\phi < 0$. This means that $|\bar{\xi}_{00}| < 1$, since, at $|\bar{\xi}_{00}|$, the angle $\phi = 0$. At $|\bar{\xi}| = 2$, there are five solutions for $M = M_{0C}$ and $\phi > 0$, so that we conclude that $|\bar{\xi}_{01}| < 2$. (These observations are confirmed by the numerical results given above in Tbl. 2.4.). When $|\bar{\xi}| = 4$, data from Figure 2.15 show again that $\phi < 0$; at that level, there are nine solutions at $M = M_{0C}$; so we deduce $|\bar{\xi}_{03}| > 4$.

Similar results can be obtained for moveable hinged caps (see Fig. 2.9). Taking into account the skew-symmetry of the deformation curve, and assuming that $M_0^+ \geq M_0^-$, it is possible to establish that the ordinate M_{0C} of the center of symmetry C is the lower (upper) boundary of the upper (lower) critical loading (see (2.41)).

$$M_0^- \leq M_{0_C} = -2(1+\nu)\bar{\xi} \leq M_0^+ . \qquad (2.46)$$

These inequalities are valid for all $|\bar{\xi}|$ greater than $|\bar{\xi}_{00}|$, independent of the number of pairs of extremal points on the deformation curve. This conclusion is also reached in the investigation of the stability of the equilibrium states, which will be discussed in detail in Chapter 6. The existence of the spectrum $|\bar{\xi}_{0i}|$, with its underlying phenomena and inequality (2.46), constitutes Property 4.

With the help of Figure 2.19, we can draw an important conclusion from this property. During the loss of stability by snapping, the shell passes from an unstable state to a new stable state of a reduced level of potential energy. The snapping process frees a large amount of the membrane energy U_m (see (2.31) and (2.32)). But the magnitude of U_{m2} of the post-snapping state is equal to the U_{m1} of its reciprocal state (point 1), the loading level of which is less than M_0^+; that is, $M_1 < M_{0C} < M_0^+$. Consequently, U_{m1} is less than U_m in the pre-snapping state (point A) at $M_0 = M_0^+$.

Thus, the transition from state A to state 2 takes place with the release of an essential part of U_m. This is particularly clear in the special case for which $M_0^+ = 2M_{0C}$. Point 1, then, coincides with the origin, where $M_1 = 0$, so that $U_{m2} = U_{m1} = 0$ and the energy release would then be equal to U_m in point A.

A similar release occurs during the process of snap-back from point B to point 1. The qualitative picture sketched above is also valid for more complicated shapes of deformation curves, independent of the number of pairs of extremal points or of the existence of isolas (closed loops). Some of these complicated deformation curves are given in Chapter 3, where specific numerical results are discussed.

We can now see how important the role of the membrane stresses is in the process of shell-buckling. This role is independent of the question of whether the system is reciprocal or not. Interesting results in this connection are given by Volmir [1963] for the specific case of an immovable, clamped, shallow cap (with $\bar{\xi} = -4$) subject to external pressure q. At the first limit point, the ratio $U_m/(U_m + U_b)$ is equal to 90%, while at the post-snapping point, it is only about 10%. A series of such numerical data were produced by Valishvili [1976] and are presented in Table 2.3. In each entry of this table, the ratio of U_m/U_b is given at the limit points of spherical caps subjected to external uniform pressure.

Table 2.3
$\nu = 0.3$

$\lvert\xi\rvert$	2.424	5.454	9.696	15.150	21.816
Movable hinge	1.20	1.63	1.59	1.59	1.59
Immovable hinge	6.23	3.45	3.89	8.66	2.98
Movable clamping	1.88*	1.56	1.57	1.57	
Immovable clamping	2.63	4.01	3.94	13.19	9.27

* no buckling

Determination of the spectrum $\lvert\bar{\xi}_{0i}\rvert$. The values $\lvert\bar{\xi}_{0i}\rvert$ are the values of $\lvert\xi\rvert$ at which the deformation curve has a horizontal tangent ($\phi = 0$) at the center of symmetry C. At this point C, the governing equations (2.35) and (2.36) possess the solution (2.43). To make use of the condition $\phi_C = 0$, it is necessary to be able to determine $\phi_C = \dfrac{dM_{0C}}{d\xi}$ at the center C. To do this, we consider an increment δM_0 of M_{0C} without changing the other parameters and boundary conditions. The perturbed state $\theta + \delta\theta$, $\omega + \delta\omega$ is then the solution of the basic equations (2.35) and (2.36) for the boundary moment $M_{0C} + \delta M_0$. Substitution of the perturbed solution in these equations, and taking into account that, at the same time, θ and ω are also solutions of the same equations, we obtain the following relationships for the increments $\delta\theta$ and $\delta\omega$.

$$\mathrm{L}(\delta\omega) = -\frac{1}{2\rho}\left[2\delta\theta\left(\theta+\bar{\theta}\right)+(\delta\theta)^2\right];\quad \mathrm{L}(\delta\theta) = -\frac{m}{\rho}\left[(\omega+\delta\omega)\delta\theta+\left(\theta+\bar{\theta}\right)\delta\omega\right]. \tag{2.47}$$

Because $\delta\theta$ and $\delta\omega$ are taken arbitrarily small, these equations can be linearized. Substituting the expressions (2.43) for θ and ω in (2.47), we find after linearization:

$$\mathrm{L}(\delta\omega) \equiv 0;\quad L(\delta\theta) - m\left[A + \frac{1}{4}\bar{\xi}^2\rho^2\right]\delta\theta = 0;\quad m = 12\left(1-\nu^2\right). \tag{2.48}$$

The constant $A = \omega'(0)$ can be obtained from the boundary condition of ω (for example, $\omega(1) = 0$). The boundary conditions of $\delta\omega$ are homogeneous and, because of this, the solution of the first equation (2.48) is $\delta\omega \equiv 0$.

The boundary conditions for $\delta\theta$ are given by:

$$\delta\theta(0) = 0;\quad \delta\theta'(1) + \nu\delta\theta(1) = \delta M_{0C}, \tag{2.49}$$

because the boundary conditions for θ are $\theta(0) = 0$; $\theta'(1) + \nu\theta(1) = M$, in conformity with

Table 1.2 (hinged support) for $K = 1$, $d_r = 1$, and $M_T = 0$. The solution $\delta\theta$ of (2.48) can be found in the form of a series.

$$\delta\theta(\rho) = \sum_{n=1}^{\infty} \alpha_{2n-1}\rho^{2n-1}; \quad (0 \le \rho \le 1). \tag{2.50}$$

It follows from the physics of the problem that $\delta\theta$, as θ, must be an odd function of ρ [Shilkrut, 1974, 1983b]. Thus,

$$\left[(2n-1)^2 - 1\right]\alpha_{2n-1} - S\left[\gamma\alpha_{2n-3} + \frac{1}{4}\alpha_{2n-5}\right] = 0; \quad (n = 1,2,3,...), \tag{2.51}$$

where $S = m\overline{\xi}^2$; $\gamma = \omega'(0)/\overline{\xi}^2$, and $\alpha_i = 0$ if $i < 0$. Satisfying conditions (2.49) and evaluating $\delta\xi = \int_0^1 \delta\theta d\rho$, we can write for the unknown angle ϕ_C:

$$\tan\phi_C = \frac{\delta M_{0C}}{\delta\xi} = \frac{\sum_{n=1}^{\infty}[2n-1+\nu]\alpha_{2n-1}}{\sum_{n=1}^{\infty}\frac{1}{2n}\alpha_{2n-1}}. \tag{2.52}$$

Thus, the condition $\phi_C = 0$ yields

$$\sum_{n=1}^{\infty}[2n-1+\nu]\alpha_{2n-1} = 0. \tag{2.53}$$

Equation (2.53) is a solution of the eigenvalue problem of equation (2.48) for conditions (2.49). Tricomi [1961] has shown that there exists an infinite discrete spectrum of these eigenvalues. Thus, the transcendental equation (2.53) has an infinite number of roots, which offers the possibility of obtaining the spectrum $|\xi_{oj}|(j = 0,1,...)$. In order to find these roots, it is convenient to express all α_j (j = 3,5,...) in terms of α_1 by means of the recurrence relations (2.51). Thus,

$$\alpha_j = K_j\alpha_1; \quad (j = 1,3,5,...), \tag{2.54}$$

where K_j are known constants which depend on $\overline{\xi}^2$. For example,

$$K_1 = 1; \quad K_3 = S\gamma/8; \quad K_5 = S\left(\frac{S\gamma^2}{8} + \frac{1}{4}\right)/24;... \tag{2.55}$$

Then, substituting α_j from (2.54) into (2.53), we obtain the equation

$$\alpha_1 \sum_{n=1}^{\infty} [(2n-1)+\nu] K_{2n-1} = 0; \quad \alpha_1 \neq 0 \tag{2.56}$$

for the spectrum $|\bar{\xi}_{0n}|(n = 0,1,2,...)$; α_1 remains undetermined because of the homogeneity of the problem.

Table 2.4 gives the first two members $(|\bar{\xi}_{00}|, |\bar{\xi}_{01}|)$ of the spectrum for two types of hinged supports of spherical caps. They were calculated for $\nu = 0.3$ by solving (2.56) with about a thousand terms of the series (2.56). (A thousand terms were more than sufficient for an accurate numerical solution.)

Table 2.4

$\nu = 0.3$

	$\lvert\bar{\xi}_{00}\rvert$	$\lvert\bar{\xi}_{01}\rvert$
Immovable hinge	0.69	1.77
Movable hinge when boundary membrane force $N_r(1) = 0$	1.92	4.40

These results, and other, similar data for a number of reciprocal systems of different shapes, are given in [Shilkrut, 1967, 1974, 1980; Shilkrut & Gerlaku, 1969]. For example, for the combination of uniform pressure and a clamped edge, the corresponding reciprocal system is described by the initial shape $\bar{\theta}(\rho) = 4\bar{\xi}(\rho - \rho^3)$ and the corresponding values of $|\bar{\xi}_{00}|, |\bar{\xi}_{01}|$ are: $|\bar{\xi}_{00}|$ = 1.35, $|\bar{\xi}_{01}|$ = 2.47 for immovably clamped, and $|\bar{\xi}_{00}|$ = 2.24, $|\bar{\xi}_{01}|$ = 4.19 for movably clamped. If $N_r(1) \neq 0$, the parameter $|\bar{\xi}_{0i}|$ takes other values. Note that the effect of $N_r(1) \neq 0$ on $|\bar{\xi}_{0i}|$ is introduced through $\omega'(0)$.

There is a very important qualitative conclusion that can be deduced from these results. The spectrum $|\bar{\xi}_{0j}|$ is infinite, as was shown above. Each member of the spectrum $|\bar{\xi}_{0j}|$ introduces an additional pair of solutions, and when $|\bar{\xi}_0|$ goes to infinity, the number of solutions of the basic equations also goes to infinity. In other words, when the height of the shell increases (the cap becomes deeper), or when the thickness of the shell body decreases, the number of equilibrium states for a given value of M increases beyond bounds; i.e., the degree of non-uniqueness of the corresponding nonlinear boundary value problem increases. Note that it is better to connect this phenomenon with a decrease of the wall thickness of the cap rather than with an increase of the depth of the shell; this in view of the shallow shell assumptions.

The same behavior, in a qualitative sense, is also prevalent with general non-shallow shells, as will be shown in Chapter 5. Consequently, the existence of an infinite spectrum $|\xi_{0j}|$ presents an argument for the ever-increasing number of possible equilibrium states at a fixed value of the loading parameter. We have given here a mathematical explanation of this very important property of the solutions of geometrically nonlinear shells. The physical explanation of this property will be discussed in Section 2.11.

Property 5. Property 2 remains valid for the case of nonsymmetric deformation states of reciprocal systems — deformation states that can occur under symmetric loading conditions and which are triggered by a bifurcation (buckling) mechanism (see the corresponding part of the Section 2.1). (Some recent investigations of these phenomena are given in Shilkrut and Kochavi [1986], and many specific cases of this type of deformations are discussed in Chapters 3 and 4.) These buckling states are (loosely) called nonsymmetric (to contrast them with the axisymmetric states), but they are actually symmetric with respect to certain rotations (smaller than 2π) around the axis of the cap. From each bifurcation point, one or more paths that correspond to these nonsymmetric deformation states branch out from the pre-buckling path of symmetric states. (This phenomenon is depicted qualitatively in Figure 2.2.) According to Property 5, the distribution of the bifurcation points and its branches are reciprocal relative to the same center of symmetry C of the characteristic curves of axisymmetric states described by Property 2. The reciprocity property holds for bifurcations in compression as well as in tension.

The geometrical features mentioned above will be studied in detail in Chapters 3 and 4, where it will also be shown that Property 5 is of crucial importance for gaining an understanding of the nature of the loss of stability of axisymmetric equilibrium states when this loss takes place at the bifurcation points along the axisymmetric pre-buckling states.

Property 6. Property 6 deals with the features of the transient behavior of reciprocal systems under a static load. Consider a reciprocal pair of transient states which satisfy the equalities (2.1) at each instant of time. In this case, all functions in (2.1) except the initial geometry $\overline{w}$ also depend on the time t, which means that we must replace equations (2.1) by

$$w_{1,t} + w_{2,t} \equiv w_{1,tt} + w_{2,tt} \equiv 0, \tag{2.57}$$

$$F_{1,t} \equiv F_{2,t}, \tag{2.58}$$

where a comma followed by a t indicates differentiation with respect to t.

It is easy to show that the basic condition (2.5) for the existence of reciprocal systems also remains valid in the dynamic case. The basic equations (1.74) and (1.76) must now be changed by replacing q by $q - sw,_{tt} \equiv q - s(\overline{w} + w),_{tt}$, where s is the mass per unit of area of the middle surface. The boundary conditions, which, together with (2.5), ensure that the system is reciprocal, are the same for both the transient and the static cases (see (2.29) and (2.34)), because they involve edge conditions that do not depend on the time t. It is necessary, however, to add two initial conditions for each member of the reciprocal pair. These initial conditions must agree with the relations of the reciprocal (2.1) (2.57) and (2.58); that is, the initial reciprocal states must satisfy the coupling condition (2.1) and the initial reciprocal velocities must satisfy the first and third relations of (2.57) and (2.58).

Consider two different transient states of shallow, spherical, hinged caps which satisfy (2.34) and which are generated by a boundary moment loading $M_{1|C}$ and $M_{2|C}$, respectively. Let the initial conditions for both states be consistent with (2.1) and (2.57) and (2.58).

$$w_1(\rho,0) + w_2(\rho,0) \equiv -2\overline{w}(\rho); \quad F_1(\rho,0) \equiv F_2(\rho,0);$$

$$w_{1,t}(\rho,0) + w_{2,t}(\rho,0) \equiv 0; \quad F_{1,t}(\rho,0) \equiv F_{2,t}(\rho,0). \tag{2.59}$$

States 1 an 2 are then reciprocal in the above sense; i.e., they are coupled through (2.1) so that all relations and consequences that result from (2.1) follow. Naturally, the terms of all these relations are now also functions of the time t. From a physical point of view, this means that the twin states are symmetric with respect to the base plane (as shown in Fig. 2.18) at each instant of time, and that their corresponding velocity and acceleration distributions are equal but of opposite signs at all times. A consequence of this property is that the proper modes of free vibrations around reciprocal equilibrium states are also reciprocal while they are found at the same natural frequencies. This holds both for axisymmetric and nonsymmetric equilibrium states. The distribution of the natural frequencies and the question whether they are real or imaginary play a very important role in the investigation of the stability of the equilibrium states of the cap, as will be explained in Chapter 6.

Property 7. The circular plate ($\overline{w} \equiv 0$) is always a reciprocal system independent of the type of loading and boundary conditions. For each pair w_1, w_2 of reciprocal states, the following conditions hold. The transverse loads that generate these states are equal but of opposite signs. The external tangential forces (including the forces acting along the edge) are equal for both states. Both these conditions hold at all times.

The first result follows immediately from (2.5), which possesses a trivial solution ($\overline{w} \equiv 0$) when $q_1 + q_2 = 0$. The second follows for axisymmetric deformations (see (2.35),

(2.36)) if $q_{t1} \equiv q_{t2}$, because, for a plate, $\theta_1 + \theta_2 \equiv 0$ (see (2.30)). The center of symmetry C coincides with the original point of the plane (ξ, p), where p denotes any parameter of the external transverse loading.

2.4 Nonreciprocal shallow systems.

In the previous two Sections, we have shown that when thermal stresses are neglected, the shallow, isotropic, elastic spherical cap with a uniform stiffness distribution is a reciprocal system when its edge is hinged and its transverse loading is limited to an edge-bending moment M only. If one of these conditions is violated, the system ceases to be a reciprocal system. But it turns out that the most important properties of reciprocal systems can, in many cases, be extended to non-reciprocal systems. In order to show this, we shall assume that the load system consists of a pressure distribution q together with an uniformly distributed edge moment M. It can easily be verified that, in this case, as in the case of reciprocal systems, all the possible states are composed of pairs of reciprocal modes generated by loading para-meters M_1, q_1 and M_2, q_2, which satisfy the relation (2.29) or (2.34) and

$$q_1 + q_2 \equiv 0. \tag{2.60}$$

However, the transverse loading need not be axisymmetric; that is, q in (2.60), can be expressed in the form $q = q_0\lambda(r, \phi)$, where q_0 is the loading parameter and the function $\lambda(r, \phi)$ describes the distribution of q. The boundary conditions for reciprocal systems remain satisfied in this more general case.

The case considered here is thus quite general, and this is an indication that the features of reciprocal systems can be extended to a wider class of problems. The proof of the properties of this particular case can be given in the same way as the proof of Property 6 for the transient solutions given in the previous section. No restriction on the type of deformations is necessary here, so that these properties are valid for both axisymmetric and nonsymmetric deformations.

The are some important consequences that result from characteristics of the solutions discussed above. We summarize these below.

1) Contrary to a reciprocal system, the deformation curve (M versus ξ) for given value $q_0 \neq 0$ of the system discussed above, is not a skew-symmetric curve (see Property 2, Section 2.3). It does not pass through a center of symmetry C; this occurs only with reciprocal systems (see Property 3 of the reciprocal systems). But according to the property under discussion here, two deformation curves $M(\xi)$, one of which is obtained

for some $q_0 = a \neq 0$ and the second for $q_0 = -a$, are symmetric curves with respect to the center of symmetry C that corresponds to the case of $q_0 = 0$ (see the curves α and α_2 in Fig. 2.20, where, for example, a pair of symmetric modes are indicated by 1′, 2′).

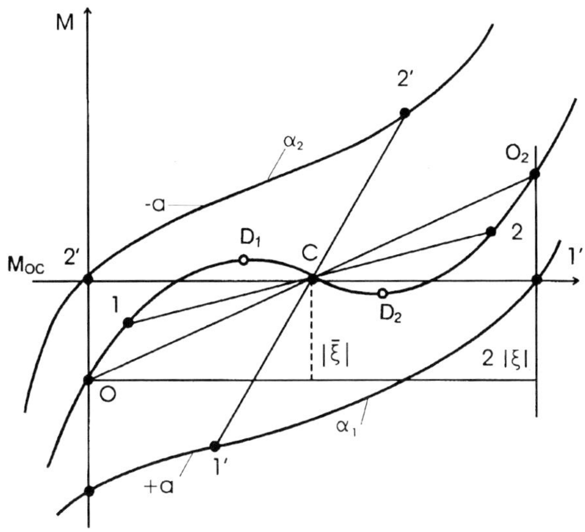

Figure 2.20 Example of a shallow system that is no longer reciprocal.

2) Proceeding from the previous observation, it can be concluded that, from the point of view of stability, the qualitative behavior of the cap for $q_0 > 0$ coincides with the behavior for $q_0 > 0$; that is, if, for example, for a given value of $q_0 = a > 0$, no snapping occurs because the corresponding deformation curve is monotonic, this feature also exists for $q_0 = -a$. In the other case (the deformation curve $M(\xi)$ is not monotonic), the relation (2.41) remains valid, with the difference that one of the two values, M^+, M^-, is now related to $q_0 = a > 0$ and the other to $q_0 = -a$, and vice versa. The same feature is also present when, instead of the pressure loading q, a combination of a pressure load q and a concentrated force P at the pole, or a radially distributed line load P_1, is acting on the shell. An example is given in Figure 2.10, which depicts a family of deformation curves $M(\xi)$, each of which corresponds to a constant value of the pressure q. The edge conditions of the cap correspond here to movable hinges and the apex rise of the cap is given by $|\xi| = 4$. The center C is located along the solution curve for $q = 0$. The set of solution curves consists of pairs of symmetric (relative to C) curves (see, e.g., the pairs of curves for $q_0 = \pm 20$; $q_0 = \pm 40$...).

For the same cap, another similar set of deformation curves is shown in Figure 2.12, where an additional (membrane) boundary force, $N = -5$, is applied. In the last figure, only half of the set of curves is drawn ($q_0 \geq 0$); the other half is symmetric relative to the point $C\left(\xi = -\overline{\xi} = 4;\ M = -2\overline{\xi}(1+\nu) = 10.4\right)$. The case of a movable hinge with $\overline{\xi} = -7$ is shown in Figure 2.21.

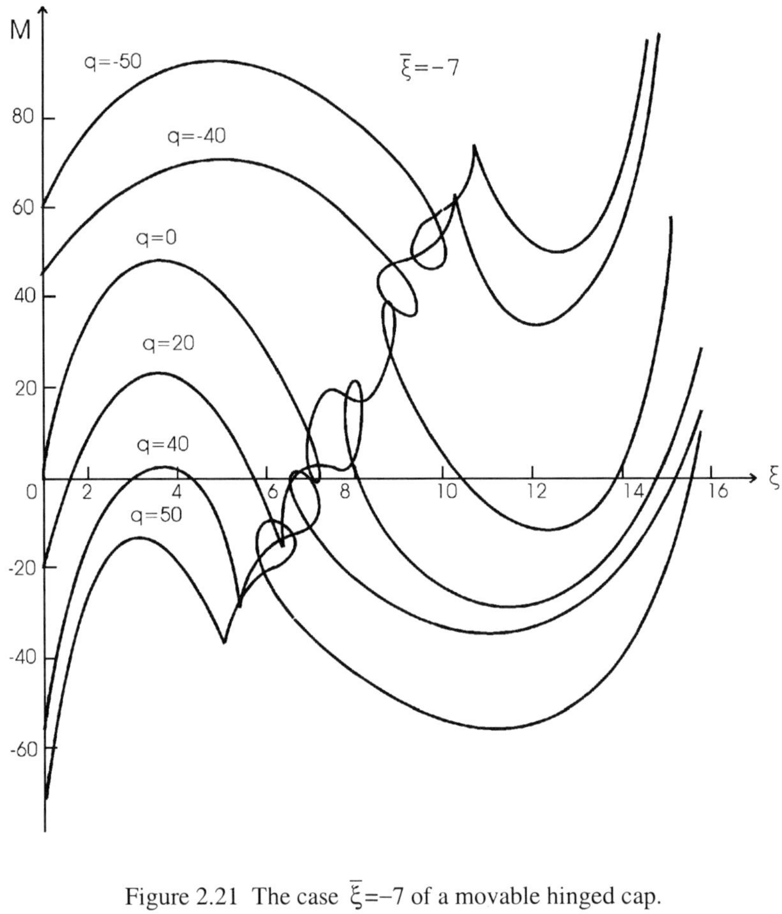

Figure 2.21 The case $\overline{\xi}$=−7 of a movable hinged cap.

Thus, it is possible, to shorten the investigation of the behavior of the cap subjected to a multi-parameter loading by making use of the symmetry properties of the solution curves. This is particularly useful in cases when, in addition to M, the loading q or other transverse forces change their sign. Please note that this feature is not trivial, because $q > 0$ means external pressure, while $q < 0$ corresponds to internal pressure. It is self-evident that these two cases represent a completely different situation.

So far, the plane (ξ, M) was used to display the results, but the same effect can also be achieved by using the plane (ξ, q). This means that the deformation curves will be given by the parameter representation $q = q(\xi)$ where M is kept constant. In this case, keeping in mind the relations (2.60) and the last two expressions (2.30), it can be seen that such a family of curves possesses a center of symmetry S with coordinates $\xi = -\overline{\xi};\ q = 0$.

Thus, a point with the coordinates (q_1, ξ_1) of the equilibrium curve for $M = M_1$ is the mirror image of the point with coordinates $(q_2, \xi_2) = (-q_1, -2\overline{\xi} - \xi_1)$ that belongs to the second equilibrium curve in the sense that the point (q_2, ξ_2) belongs to a straight line through the point (q_1, ξ_1) and S, and is located at a distance measured from S that is equal to the distance between (q_1, ξ_1) and S. An example of this characteristic is given in Figure 2.11, where half of the curves shown in Figure 2.10 were transferred from the (ξ, M) plane for q = const. to plane (ξ, q) for M = const. The second half of Figure 2.11 can be reconstructed using the symmetry relative to S.

This particular property also extends itself to the bifurcation points. This means that the bifurcation points occur in pairs, symmetrically placed relative to the centers of symmetry C or S. The corresponding equilibrium states for q_1 and $-q_1$ (or other loading parameters) are mutually symmetric in the sense described above.

All the properties related to non-reciprocal systems that are described in this section can be extended to a larger class of shells, the load systems of which include tangential and thermal stresses. For example, consider an isotropic shell with uniform rigidity which undergoes axisymmetric thermo-elastic deformations. The governing equations that apply in this special case are derived from the general formulation (1.58) and (1.59) for $K^2 = \lambda^2 = 1;\ \overline{\theta} = 2\overline{\xi}\rho$. It turns out that the features of the solutions discussed above still exist if the following relationships are fulfilled:

$$q_{t1} \equiv q_{t2};\ \frac{\partial T}{\partial \rho} \equiv 0, \tag{2.61}$$

where q_t denotes the distribution of the tangential forces. In this case, the thermal part of the loading is absent in the governing equations, but it is still present in the boundary conditions. In this case, all boundary conditions must be rewritten, including the corresponding thermal terms expressed by M_T and N_T (see Table 1.1 and (1.62), (1.63)). If the temperature distribution T depends on ρ and is an even function of ξ, the condition (2.61) for T can be replaced by the first condition of:

$$T_1(\rho,\xi) \equiv T_2(\rho,\xi);\ T_1(\rho,\xi) \equiv -T_2(\rho,\xi);\ \ (0 \le \rho \le 1);\ \left(-\frac{h}{2} \le \xi \le \frac{h}{2}\right) \tag{2.62}$$

If T is an odd function of ξ, it must satisfy the second equality of (2.62) instead of (2.61). Naturally, in these particular cases, the temperature dependent terms must be included in all the relevant relationships and equations. These cases can describe, for example, the deformation of a hinged spherical cap subjected to forces q, M, or to a temperature field T that varies with ξ. They also describe the case when the cap is loaded by a centrifugal force q_t. The boundary conditions for ω do not influence the features of solutions related to symmetry.

Thus, by generalizing the symmetry concept, practically all cases of hinged spherical caps that are subjected to multi-parameter loads can be analyzed using properties the existence of which can be rigorously proven. The general theory of reciprocal systems, which covers shells of arbitrary shape, were given by Shilkrut [1967, 1969a,b, 1980].

2.5 A class of reciprocal systems based on spherical caps subjected to nonsymmetric loading.

The majority of the reciprocal systems that we examined in the previous sections are axisymmetrically loaded spherical caps. In these cases (as mentioned in section 2.1), nonsymmetric deformations (buckled states) can occur that possess symmetry with respect to certain rotations around the axis of the shell; that is, the buckling modes in terms of $w(r, \phi)$ and $F(r, \phi)$ are periodic functions of ϕ, with a period $2\pi/n$, where $n \geq 1$ is an integer. This is the first group of problems where nonsymmetric deformations can occur. A second group exists where nonsymmetric deformation states arise when the external loading is not axisymmetric.

We shall now see that the properties of reciprocal systems can be generalized to hinged spherical caps loaded by nonsymmetric systems of forces without any restrictions. This generalization can also be carried out when the support conditions are not axisymmetric. Using the governing equations for nonsymmetric deformations of isotropic shallow caps of revolution, which were specified for spherical caps in Section 1.3 (see (2.17)), the following relations can be obtained for pairs of reciprocal solutions (of general shape).

$$w_1(r,\phi) + w_2(r,\phi) = -2\overline{w}(r); \quad u_1 \equiv u_2; \; v_1 \equiv v_2; \; F_1 \equiv F_2. \tag{2.63}$$

$$N_{r1} \equiv N_{r2}; \; N_{\phi 1} \equiv N_{\phi 2}; \; N_{r\phi 1} \equiv N_{r\phi 2}. \tag{2.64}$$

$$\chi_{r1} + \chi_{r2} \equiv \chi_{\phi 1} + \chi_{\phi 2} \equiv 4H / a^2; \; \gamma_1 + \gamma_2 \equiv 0. \tag{2.65}$$

$$M_{r1} + M_{r2} \equiv M_{\phi 1} + M_{\phi 2} \equiv D(1+\nu)4H / a^2; \quad M_{r\phi 1} + M_{r\phi 2} \equiv 0. \tag{2.66}$$

$$Q_{r1} + Q_{r2} \equiv Q_{\phi 1} + Q_{\phi 2} \equiv 0;\ q_1 + q_2 \equiv 0. \tag{2.67}$$

$$U_{m1} \equiv U_{m2};\ \ (0 \le r \le a);\ \ (0 \le \phi \le 2\pi);$$

$$U_{b2} = U_{b1} + \frac{4(1+\nu)HD}{a^2}\left[4\pi H - \int_0^a \int_0^{2\pi} \left(\chi_{r1} + \chi_{\phi 1}\right) r dr d\phi\right]. \tag{2.68}$$

The boundary conditions for the second solution can be derived from the boundary conditions for the first with the help of relations (2.63)-(2.68) and (2.17). Thus, the hinged shallow spherical cap is a generalized reciprocal system for nonsymmetric equilibrium states, and the characteristic properties described earlier also apply in the present case. For instance, if only a boundary moment $M(\phi) = \lambda f(\phi)$ is acting (where λ is the loading parameter, and $f(\phi)$ is a given function that describes the distribution of M along the cap's edge), then the solution is symmetric relative to center C, with coordinates w_C, λ_C

$$w_C = w\left(\bar{r}, \bar{\phi}\right) = -\overline{w}(\bar{r});\ \ \lambda_C = D(1+\nu)2H / a^2 f\left(\bar{\phi}\right), \tag{2.69}$$

where: $\bar{r}$ and $\bar{\phi}$ are the coordinates of an arbitrary point of the interior of the shell (and thus not of the edge); and w denotes the selected displacement to describe the relation $\lambda(w)$. If a nonsymmetric, distributed transverse loading q is added, the corresponding curves for the loads $q_2(r,\phi) \equiv -q_1(r,\phi)$ are symmetric about the same center as in the previous case.

2.6 Some general remarks concerning the properties of reciprocal systems.

Despite the fact that reciprocal systems constitute a specific group of shells with special external conditions, many of their principal properties extend to all geometrically nonlinear shells, at least from a qualitative point of view. We can summarize these observations in the following way:

1) When the rise parameter $|\xi|$ increases, the number of possible equilibrium states n that correspond to a given value of the loading parameter (the degree of non-uniqueness of solutions of corresponding boundary value problem) also increases. In the case of hinged supports, it increases beyond bounds if $|\xi| \to \infty$, as is proven in the theory of reciprocal systems (see Property 4). This phenomenon has a geometric background that can be explained by the so-called "nonregular isometric transformation" of surfaces, whose role in the theory of shells is considered in Section 2.11.

The analysis of the case of a clamped edge follows the same pattern as that of the case of a hinged edge, although the behavior of the degree of non-uniqueness of the solutions n when $|\xi| \to \infty$ has not yet been established. In order to do this, it is necessary to determine the class of initial shapes $\overline{w}$ that lead to reciprocal solutions for the case where the edges are clamped.(2.19)-(2.21). An example of such a case, where the shell is loaded by external pressure, is presented by formulae (2.19)-(2.21).

2) It follows from Property 4 of the reciprocal systems that the loss of stability by snap-through is accompanied by the release of an essential part of the membrane energy that belongs to the prebuckling state. This part is transformed into kinetic energy, which illustrates that, in principle, snapping is a purely transient process. This characteristic clarifies the crucial role the membrane stresses play in the snap-through process of a shell.

3) Property 5 of the reciprocal systems shows that the spectrum of bifurcation points is distributed symmetrically relative to the center C of the deformation curves. Thus, bifurcation points exist along all parts (stable or unstable) of the equilibrium curves. This observation is confirmed by many numerical solutions (see, for example, Section 2.1, and Figs. 2.14 and 2.17).

The search for qualitative properties — which, as a rule, are important because of their general character — continues in the following sections.

2.7 Reduction of some thermo-elastic problems of shallow shells to equivalent elastic problems.

When temperature dependent terms are absent in equation (1.58), and we exclude the special case $M_T \neq C\rho$ (C is an arbitrary constant), each thermo-elastic problem is equivalent to a pure elastic problem whereby the temperature-dependent terms of equation (1.59) can be interpreted as an externally applied transverse load. The boundary conditions of the equivalent elastic problem are given in Table 1.1. Examples follow below:

1) A thermo-elastic isotropic shell ($K = \lambda = 1$) of uniform rigidity, where the temperature distribution T is a function of ξ alone. Then $N_T \equiv$ const., $M_T \equiv$ const., and all the temperature dependent terms disappear from both equations (1.58) and (1.59). The influence of the temperature field on the governing functions ω and θ is thus induced by boundary conditions (see Table 1.1). In this case, the role of the corresponding equivalent loading are the terms explicitly dependent on T present in the boundary conditions.

2) An orthotropic shell with uniform rigidity that is thermally isotropic ($K \neq 1$, $\lambda = 1$); T is an even function of ξ and does not depend upon ρ. Under such conditions, $N_T =$ const., $M_T = 0$, and the case is similar to the preceding one.

3) An orthotropic shell with T as an odd function of ξ. In this case, $N_T \cong 0$; M_T depends on ρ; and if $M_T \neq C\rho$, the temperature dependent terms are equivalent to an externally applied transverse loading, as can be seen in equation (1.59).

2.8 Some general properties of elastic, geometrically nonlinear, axisymmetrically deformed shallow caps of revolution.

We shall begin the discussion of the general features of solutions to equations that describe the deformation and stability of spherical caps for the simplest case; i.e. for an isotropic elastic cap of uniform rigidity that is loaded by uniformly distributed pressure in combination with uniformly distributed edge moments and forces. Although we consider at first only the axisymmetric equilibrium states, the study of these states turns out to be of crucial importance to the understanding of the behavior of geometrically nonlinear shells in general, as we will proceed to show. These features can be revealed by investigating the qualitative properties of solutions of the corresponding boundary value problems formulated in its general integral form.

The governing equations of the specific cases investigated here are (2.35) and (2.36) without the terms containing q_t and P. Each of these equations can be rewritten in the generalized form (2.70).

$$L(\eta) = f(\rho); \quad \rho \subset R, \tag{2.70}$$

where $\eta(\rho)$ denotes the unknown variable; ρ is the argument; L() is the linear operator (2.37), which is, in particular, the operator of the Sophie Germain-Lagrange equation for axisymmetric bending of a linearized annular plate problem; $f(\rho)$ denotes, in a very generalized way, the right-hand side containing the nonlinear terms of the investigated governing equation; and R denotes the interval ($0 \leq \rho \leq 1$). The general solution of equation (2.70) can be expressed in the form:

$$\eta(\rho) = A\rho + \frac{B}{\rho} + \frac{1}{2}\int_0^\rho \left(\rho - \frac{\lambda^2}{\rho}\right) f(\lambda) d\lambda; \quad in\ R. \tag{2.71}$$

The solution of the boundary problem

$$\eta(0) = 0; \;\; \alpha\eta'(1) + \beta\eta(1) = \gamma, \tag{2.72}$$

is given by formulae (2.73) and (2.74) below:

$$A = \eta'(0) = -\frac{1}{2}\int_0^1 f(\lambda)d\lambda + \frac{1}{\alpha+\beta}\left[\gamma - \frac{\alpha-\beta}{2}\int_0^1 \lambda^2 f(\lambda)d\lambda\right]; \;\; B{=}0. \tag{2.73}$$

$$\eta(\rho) = \frac{\gamma}{\alpha+\beta}\rho - \frac{1}{2}\left\{\rho\int_\rho^1 f(\lambda)d\lambda + \frac{\alpha-\beta}{\alpha+\beta}\rho\int_0^1 \lambda^2 f(\lambda)d\lambda + \frac{1}{\rho}\int_0^\rho \lambda^2 f(\lambda)d\lambda\right\}, \tag{2.74}$$

where α, β, and γ are given constants and $\alpha + \beta \neq 0$. When $\alpha + \beta = 0$, we deal with a particular case, and its solution exists only if the additional condition,

$$\alpha\int_0^1 \lambda^2 f(\lambda)d\lambda = \gamma, \tag{2.75}$$

holds; a condition that ensures that $\eta'(0)$ remains bound. In the latter case, the solution becomes the form (2.76) presented below, which results from the theory of self-adjoint linear differential equations [see, e.g., Courant & Hilbert, 1953]:

$$\eta(\rho) = \frac{1}{2}\left[-\rho\int_0^1\left(1+2\lambda^3\right)f(\lambda)d\lambda + \int_0^\rho\left(\rho - \frac{\lambda^2}{\rho}\right)f(\lambda)d\lambda\right]; \;\; \alpha+\beta = 0. \tag{2.76}$$

Because this case is so particular, we will not discuss it further.

The relationship (2.74) above is in fact an integral equation for η, because the function f depends on the unknown η. This makes it possible to investigate the qualitative features of solutions for η, because an integral equation, as is well-known, is, as a rule, much simpler to analyze than a differential equation.

We will now list some equalities and identities which serve as mathematical tools for the qualitative analysis of the solutions of (2.70). These solutions can be rewritten in the form (2.77) given below and which, after integration, lead to the integral expression (2.78).

$$\left(\rho\eta\eta'\right)' = \rho\eta'^2 + \frac{1}{\rho}\eta^2 + \rho\eta f. \tag{2.77}$$

$$\rho\eta(\rho)\eta'(\rho)\Big|_{\rho_0}^{\rho} = \int_{\rho_0}^{\rho}\left[\rho\eta'^2 + \eta^2/\rho + \rho\eta f\right]d\rho; \quad \rho_0 \subset R. \tag{2.78}$$

The next two identities are then easily derived:

$$\int_{\rho_0}^{\rho}\rho^n f d\rho = \int_{\rho_0}^{\rho}\rho^n \mathrm{L}(\eta)d\rho = \rho^{n-1}[\rho\eta' - (n-1)]\eta\Big|_{\rho_0}^{\rho} + n(n-2)\int_{\rho_0}^{\rho}\rho^{n-2}\eta d\rho. \tag{2.79}$$

$$\int_{\rho_0}^{\rho}\eta\mathrm{L}(\eta)d\rho = \left[\eta\eta' + \eta^2/(2\rho)\right]\Big|_{\rho_0}^{\rho} - \int_{\rho_0}^{\rho}\left[\eta'^2 + \eta^2/\left(2\rho^2\right)\right]d\rho. \tag{2.80}$$

Expression (2.80) is a direct consequence of the identity

$$\int \eta\eta' \frac{d\rho}{\rho} = \frac{1}{2}\left[\frac{1}{2}\eta^2 + \int \frac{1}{\rho^2}\eta^2 d\rho\right]. \tag{2.81}$$

We shall also add Green's identity, which holds for self-adjoint linear operators, such as the operator $\rho\mathrm{L}()$, which we need in our case [Courant & Hilbert, 1953]. This identity is given by

$$\int_{\rho_0}^{\rho}[v\rho\mathrm{L}(u) - u\rho\mathrm{L}(v)]d\rho = [\rho(u'v - uv')]\Big|_{\rho_0}^{\rho}, \tag{2.82}$$

where u, v are arbitrary functions of ρ that permit the operations indicated in this identity.

a) ***Properties of solutions of equation*** $\mathbf{L(\eta) = f}$.

In this section, we will investigate the main qualitative properties of the solutions of equation (2.70). On the basis of these results, one can obtain the corresponding properties of solutions of shell equations. The treatment is very simple from the formal mathematical point of view, but the results are very useful for applications.

Property 1. If $f(\rho)$ is an odd function of ρ, then $\eta(\rho)$ is also an odd function, and vice versa. The first assertion is a simple consequence of (2.71), while the second follows directly from equation (2.70). What this property shows, in particular, is the structure of the solutions $\eta(\rho)$

in the neighborhood of the apex $\rho = 0$. The latter information is essential, especially for numerical solution methods in connection with the removal of the singular condition of the operator (2.37) $\rho = 0^+$.

Property 2. (Positivity of the operator L()). If $f(\rho) \geq 0$ (≤ 0) in the interval $R(0 \leq \rho \leq 1)$, and $\eta'(0) \geq 0$ (≤ 0), then $\eta(\rho)$ is a monotonic function in the interval R, and its sign coincides with the sign of $f(\rho)$. The property concerning the sign of $\eta(\rho)$ follows directly from (2.71), taking into account that $B = 0$; $A = \eta'(0)$, and that the kernel of the integral $\rho - \dfrac{\lambda^2}{\rho} \geq 0$ is always non negative ($\lambda \leq \rho$, and $0 \leq \rho \leq 1$). The behavior of the sign of $\eta'(\rho)$ is a consequence of

$$\eta'(\rho) = \eta'(0) + \frac{1}{2}\int_0^\rho \left(1 + \frac{\lambda^2}{\rho^2}\right) f(\lambda)d\lambda, \tag{2.83}$$

which is obtained by differentiation of (2.71).

Property 3. If $f(\rho) \geq 0$ (≤ 0) in R, then:

1) $\eta(\rho)/\rho$ is a monotonic function in R and sign $(\eta/\rho)' = sign\, f$.

2) The following three inequalities hold in R:

$$\eta'(\rho) - \frac{1}{\rho}\eta(\rho) \geq 0(\leq 0). \tag{2.84}$$

$$\eta(\rho) \geq (\leq)\eta'(0)\rho. \tag{2.85}$$

$$\eta'(\rho) \geq (\leq)\eta'(0). \tag{2.86}$$

Point 1 of this property is a direct consequence of identity (2.87),

$$\left(\frac{\eta}{\rho}\right)' = \frac{\eta'}{\rho} - \frac{\eta}{\rho^2} = \frac{1}{\rho^3}\int_0^\rho \lambda^2 f d\lambda. \tag{2.87}$$

which is a particular case of (2.79) for n = 2. Equation (2.84) also results from (2.87). Inequality (2.85) is determined by (2.71) and (2.86) using (2.83).

Property 4. If $\eta(\rho)$ is a monotonic function in the interval R, then it is of constant sign in R, and this sign coincides with that of $\eta'(\rho)$. This is a consequence of $\eta(\rho)$ being a monotonic function, and that $\eta(0)=0$.

Property 5. In any point $\rho = \bar{\rho}$ in R, where $\eta'(\bar{\rho})=0$ and $\eta(\bar{\rho})$ has a positive maximum (negative minimum) or an inflection point, the sign of $f(\bar{\rho})$ is opposite to that of $\eta(\bar{\rho})$. Indeed, if for example, at $\rho = \bar{\rho}$, the function $\eta(\rho)$ reaches a positive maximum, then, at this point, $\eta \geq 0$; $\eta' = 0$, and $\eta'' \leq 0$. As a consequence, the following inequality is implied by equation (2.70):

$$f(\bar{\rho}) = \eta''(\bar{\rho}) + \eta'(\bar{\rho})/\rho - \eta(\bar{\rho})/\bar{\rho}^2 < 0. \tag{2.88}$$

At an inflection point $\eta'' = 0$ and thus the property holds for this condition as well. As a further consequence of this property, it follows that, if η is not a monotonic function, the signs of η and f cannot be the same in the entire interval R.

The properties discussed above concern the initial value problem (Cauchy's problem) of equation (2.70). The properties of the solutions of the boundary value problem (2.72) of (2.70) are still to follow.

Property 6. If: a) sign f = const.; b) sign $f = -\text{sign}\gamma/(\alpha + \beta)$ or $\gamma = 0$; c) const. $æ = (\alpha - \beta)/(\alpha + \beta) \geq -1$; then, $\text{sign}\eta(\rho)$ = const. = $-\text{sign}f$; and $\eta'(0) \neq 0$. The first assertion of the present property results from relationship (2.74) if the expression between the braces is rewritten as follows:

$$\rho\int_{\rho}^{1}\left(1+æ\lambda^2\right)f(\lambda)d\lambda + \rho\left(æ+\frac{1}{\rho^2}\right)\int_{0}^{\rho}\lambda^2 f(\lambda)d\lambda.$$

The feature $\eta'(0) \neq 0$ results directly from (2.73).

Condition a) for the existence of this property is sufficient, but not necessary. For example, let $f(\rho) = 2\rho^3 - 2.5\rho$, which does not have the same sign in the interval. The corresponding function $\eta(\rho)$ is then given by

$$\eta(\rho) = \frac{\gamma}{\alpha+\beta}\rho + \frac{\rho^3}{8}\left(\rho^2 - \frac{5}{2}\right) \leq 0 \text{ if } \gamma/(\alpha+\beta) < 0.$$

Hence, the property of reciprocity does not always exist. That f will have a constant sign does not follow from the constant sign of η.

Property 7. Let $\eta(\rho)$ satisfy any one of the two boundary conditions.

$$\eta(1)=0; \quad \text{or } \eta'(1)+\beta\eta(1)=0. \tag{2.89}$$

In that case, both inequalities

$$\int_0^1 \eta L(\eta)d\rho \le 0; \quad \int_0^1 \rho\eta L(\eta)d\rho \le 0, \tag{2.90}$$

hold as long as $\beta \ge 0$. But, if $-0.5 \le \beta \le 0$, only the first inequality is satisfied.

When the first condition of (2.89) holds, the first part of this property is a direct consequence of (2.80), keeping in mind that $\eta(0)$ is always equal to 0. Using (2.80), the second condition of (2.89), can be transformed into

$$\int_0^1 \eta L(\eta)d\rho = \eta(1)\eta'(1)+\eta^2(1)/2-\int_0^1\left[\eta'^2+\eta^2/\left(2\rho^2\right)\right]d\rho = \eta^2(1)\left(\frac{1}{2}-\beta\right)$$

$$-\int_0^1\left[\eta'^2+\eta^2/\left(2\rho^2\right)\right]d\rho \le \eta^2(1)\left(\frac{1}{2}-\beta\right)-\eta^2(1)-\int_0^1 \eta^2/\left(2\rho^2\right)d\rho$$

$$=-\eta^2(1)\left(\frac{1}{2}+\beta\right)-\int_0^1 \eta^2/\left(2\rho^2\right)d\rho. \tag{2.91}$$

The previous derivation makes use of the inequality $-\int_0^1 \eta'^2 d\rho \le -\eta^2(1)$, which is actually derived from Schwarz's inequality [Courant & Hilbert, 1953],

$$\left[\int \phi F d\rho\right]^2 \le \int \phi^2 d\rho \int F^2 d\rho, \tag{2.92}$$

setting $\phi = \eta'$; and $F = 1$.

The relation (2.91) proves the validity of the first inequality of (2.90) under the given conditions. The second inequality of (2.90), for $\beta \ge 0$, can be obtained by integration of parts of

$$\int_0^1 \rho\eta L(\eta)d\rho = \int_0^1 \rho\eta\left(\eta''+\eta'/\rho-\eta/\rho^2\right)d\rho = \rho\eta\eta'\Big|_0^1-\int_0^1 \rho\eta'^2 d\rho$$

$$-\int_0^1 \eta^2/\rho d\rho = -\beta\eta^2(1) - \int_0^1 \left(\rho\eta'^2 + \eta^2/\rho\right)d\rho \le 0$$

The inequality given above is derived for the case that η satisfies the second boundary condition of (2.89). When η satisfies the first condition of (2.89), one gets a similar result given by:

$$-\int_0^1 \left(\rho\eta'^2 + \frac{\eta^2}{\rho}\right) \le 0.$$

So far, the properties of η have been investigated in the entire interval R. It turns out that many of the properties that hold in the whole interval R can also be formulated for a part of R. This is done in the next proposition.

Property 8. Property 8 is concerned with the features of $\eta(\rho)$ in an interval $R_1(0 \le \rho \le \rho_1)$, where the function $f(\rho)$ does not change sign before $\rho_1 \ne 0 \subset R$.

1) The Properties 2 and 3 remain valid in R_1. This is obvious and does not require any additional proof.

2) If at some point $\rho_2 \le \rho_1$, the function η satisfies

$$\alpha\eta'(\rho_2) + \frac{\beta}{\rho_2}\eta(\rho_2) = \gamma, \tag{2.93}$$

the constant $\gamma/(\alpha + \beta)$ and f are of opposite sign (or $\gamma = 0$) and $æ = \dfrac{\alpha - \beta}{\alpha + \beta} \ge -1$, then η keeps its sign in the interval $(0 \le \rho \le \rho_2)$ and it is opposite to the sign of $f(\rho)$.

The idea that leads to the proof of the last assertion can be given as follows. Using the general expression (2.71), one can obtain the solution of the boundary value problem in the interval $(0 \le \rho \le \rho_2)$ for conditions $\eta(0) = 0$, and (2.93). This solution has a form analogous to (2.74).

$$\eta(\rho) = \frac{\gamma}{\alpha+\beta}\rho - \frac{1}{2}\int_0^{\rho_2}\left(\frac{1}{\rho} + \frac{æ\rho}{\rho_2^2}\right)\lambda^2 f(\lambda)d\lambda - \frac{1}{2}\int_\rho^{\rho_2}\left(\rho - \frac{\lambda^2}{\rho}\right)f(\lambda)d\lambda;$$

$$\left(0 \le \rho \le \rho_2\right);\ \alpha + \beta \ne 0. \tag{2.94}$$

The second part of the property is then obtained by a simple transformation of this expression.

The analysis of the properties of the solutions η is not exhausted by what has been described so far, but the results that we obtained here suffice to establish the main qualitative characteristics of the behavior of the spherical cap.

b) Properties of solutions of governing equations of the investigated cap.

The properties of η described above provide the means to investigate the qualitative behavior of the cap that we consider here. (The previous categorization of the features is continued.)

Property 9. The initial (i.e., undeformed) configuration $\overline{w}(\rho)$ of the cap and the displacement $w(\rho)$ are even functions of ρ because the initial geometry of the shell and its ensuing deformation are assumed to be axisymmetric. It then follows that the slopes $\theta(\rho)$ and $\overline{\theta}(\rho)$, which are the first derivatives of $\overline{w}$ and w, respectively, must be odd functions of ρ. Hence, the membrane stresses function $\omega(\rho)$ is an odd function too. This is easy to see if one applies Property 1 to the equation (2.35) (for $q_t \equiv 0$). The relations (1.61), (1.62), and (1.63) (for this particular case) then show that the parameters of the problem — i.e., $\sigma_r(\rho)$, $\sigma_\phi(\rho)$, $M_r(\rho)$, $M_\phi(\rho)$, $N_r(\rho)$, $N_\phi(\rho)$, $\varepsilon_r(\rho)$, $\varepsilon_\phi(\rho)$ — are even functions of ρ, while the shear force $Q(\rho)$ and horizontal displacement $u(\rho)$ are odd functions of ρ. Of course, this property is absolutely transparent from the physical point of view.

Because θ and ω are odd functions, the following relationships for the initial values of changes of curvature and membrane stresses can be derived:

$$\sigma_r(0) = \sigma_\phi(0) = \omega'(0);\;\; \varepsilon_r(0) = \varepsilon_\phi(0) = \frac{(1-\nu)h^2}{a^2}\omega'(0);$$

$$\chi_r(0) = \chi_\phi(0) = \theta'(0);\;\; M_r(0) = M_\phi(0) = (1+\nu)\theta'(0). \tag{2.95}$$

Property 10. This property can be split into two parts: a) When the edge of the cap is free to move in its base plane, $\sigma_\phi(\rho) = \omega'(\rho)$ is a function of variable sign in R, independently of the features of $\sigma_r = \omega/\rho$; b) If σ_ϕ does not change sign in R, then $\sigma_r(\rho)$ does not change sign either, so that both stress components are of the same sign.

The proofs can be given as follows.

1) When the edge is free to move, boundary conditions (2.96) hold; i.e.,

$$\omega(0) = \omega(1) = 0. \tag{2.96}$$

Hence, the derivative $\omega'(\rho)$ changes its sign in R, and thus the hoop stress σ_ϕ is a function of variable sign in R.

2) Taking into account the first relationship of (2.95), we can show the validity of the second part (b) of Property 10.

These two, simple features are important for the understanding of the buckling phenomena that can occur when the cap is loaded by internal pressure (see Section 2.1, part b). In bifurcation problems of this type, a buckling pattern is formed in a circular, belt-like region of the cap's surface, which is multi-lobed in the circumferential direction. Indeed, since, under the given conditions, the hoop stress σ_ϕ is a function of variable sign, there exists an interval $\rho_1 < \rho < \rho_2$, where $\sigma_\phi < 0$. This region cannot be close to the apex of the shell because the internal pressure creates a zone of stretching around the apex so that, at the apex itself, $\sigma_\phi(0) = \sigma_r(0) > 0$ (see (2.95)). The positive sign of $\sigma_r(0)$ follows directly from inequalities (2.104) for a plate and (2.134) for a cap.

Because the hoop stress σ_ϕ can become negative in an intermediate region between the apex and the edge, this kind of membrane stress distribution creates the necessary condition for forming a buckling pattern. It is a type of buckling that we already encountered in Section 2.1, where we categorized it "buckling in tension." Analyses that deal with it can be found in, for example, Shilkrut [1967, 1974, 1983a], Biezeno and Grammel [1939], Bushnell [1967a], Gerlaku and Shilkrut [1970], Morozov [1961], and Panov and Feodosiev [1948]. The results of these investigations will be discussed in Sections 3.2 and 3.7.

It is obvious that, if a sufficiently large positive load N is applied at the movable edge, the stress distribution $\sigma_r(\rho)$ will be turned into a monotonically increasing, positive function in R (the boundary condition will then be $\omega(1) = N > 0$). In that case, the hoop stress σ_ϕ will also become positive everywhere in R. In this way, buckling can be avoided. As will be shown in Property 11, the condition $\sigma_r > 0$ in R holds almost always when we deal with a circular plate.

Property 11. This property concerns the principal characteristics of the behavior of circular plates, which undergo axisymmetric deformations.

1) Independently of the type of external loading and edge supports, any axisymmetric deformation of a plate is characterized by a stress distribution where $\sigma_r(\rho)$ is a monotonically decreasing function, and $\sigma_\phi(\rho)$ obeys

$$\sigma\phi(\rho) \leq \sigma\rho(\rho) \text{ in } R. \tag{2.97}$$

This property is a direct consequence of Property 3 (Section 2.8) if the latter is applied to equation (2.98) for plates.

$$L(\omega) = -\theta^2/2\rho. \tag{2.98}$$

2) If $\omega(\rho)$ satisfies the generalized the boundary condition

$$\alpha\omega'(1) + \beta\omega(1) = \gamma; \quad (\alpha + \beta \neq 0), \tag{2.99}$$

where

$$æ = \frac{\alpha - \beta}{\alpha + \beta} \geq -1, \tag{2.100}$$

the following inequalities hold

$$\frac{\gamma}{\alpha + \beta}\rho \leq \omega(\rho) \leq \omega'(0)\rho \text{ in } R. \tag{2.101}$$

independently of the kind of axisymmetric loading and boundary conditions for $\theta(\rho)$.

Actually, the second inequality (2.101), which is a direct consequence of (2.97), always holds, and not only for the set of conditions (2.99). The first inequality (2.101) can be proved by splitting $\omega(\rho)$ in two parts

$$\omega(\rho) = \overline{\omega}(\rho) + \omega_\Delta(\rho), \tag{2.102}$$

where $\overline{\omega}$ satisfies the homogenous equation $L(\overline{\omega}) = 0$ and boundary condition (2.99); i.e., $\overline{\omega}(\rho) = \gamma\rho/(\alpha + \beta)$. Then $\omega_\Delta(\rho)$ satisfies equation (2.98) and the conditions ($\gamma = 0$) (2.99) and (2.100). Applying Property 6 to $\omega_\Delta(\rho)$, one must arrive at the conclusion that $\omega_\Delta(\rho) \geq 0$ in R. This fact, together with the expression (2.102), proves the inequality considered here.

As a consequence of the above, the following bounds on the values of $\sigma_r(\rho)$ are obtained:

$$\frac{\gamma}{\alpha + \beta} \leq \sigma_r(\rho) \leq \sigma'(0). \tag{2.103}$$

In particular, $\omega(\rho) \geq 0$ and $\sigma_r(\rho) \geq 0$ in R if $\gamma/(\alpha + \beta) \geq 0$. This follows for immovable supports (where $\alpha = 1$, $\beta = -\nu$, $\gamma = 0$) and elastic supports, for which the boundary condition can be written in the form

$$\omega(1) = N = -Cu(1),$$

or

$$C\omega'(1) + (1 - C\nu)\omega(1) = 0; \; C \geq 0 \tag{2.104}$$

The case of the elastic support $C \neq 0$ is an intermediate case between movable and immovable supports, and it can be interpreted as an imperfect boundary condition. However this case does not change the qualitative behavior of a symmetrically deformed plate. The inequality $\sigma_r(\rho) \geq 0$ also holds in the case of movable, simply supported edges where a positive membrane force $N \geq 0$ is applied ($\omega(1) = N \geq 0$). That the stress component $\sigma_r(\rho)$ is positive definite in R for simply supported plates was probably proved for the first time by Shilkrut [1967].

It is noted that numerous numerical results for different load and support conditions confirm the existence of these properties of the solutions (see, for example, the very informative book by Volmir [1963]).

3) If the edge load N is compressive (i.e., $N < 0$, the function $\omega(\rho)$ is either negative everywhere in R or $\omega(\rho)$ is of a variable sign (because $\omega(\rho)$ is a monotonically decreasing function). In the latter case, there can be only two regions where $\omega(\rho)$ differs in sign. In the first region, which contains the point $\rho = 0$, $\omega(\rho)$ can only be positive, so that, because it is monotonic, it must be negative in the second.

Indeed, if $\omega'(0) < 0$, then, in conformity with Property 2 applied to equation (2.98), $\omega(\rho)$ is a monotonically decreasing function and $\omega(\rho) \leq 0$ everywhere in R. If $\omega'(0) > 0$, then in an interval $0 \leq \rho \leq \tilde{\rho}$, $\omega(\rho) \geq 0$. At $\rho = \tilde{\rho}$, $\omega = 0$, and for $\rho > \tilde{\rho}$, ω will be negative and monotonically decreasing on the basis of Property 5. If $\omega(\rho) < 0$ and $\omega(\rho)$ is not a monotonically decreasing function, $\omega(\rho)$ would possess a negative minimum. According to Property 5, the left-hand side of (2.98) should be positive at this point, which is clearly impossible. This contradiction shows that the assertion holds.

There are many numerical solutions which confirm the features of $\omega(\rho)$ described above. For example, in Figure 2.22a [Shilkrut, 1983b], plots of $\omega(\rho)$ are presented for different equilibrium states of a plate for $N < 0$ and, in addition, a bending moment is acting along the edge. The corresponding deformation curves are presented in Figure 2.22b. As can be observed, both features of the solutions for $\omega(\rho)$ described above are clearly present.

The graphs $\omega'(\rho)$, which correspond to $\omega(\rho)$ depicted in Figure 2.22a, are given in Figure 2.23. As can be seen in both of these figures, close to the edge, regions exist where σ_r and σ_ϕ attain very large negative values (this is particularly clear in case 5). The negative sign of σ_r and σ_ϕ provide the necessary conditions for a bifurcation phenomenon to occur that is associated with a nonsymmetric buckling pattern. It is interesting to note that the same features of ω are also present in the case of orthotropic plates (see Chap. 4, Figs. 4.21, 4.22).

theorem requires restrictions on the external forces and initial shape of the membrane cap, which are of the following type:

$$m_1\rho^2 \le \int_0^1 \rho q d\rho \le m_2\rho^2; \quad \bar{\theta} \le m_3\rho, \tag{2.151}$$

where m_1, m_2, and m_3 are positive constants. For these restrictions, the theorem of the existence of the solutions in question was also proved by Srubtchik and Yudovitch [1962].

Combining the three features of membrane caps described above, we can arrive at the following important conclusion. The membrane cap for any $|\xi| \neq 0$ can lose its stability by jumping in the space of symmetric deformation states, where the upper and lower critical loads (q_1 and q_2) are related to each other by the first formula (2.150). Here we face the cardinal differences in the behavior of caps with bending rigidity and membrane caps. The former are able to lose their stability by jumping only when their geometric parameter $|\xi|$ is greater than some non-zero number, the magnitude of which depends on the boundary conditions and the kind of external loading (see Section 2.1). We shall encounter this phenomenon again in the next chapter with the discussion of various numerical solutions.

In contrast, membranes can lose their stability by jumping independently of the value of $|\xi|$ as long it is non-zero. The membrane stresses, in particular σ_r, play the crucial role in the jumping mechanism because there are no bending stresses in such structures. These jumping mechanisms are in essence similar to that of the von Mises frame, which is shown in Figure 2.26. Its load-deformation curve, depicted in Figure 2.27, is of the same qualitative type as that of the membrane cap with homogenous support conditions. This correspondence exists because both elements of this frame undergo only compression or stretching; that is, there are only "membrane" forces. This results from the fact that all nodes are hinged and P acts on the central hinge.

If an arch is taken instead of the frame, the picture of its behavior will be similar to that of a cap with bending rigidity; i.e., the loss of stability by jumping will occur only after some value of $|\xi|$ is exceeded. Bending stresses are then induced in the arch in addition to membrane forces, and this avoids the possibility of the jumping of arches for small $|\xi|$. When $|\xi|$ increases (that is, the height of the cap or arch increases, or the thickness of their bodies decreases), the membrane stresses become larger and larger, so that the jumping process becomes more and more violent. (This effect can easily be understood considering the von Mises frame; when the initial angle α_0 (see Figs. 2.26 and 2.27) increases, the initial height increases, so that the axial forces acting in the frame's members increase as well.)

From this qualitative consideration, it is clear that the jumping process can be eliminated by the addition of meridional stiffeners, which diminish the membrane stress σ_r.

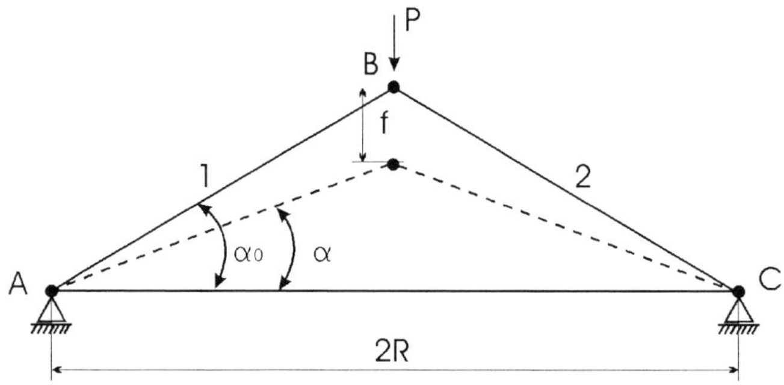

Figure 2.26 The von Mises frame.

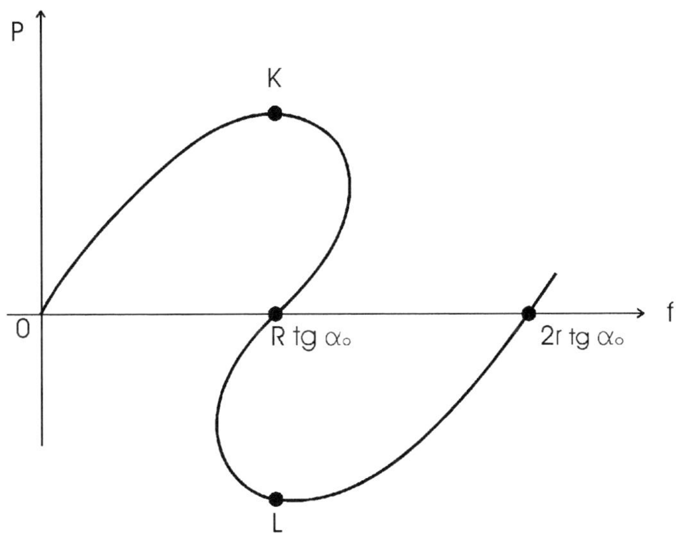

Figure 2.27 Response curve for the von Mises frame.

4) All the features of $\omega(\rho)$ given above that are independent of $\theta(\rho)$ for shells with bending rigidity remain valid for membranes. This conclusion follows from the fact that the equation for ω, (2.147), for the membrane is the same as for the shell with bending stiffness. For example, Properties 12, 13, and 16 are valid for both these cases. In particular, inequality (2.138), which eliminates the danger of loss of stability by snapping, holds for a membrane cap as well.

Property 17. About the nature of bifurcation phenomena. In the previous discussion, we established the importance of the role that $\omega(\rho)$ and σ_r play in the snap-through behavior of caps when only axisymmetric deformations are considered. As was stated earlier (see Sections 2.1 and 2.3), besides this type of stability loss, a bifurcation phenomenon can take place whereby the axisymmetric state goes over into a nonsymmetric state. The changeover takes place at a certain value of the external load (which depends on particular conditions of the problem), called the "bifurcation load," while the axisymmetric equilibrium state that corresponds to this load is called a "bifurcation point."

The geometrical significance of this phenomenon is that branch I of axisymmetric deformation states is crossed by a branch II of nonsymmetric deformation states at this particular point. When the actual cap is loaded up to and beyond the bifurcation load along I, two different kinds of behavior can be observed. If the bifurcating branch of nonsymmetric states is stable (in the small) in the neighborhood of the bifurcation point, the cap will transfer smoothly into the nonsymmetric form when the load is further increased. However, when the bifurcating branch is unstable, loss of stability occurs at the bifurcation load by a snap-through process (called snap-buckling), which comes to an end at a new stable state of the cap. In this case, the bifurcation phenomenon plays the role of a trigger of the jump from the "initially" basic symmetric equilibrium state. The new postbifurcation (post-jumping) state can either be symmetric or nonsymmetric. What the outcome will be in this case can only be determined by an investigation using a transient integration method, because the snap-through process is essentially a time-dependent phenomenon.

It is also possible that more than one nonsymmetric mode will develop at the bifurcation point. The bifurcation phenomenon is then more complicated, but the actual behavior of the cap will still follow the description given above.

In Sections 2.1 and 2.3, two different kinds of bifurcation phenomena were identified: bifurcation in "tension" and bifurcation in "compression." Apart from some additional conditions, "bifurcation in tension" can take place for any value of $\left|\bar{\xi}\right|$ (including $\left|\bar{\xi}\right| = 0$) if the transverse loading is acting from the inside out; e.g., internal pressure. Characteristic for this type of bifurcation is that the meridian is stretched everywhere in R in the pre-buckling state. On the other hand, "bifurcation in compression" occurs when the apex rise satisfies $\left|\bar{\xi}\right| > \xi_0 > 0$ and the cap is loaded by external transverse loads that are pointing inward. The value of ξ_0 must be greater than ξ_{00}, where the latter corresponds to the magnitude of $\left|\bar{\xi}\right|$, at which loss of stability by snap-through in the subspace of axisymmetric deformation occurs (see Property 4, Section 2.3). This means that, in the actual physical situation, buckling in compression does not occur for plates or for very shallow shells. By considering Property 17, we shall now investigate the role of membrane stresses in these cases.

It is clear that the waviness in the circumferential direction of the nonsymmetric mode that develops after bifurcation is caused by the membrane (hoop) stress σ_ϕ and its interaction with σ_r, and that, in the circumferential region where the buckles occur, σ_ϕ must be compressive. In addition, it is necessary that meridional stress σ_r does not prevent the formation of these buckles. It follows from these observations that a necessary condition for the development of a buckling pattern is given by

$$\sigma_\phi < 0, \text{ and } |\sigma_r| \le |\sigma_\phi| \text{ in } B \subset R, \tag{2.152}$$

where B denotes the width of the band of buckles. Conditions (2.152) hold for both positive and negative σ_r, but they can only be appreciated as necessary conditions, not as sufficient conditions.

For the two cases that are investigated here (movable and immovable supports in the radial direction), the property presently under discussion can now be summarized in the following way.

1) If $|\xi|$ is small (enough), bifurcation in "compression" cannot occur in either of the two cases of support. It can only occur when $|\xi|$ exceeds a certain, configuration dependent value.

2) In contrast, bifurcation in "tension" can happen when the support is movable, but it cannot happen when the support is immovable.

If $|\xi|$ is small and the external loading $q > 0$ (which is acting from the outside of the cap) is also small, then it can be assumed that $0 \le \theta(\rho) < |\overline{\theta}(\rho)|$; that is, the possible equilibrium states with the smallest ξ (a number of equilibrium states with different ξ can exist for the given q) are of the type shown in Figure 2.28a. In this case, this is the right-hand side of the equation of ω (see (2.147) or (2.35) when $q_t \equiv 0$) is positive, and then, in conformity with the Property 6, $\omega(\rho)$ is necessarily negative. $\sigma_r(\rho) \le 0$ is a monotonically increasing function, and, because of Property 3, the following relation holds:

$$\sigma_r(\rho) \le \sigma_\phi(\rho) \text{ in } R. \tag{2.153}$$

When the support is movable, σ_ϕ can be both positive and negative (Property 10). In the interval, where σ_ϕ is positive, inequalities (2.152) are naturally not fulfilled. But in the vicinity of the apex, an interval exists where both membrane stresses are negative simultaneously (see the graphs of these stresses, depicted qualitatively in Fig. 2.29). The necessary conditions (2.152) are again not fulfilled in this interval because of (2.153). When the edge is immovable, the same conclusion holds because of (2.153) in R, despite the fact that both σ_r and σ_ϕ can be negative in a part of R. This means that the first part of the considered property is proved.

can be written on the basis of (2.74) and (2.156). This means that $\sigma_r(\rho) > |\sigma_\phi(\rho)|$ in R, so that condition (2.152) is violated, thus confirming the validity of the second part of the content of Property 17.

The same features of bifurcation in tension hold when $\theta(\rho) \geq -2\left|\overline{\theta}(\rho)\right| \geq 0$; i.e., when the deformed shape is of the type depicted in Figure 2.28c and $q > 0$ corresponds to external pressure. In this case, the geometry of the equilibrium state is comparable to the inverted initial geometry of the cap and the pressure q can now be conceived as an internal pressure for the deformed cap.

It is useful to consider the case of elastic supports with boundary conditions,

$$\omega(1) = N = -Ku(1),$$

which can be transformed to

$$\omega'(1) + \left(\frac{a}{Kh} - \nu\right)\omega(1) = 0, \tag{2.158}$$

where the constant $K > 0$ is the rigidity of the elastic support, $u(1)$ is the horizontal displacement of the edge (which is given by corresponding formula (1.63); $K = 0$ means a movable support, and $K = \infty$ for immovable support), a is the radius of the base of the shell, and h is its thickness. The investigation of this case, which can be carried out in the same way as was described above for the other cases, shows that bifurcation in compression cannot occur for small $|\xi|$, as was observed for the support conditions previously considered.

Bifurcation in tension is possible if

$$0 \leq K \leq \frac{a}{h(1+\nu)}. \tag{2.159}$$

The range of values of K satisfying (2.159) is rather large, considering the values of a/h that are usually encountered in practice. Property 17 classifies the most important cases of bifurcation in tension. However, one should note that this phenomenon cannot occur for immovable supports, a fact that has been observed in practice by the author [see Shilkrut, 1983a].

In order to avoid the jumping process in the subspace of axisymmetric deformation, meridian stiffeners can be used, while the bifurcation phenomena can be prevented by the application of ring stiffeners, which are able to reduce the compressive hoop stresses σ_ϕ. As a result, application of these two types of stiffeners leads to a circular orthotropic shell. From the point of view of stability, this optimization of the shell structure is connected with the introduction of anisotropy. The shell (or plate) equipped with a set of stiffeners in fact represents an anisotropy of a type called "structural" in contrast to the anisotropy of the material of the shell wall.

There are many examples of investigations which deal with plates and shells with structural anisotropy. A survey of the literature published before 1975 dealing with the stability of such shells is presented by Nemirovsky [1976]. A series of interesting results about the stability of structurally anisotropic, geometrically nonlinear spherical shells can be found in Simitses and Blackmon [1975] and Simitses and Cole [1971]. Some particular problems of stability of circular plates with structural anisotropy are in Vainberg's [1973] reference book.

Property 18. Some features of dynamic deformations states. The governing equations of dynamics are obtained from the static equations by replacing the transverse loading q by the expression $q - s\dfrac{\partial^2 w}{\partial \tau^2}$, where the additional term represents the inertial forces due to the acceleration of the body particles in the z direction. The inertial force components caused by the other displacements u and v are, as usual, neglected, as are the inertial moments related to the rotation of the normals to the midsurface. In the axisymmetric case, the dimensionless mass density per unit area of the middle surface is defined as $s = s^* a^4 g/Eh^4$, where s^* is the actual specific mass of the shell wall material, and g is the gravitational acceleration. Note that we introduce here the dimensionless time $\tau = t\sqrt{g}/\sqrt{h}$, where t is the realtime. All the unknown functions are now dependent on the additional dimension τ, and thus, in addition to the boundary conditions, we also need initial conditions, which are of the form

$$w|_{\tau=0} = f_1; \quad \frac{\partial w}{\partial \tau}\Big|_{\tau=0} = f_2, \tag{2.160}$$

where $f_i (i = 1.2)$ are given functions of the coordinates of the middle surface. The functions f_i should satisfy the kinematic boundary conditions, which can be time-dependent in the general case.

This concludes the formulation of the transient problem. We can now examine some of the main features of the solutions of governing equations of motion. We will assume here that the cap is isotropic, shallow, of uniform rigidity, and that there is no thermal loading and no force system with components tangential to the shell surface. In that case, the equations of motion for axisymmetric deformations are given by

$$L(\omega) = -\frac{1}{2\rho}\left[\theta^2 + 2\theta\bar{\theta}\right]; \quad L(\theta) = -\frac{m}{\rho}\int_0^\rho \rho\left[q - s\frac{\partial^2 w}{\partial \tau^2}\right]d\rho + \frac{m}{\rho}\omega\left(\theta + \bar{\theta}\right). \tag{2.161}$$

The operator L() is the same as the one that appears in the static case and does not depend on τ. Sometimes it is more convenient to rewrite the term that is dependent on w in another form so that it does not contain the displacement w explicitly. This is done as follows:

$$\frac{ms}{\rho}\frac{\partial^2}{\partial\tau^2}\int_0^\rho \rho w(\rho,\tau)d\rho = \frac{ms\partial^2}{2\rho\partial\tau^2}\left[\rho^2\int_\rho^1 \theta(\rho,\tau)d\rho + \int_0^\rho \rho^2\theta(\rho,\tau)d\rho\right]. \tag{2.162}$$

For the general deformation of shallow shells — of uniform rigidity, isotropic wall construction in the absence of tangential forces, and temperature stresses — the governing equations of motion are:

$$\frac{r}{Eh}\Delta\Delta F = \mathrm{P}[w,\overline{w}];\ \ \mathrm{D}\Delta\Delta w = \mathrm{Q}[w,\overline{w},F] + q - s\frac{\partial^2 w}{\partial\tau^2}, \tag{2.163}$$

where P and Q are the right-hand side operators of the static equations (1.76) and (1.74), respectively. The unknowns and constants in (2.163) are physical quantities.

In the transient state, all the properties of the static solutions discussed earlier remain valid within the framework in which the existence of these properties were proved, keeping in mind, of course, the existence of inertial forces and the possibility that the external forces and boundary conditions can be time-dependent. This very important affirmation is based on the fact that all the features of the static solutions previously identified are based on the properties of the operator L (in the axisymmetric case) and the bi-harmonic operator $\Delta\Delta$ (in the general case). Both these operators depend only on the geometry, and not on the time τ. As a consequence, the properties of the static solutions are also valid for the transient solutions at each fixed value of τ. Of course, the previous relations that were found for a specific distribution of q, such as the inequality (2.140), lose their validity in the transient case, because, instead of q, it is necessary to take $q - s\dfrac{\partial^2 w}{\partial\tau^2}$ for the load, the distribution of which is not known.

The features of reciprocal systems, as stated in Section 2.3, Property 6, remain valid also in the transient case.

c) Small vibrations around an equilibrium state.

This problem will first be investigated for the axisymmetric case. Let $\omega_s(\rho)$ and $\theta_s(\rho)$ be the static solution around which small vibrations happen. The vibrations can be described by small, time-dependent perturbations $\delta\omega(\rho,\tau)$, $\delta\theta(\rho,\tau)$ that are added to the basic static solution. Substituting $\omega_s + \delta\omega$ and $\theta_s + \delta\theta$ into equations (2.161) (using the form (2.162)) and linearization with respect to $\delta\omega$, $\delta\theta$ leads to

$$\rho\mathrm{L}(\delta\omega) = -\delta\theta\left(\theta_s + \overline{\theta}\right);$$

$$\rho L(\delta\theta) = m\left\{\frac{s}{2}\frac{\partial^2}{\partial\tau^2}\left[\rho^2\int_0^1 \delta\theta(\lambda)d\lambda - \int_0^\rho \left(\rho^2 - \lambda^2\right)\delta\theta(\lambda)d\lambda\right] + \delta\omega\left(\theta_s + \bar{\theta}\right) + \omega_s\delta\theta\right\};$$

$$(0 \le \rho \le 1);\ (0 \le \tau < \infty) \tag{2.164}$$

The solution of these equations is sought in the form of Fourier's series:

$$\delta\omega = \sum_{n=1}^{\infty}\Omega_n(\rho)\sin(p_n\tau + \phi_n);\quad \delta\theta = \sum_{n=1}^{\infty}T_n(\rho)\sin(p_n\tau + \phi_n), \tag{2.165}$$

the coefficients of which satisfy the relations

$$\rho L(\Omega_n) = -T_n\left(\theta_s + \bar{\theta}\right);\ 2\rho L(T_n) = 2m\left[\omega_s T_n + \Omega_n\left(\theta_s + \bar{\theta}\right)\right]$$

$$-K_n^2\left[\int_0^\rho \lambda^2 T_n(\lambda)d\lambda + \rho^2\int_\rho^1 T_n(\lambda)d\lambda\right], \tag{2.166}$$

where $p_n = \overset{*}{p}_n\sqrt{h/g}$ is the dimensionless n^{th} frequency of the "proper vibrations," or "vibration modes"; $\overset{*}{p}_n$ is the physical quantity of the frequency; and $K_n^2 = msp_n^2$ is the square of the nth dimensionless eigenvalue of the boundary value problem

$$\Omega_n(0) = T_n(0);\ \alpha_1\Omega_n'(1) + \beta_1\Omega_n(1) = 0;\quad \alpha_2 T_n'(1) + \beta_2 T_n(1) = 0. \tag{2.167}$$

Relations (2.267) result from the corresponding boundary conditions on $\omega = \omega_s + \delta\omega$ and $\theta = \theta_s + \delta\theta$.

The following identity results from (1.166):

$$\int_0^1 T_n L(T_n)d\rho = m\left[\int_0^1 T_n^2\omega_s\frac{d\rho}{\rho} + \int_0^1 T_n\Omega_n\frac{\theta_s + \bar{\theta}}{\rho}d\rho\right]$$

$$-\frac{K_n^2}{2}\int_0^1 T_n(\rho)\left[\int_0^\rho \lambda^2 T_n(\lambda)d\lambda + \rho^2\int_\rho^1 T_n(\lambda)d\lambda\right]\frac{d\rho}{\rho}, \tag{2.168}$$

which is useful for the analysis of the character of the perturbation solutions. In cases when the edge of the cap is hinged or clamped, the left hand-side of (2.168) must be negative in conformity with Property 7 (see (2.90)). For the same reason, it follows that the second integral in the right-hand side of (2.168) must be positive (keeping in mind the first equation of (2.166)). It follows that

$$\int_0^1 \mathrm{T}_n^2 \omega_s \frac{d\rho}{\rho} - \frac{K_n^2}{2m} \int_0^1 \mathrm{T}_n(\rho) \left[\int_0^\rho \lambda^2 \mathrm{T}_n(\lambda) d\lambda + \rho^2 \int_\rho^1 \mathrm{T}_n(\lambda) d\lambda \right] \frac{d\rho}{\rho} < 0, \quad (n = 1,2,\ldots). \tag{2.169}$$

From (2.169), we obtain the important conclusion that K_n^2 cannot be equal to zero if $\omega_s(\rho) \geq 0$ in R. That is, any basic static state with $\omega_s(\rho) \geq 0$ is stable in the small (within the subspace of axisymmetric deformations), because the criterion for stability in the small corresponds to the condition that K_n^2 be positive for all n (see a more detailed explanation in Chapter 6). This conclusion is in complete agreement with the uniqueness theorem considered in Properties 13 (for shells) and 11 (point 6) for plates.

Using simple algebra, it is not difficult to show the validity of the following two generalized orthogonal relations.

$$\int_0^1 \left(\Omega_j \mathrm{T}_m - \Omega_m \mathrm{T}_j\right)\left(\theta_s + \overline{\theta}\right) d\rho = 0. \tag{2.170}$$

$$\int_0^1 \int_0^\rho \lambda^2 \left[\mathrm{T}_m(\rho)\mathrm{T}_j(\lambda) + \mathrm{T}_m(\lambda)\mathrm{T}_j(\rho)\right] d\lambda d\rho = 0; \quad K_j^2 \neq K_m^2. \tag{2.171}$$

One can deduce from (2.171) that all the eigenfunctions $\mathrm{T}_n(\rho)$ must be functions of variable sign, with possibly one exception. It is noted that a number of important features of proper frequencies and eigenfunctions, which are established for self-adjoint problems of mechanics (see e.g., Collatz [1963]), apply to the problem considered here.

For shallow caps of revolution, the general equations for small vibrations around a symmetric equilibrium state, that are analogous to (2.164), are given by

$$\frac{r}{Eh} \Delta\Delta(\delta F) = -w_\Delta \delta w' - w_\Delta' \delta w''; \quad \mathrm{D}\Delta\Delta(\delta w) = \frac{F_s'}{r} \delta w'' + \left(\frac{1}{r} \delta F^{\cdot} + \frac{1}{r^2} \delta \ddot{F} \right) w_\Delta$$

$$+ \frac{1}{r} F_s'' \left(\delta w' + \frac{1}{r} \delta \ddot{w} \right) + \frac{1}{r} \delta F'' w_\Delta' + s \frac{\partial^2 \delta w}{\partial \tau^2}, \tag{2.172}$$

where the basic symmetric equilibrium state is given by the functions $F_s(r)$, $w_s(r)$; $w_\Delta(r) = w_s(r) + \overline{w}(r)$, while $\delta F(r,\phi,\tau)$ and $\delta w(r,\phi,\tau)$ describe the motion from this state. The corresponding boundary conditions are again homogenous, as in the previous case.

It is interesting to consider the specific class of solutions for which $w_\Delta \equiv 0$. In that case, the first equation (2.172) shows that $\delta F \equiv 0$ (the corresponding boundary conditions are trivial), while the vibrations are described by the equation

$$\mathrm{D}\Delta\Delta(\delta w) = F_s' \,\delta w''/r + F_s'' \,(\delta w' + \delta\ddot{w}/r)/r + s\frac{\partial^2 \delta w}{\partial \tau^2}. \qquad (2.173)$$

Expression (2.173) corresponds formally to the equation of vibrations of a linear circular plate stretched axisymmetrically by an in-plane loading generated by $F(r)$. The condition $\delta F \equiv 0$ implies that the considered vibrations take place without changes in the Gaussian curvature (inner geometry) of the initial shape (the state of the considered equilibrium state around which the vibrations take place). This observation is based on the comparison of the first equation (2.172), where $\delta F \equiv 0$, with (1.88). Thus, we are dealing with a case where the vibration mode is an infinitesimal isometric geometric transformation of the initial shape determined by $w_\Delta \equiv 0$. The isometric geometrical transformation of surfaces and its role in the theory of shells and plates will be considered and discussed at a later stage in this chapter.

In the general case, the linear system (2.172) can be solved by a standard procedure using corresponding Fourier series and obtaining the spectrum of proper frequencies. In the bifurcation and limit points, one of the frequencies becomes zero. It is this criterion, and the computations that are connected with it, that we will use for the analysis of bifurcation and snap-through phenomena [see, e.g., Archer & Famili, 1965; Famili & Archer, 1965; Shilkrut & Vyrlan, 1975, 1977].

d) A paradox connected with vibrations of an added mass attached to a geometrically nonlinear cap.

Let a point-mass of weight Q^* be attached to the apex of a geometrically nonlinear spherical cap, as shown in Figure 2.30. The mass of the cap itself is very small in comparison with the added mass. We will consider the vibrations of this system when the cap undergoes axisymmetric deformations.

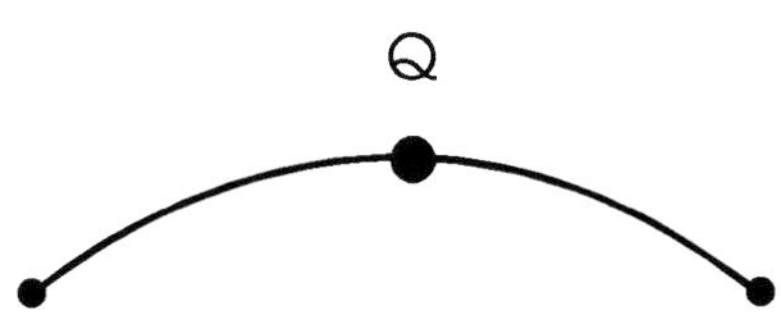

Figure 2.30 Cap with added mass.

Because the added mass is much larger than mass of the cap, we can simplify the problem by assuming that the latter can be neglected in comparison with the former. This (classical) simplification implies that the inertia of the cap itself does no longer play a role in the dynamics of the system so that the function of the cap is restricted to the role of a nonlinear, massless spring. Moreover, because we consider the added mass as a point-mass, the model of the cap with added mass can be reduced to a nonlinear system of a single degree of freedom.

The equation of motion for the point-mass can be obtained in following manner. First of all, it is necessary to obtain the solution of the static problem of axisymmetric deformation of the cap subjected to a concentrated force P at the cap's apex (see Fig. 2.31). ξ denotes the deflection of the cap's apex. In the simplest case, this static solution is given by the explicit expression (2.174)

$$P = f(\xi). \tag{2.174}$$

Then, using d'Alembert's principle — that is, substituting P by $Q^* - (Q^* / g)\ddot{\xi}$ in (2.174) — one obtains the equation of motion (2.175).

$$\frac{Q^*}{g}\ddot{\xi} + f(\xi) = Q^*, \tag{2.175}$$

where g is the constant of gravitation. A dot denotes differentiation with respect to the time t.

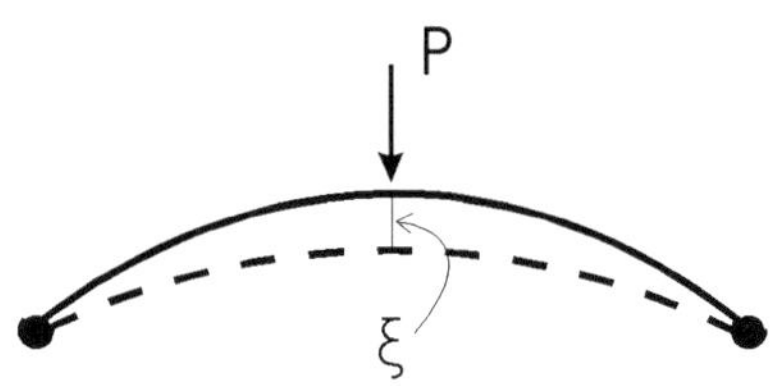

Figure 2.31 The cap as a nonlinear spring.

The properties of the autonomous, ordinary, differential equation (2.175), are well-known [e.g., Meirovitch, 1967; Stoker, 1950]. This equation admits periodic solutions, and can be integrated into

$$\frac{Q^*}{2g}\dot{\xi}^2 + \Pi(\xi) = C. \tag{2.176}$$

The period of the solutions is given by

$$T = 2\int_{\xi_1}^{\xi_2} \frac{d\xi}{\sqrt{\frac{2g}{Q^*}\left[C - \Pi(\xi)\right]}}. \tag{2.177}$$

In these expressions, the potential energy $\Pi(\xi)$ is determined by

$$\Pi(\xi) = \int f(\xi)d\xi - Q^*\xi, \tag{2.178}$$

while C follows from (2.176) and the initial conditions for $\xi(0)$, $\dot{\xi}(0)$. From (2.176) one obtains

$$\dot{\xi}(t) = \pm\sqrt{\frac{2g}{Q^*}\left[C - \Pi(\xi)\right]}. \tag{2.179}$$

The two different signs of the right side of (2.179) is an indication of the periodicity of the solution, since any periodic motion demands variable signs of the velocity. Expression (2.179) generates the second integral (2.180), below, and, in particular, the relation (2.177).

$$t - t_0 = \int_{\xi(t_0)}^{\xi(t)} \frac{d\xi}{\dot{\xi}}. \tag{2.180}$$

Parameters ξ_1, ξ_2 in (2.177) are the extreme values of ξ; that is, they are the roots of the equation $\dot{\xi}(t) = 0$, which is the same $C - \Pi(\xi) = 0$ (see (2.179)). The roots ξ_1, ξ_2 cannot be multiple; otherwise, the integral (2.177) would become meaningless.

It is very useful to illustrate the behavior of the system in the phase plane. We will consider at the beginning two different, simple, characteristic cases of $P = f(\xi)$ (curves I and II) shown in Figure 2.32.

When the height of the shell is increasing (or the thickness of it is decreasing), the curve $P(\xi)$ becomes more and more complicated in the sense that the number of the possible states of equilibrium corresponding to a given value of the loading increases (see Section 2.1). For our purposes here, however, it suffices to investigate the two cases qualitatively, as shown in Figure 2.31.

The two different phase planes of the motion of the mass described by equation (2.175) for the corresponding $f(\xi)$ given by the curves I and II are depicted qualitatively in Figures 2.33a and 2.33b, respectively.

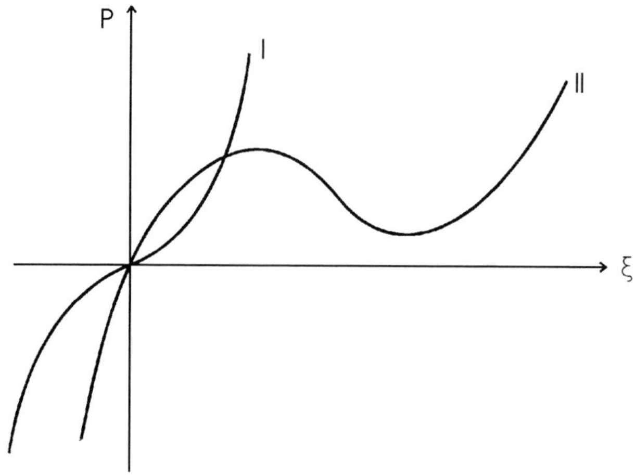

Figure 2.32 Typical responses for the cap.

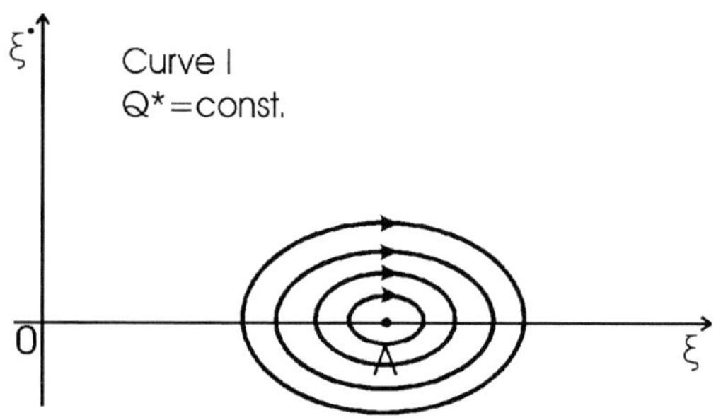

Figure 2.33a Periodic solutions in the phase plane; $P = Q^*$, case I.

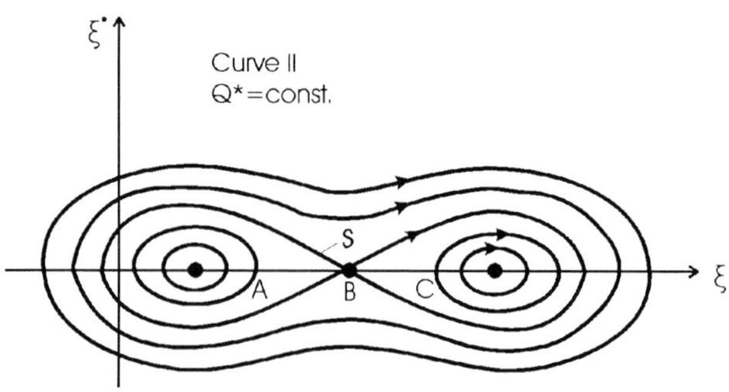

Figure 2.33b Periodic solutions in the phase plane; $P = Q^*$, case II.

The isolated singular point A in Figure 2.33a represents a unique equilibrium state of the cap, which corresponds to the load $P = Q^*$, case I. All possible motions of the mass are of a periodic character, as depicted by closed curves around A, each of which is determined by the corresponding initial conditions and the coordinates of a point through which the correspon-ding phase trajectory passes. The arrows show the directions of the motion along trajectories. Only one trajectory passes through each point of the phase plane. The equilibrium point of the type A is called a "center."

Figure 2.33b illustrates case II for a level of $P = Q^*$ where there are three equilibrium points A, B, and C, which are distinct points along the horizontal axis of the phase plane. Points A and C represent two stable equilibrium states, while B is an unstable state. The phase plane can now be divided into three distinguishing areas separated by the closed curve S (called the "separatrix"), which passes through point B, where it displays a self-intersection. Point B is called a "saddle point," while A and C, as in the previous case, are centers.

The motions of the mass are periodic around either equilibrium state A (closed trajectories within the left half of the "eight" of the separatrix S) or state C (within the right half of S), or around both states A and C (closed trajectories that are going around the whole separatrix). In addition, there is a special trajectory, represented by the separatrix S itself, that is a "periodic" motion with period $T = \infty$. This motion cannot, in fact, be realized because of the existence of various inevitable external perturbations.

This is, in general lines, a synopsis of the behavior of the structure considered above — the classical description of the behavior of an autonomous conservative system of one degree of freedom. In this model, the characteristic curves of the "spring" (flexible cap) are either of type I or II, and can be described by a single valued function $P = P(\xi)$. On the other hand, the inverse of $P = P(\xi)$, $\xi = \Xi(P)$ in the case illustrated by curve II in Figure 2.32 cannot be given in the complete interval in which $P = P(\xi)$ is defined, because, for one value of P, more than one value of ξ exists. In general, there is a very large class of geometrically nonlinear (flexible) systems that cannot be described by an explicit function $P = f(\xi)$ on the whole interval of ξ. The dependence between P and ξ can then only be given by an implicit relation

$$F(P, \xi) = 0. \tag{2.181}$$

Points at which the explicit relation $P = f(\xi)$ cannot be given are marked here by the condition $\dfrac{d\xi}{dP} = 0$, or correspondingly $\dfrac{\partial F}{\partial P} = 0$. When this occurs, the equation of the motion of the mass cannot be represented in an explicit relation (2.175) in the whole interval of ξ. (Below are some examples of geometrically nonlinear flexible systems of this type.)

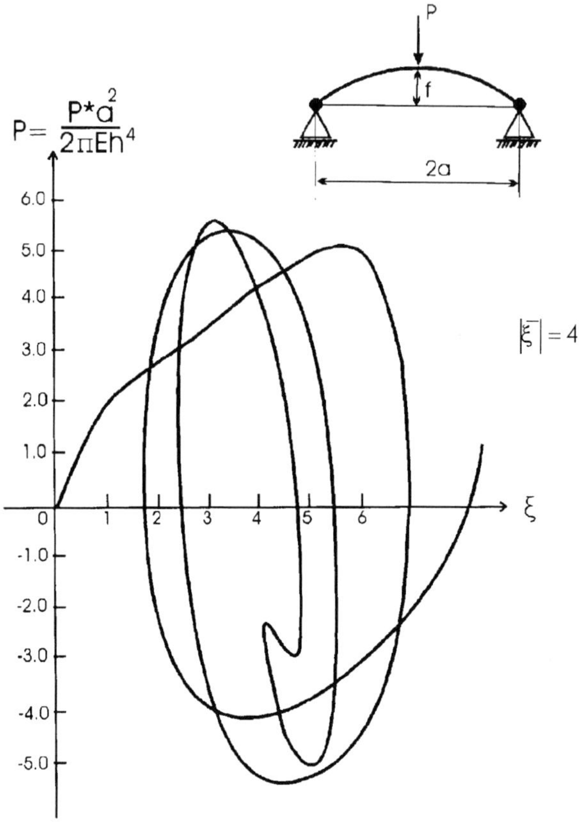

Figure 2.34 Response curve for shallow spherical cap with an immovable hinged edge; $\bar{\xi} = -4$.

Figure 2.34 shows the deformation curve for a shallow spherical cap with an immovable hinged edge for $\bar{\xi} = -4$ [see Shilkrut, Barladeanu, Vyrlan, Gerlaku, Fishberg & Tcherkezov, 1977], while Figure 2.35 shows the same for another specific geometry [see Wagner, 1985]. An analogous case is shown in Figure 2.36 for movable hinged supports and $\bar{\xi} = -12.3$ [see Mescall, 1965].

As expected, all three of these curves have similar features. The undulations are a reflection of the complicated nature of the behavior of the cap. This behavior can also be observed in the case of a shallow geometrically nonlinear hinged arch (Fig. 2.37), studied by Biezeno and Grammel, who obtained an exact analytical solution of the corresponding static problem [see Biezeno & Grammel, 1939]. But flexible structures that do not possess deformation curves with loops and self-intersections can still exhibit a behavior characterized by a multiple valued function P(ξ). For example, see the curve depicted in Figure 2.38, which describes the behavior of a curved panel [see Ramm, 1981; Wagner, 1985]. A similar curve is shown in Figure 2.39 for spherical caps of relatively small $|\bar{\xi}|$ [Shilkrut, Barladeanu, Vyrlan, Gerlaku, Fishberg & Tcherkezov, 1977].

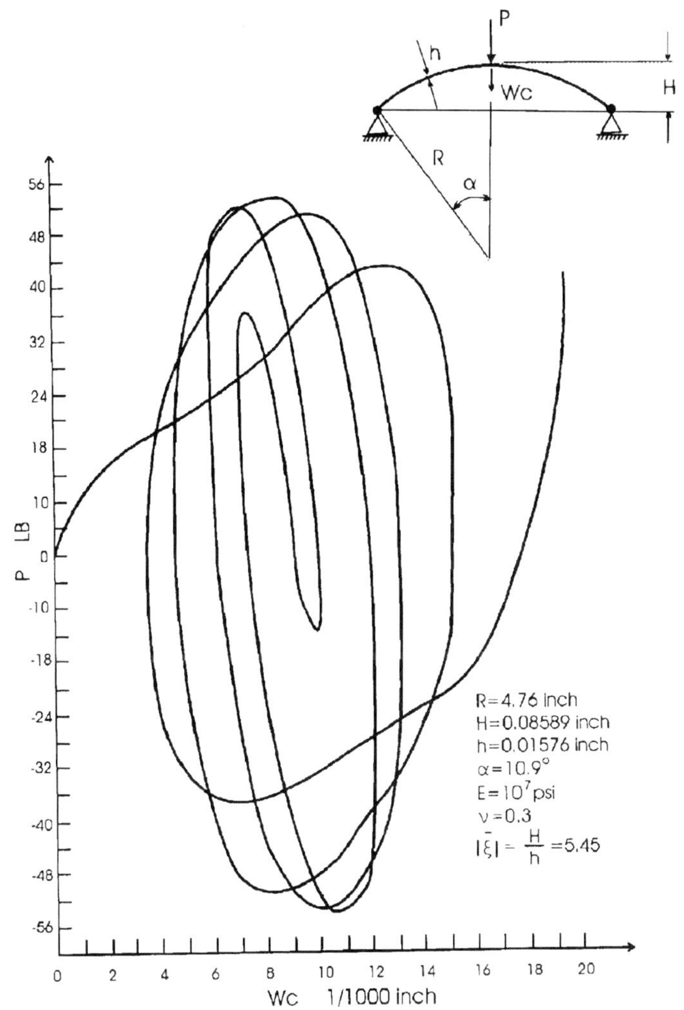

Figure 2.35 Response curve for shallow spherical cap with an immovable hinged edge, different geometry.

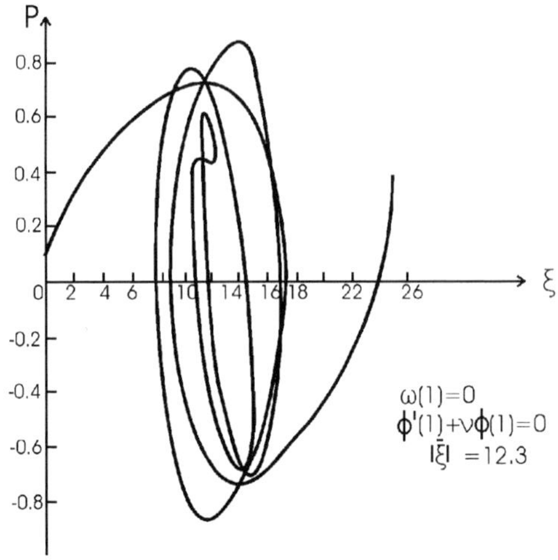

Figure 2.36 Response curve for shallow spherical cap with a movable hinged edge; $\bar{\xi} = -12.3$.

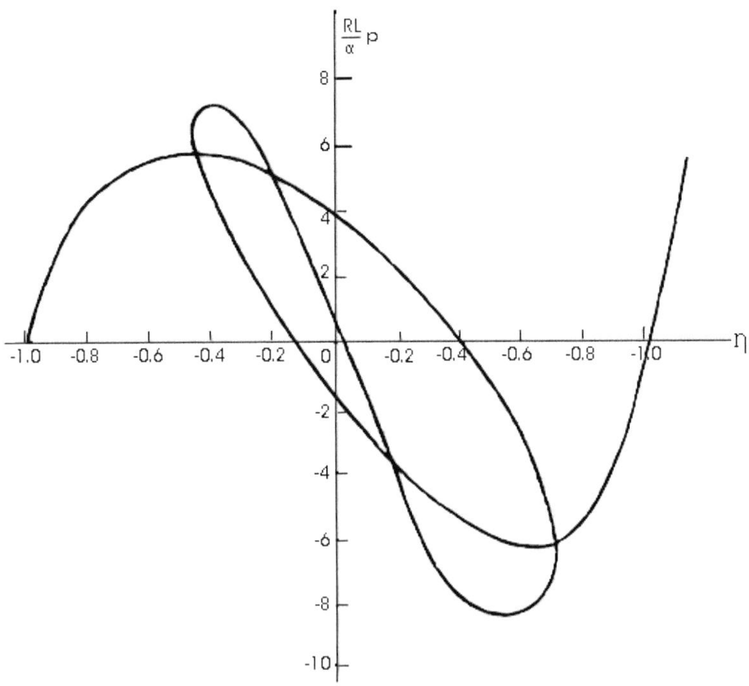

Figure 2.37 Hinged arch, Biezeno and Grammel [1939] example.

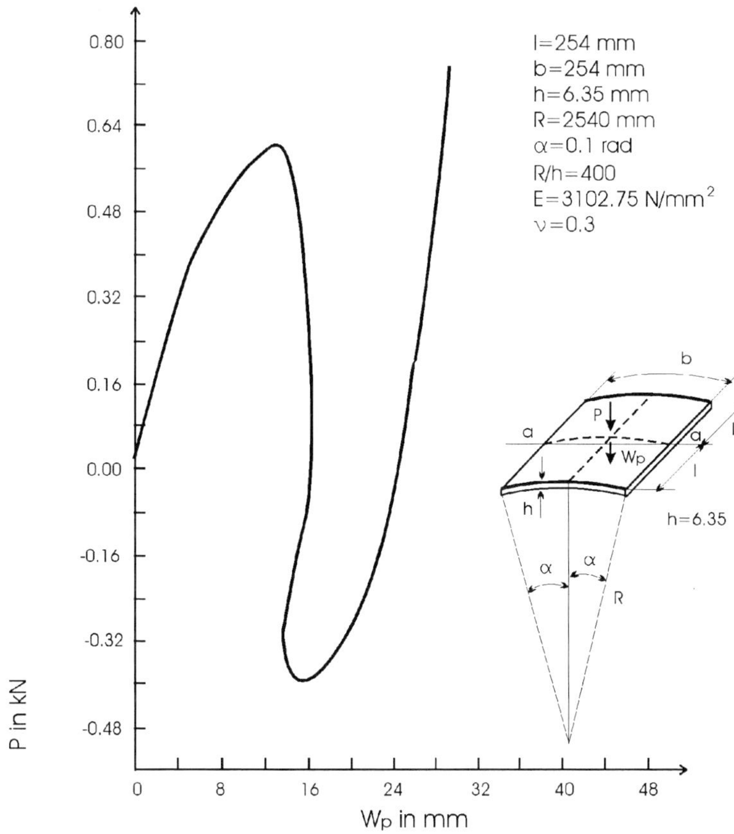

Figure 2.38 Curved panel example.

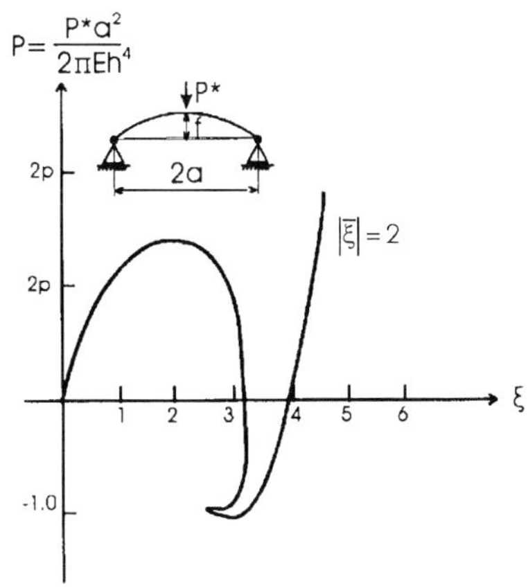

Figure 2.39 Spherical cap with $\overline{\xi} = 2$.

Thus, the other cases described above are not compatible with the classical approach, whereby the mechanical model can be reduced to a system of one degree of freedom. For these cases, the governing equation can only be obtained in implicit form,

$$F\left(Q^* - \frac{Q^*}{g}\ddot{\xi}, \xi\right) = 0, \tag{2.182}$$

or in the form (2.181).

We shall now try to construct a trajectory of (2.182) on the phase plane which corresponds to the given initial conditions

$$\xi(0) = a; \quad \dot{\xi}(0) = b. \tag{2.183}$$

Let the static deformation curve have the form that is qualitatively depicted in Figure 2.40. Points A-G denote the possible equilibrium states of the weightless spring supporting the mass, which corresponds to the static action of its weight Q^*. These states are the points ξ_A – ξ_G of the phase plane (see Fig. 2.41). (In the terminology of Poincaré, these points are called "isolated singular points.")

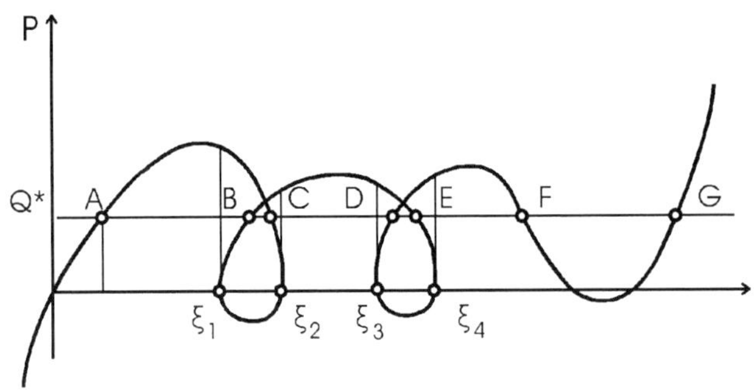

Figure 2.40 A spring characteristic of general form.

We shall approximate the trajectory in question (with initial conditions (2.183)) using Euler's integration. This approximative approach, as will be seen below, is adequate enough to establish the qualitative behavior of the system. Euler's method requires the computation of the field of the gradient $d\dot{\xi}/d\xi$ in the phase plane. This can be done using the expression

$$\frac{d\dot{\xi}}{d\xi} = \frac{\ddot{\xi}}{\dot{\xi}} = \frac{[Q^* - P(\xi)]g}{Q^* \dot{\xi}}. \tag{2.184}$$

The first equality (2.184) is a natural identity, while the second results from d'Alembert's principle. The gradient $\frac{\mathrm{d}\dot{\xi}}{\mathrm{d}\xi}$ is determined at $t = 0$ by the initial conditions (2.183)

$$\left.\frac{\mathrm{d}\dot{\xi}}{\mathrm{d}\xi}\right|_{t=0} = \frac{[Q^* - P(a)]g}{Q^* \, b}. \tag{2.185}$$

Consequently, the first step of the trajectory is obtained and this is shown in the phase plane (see Fig. 2.41).

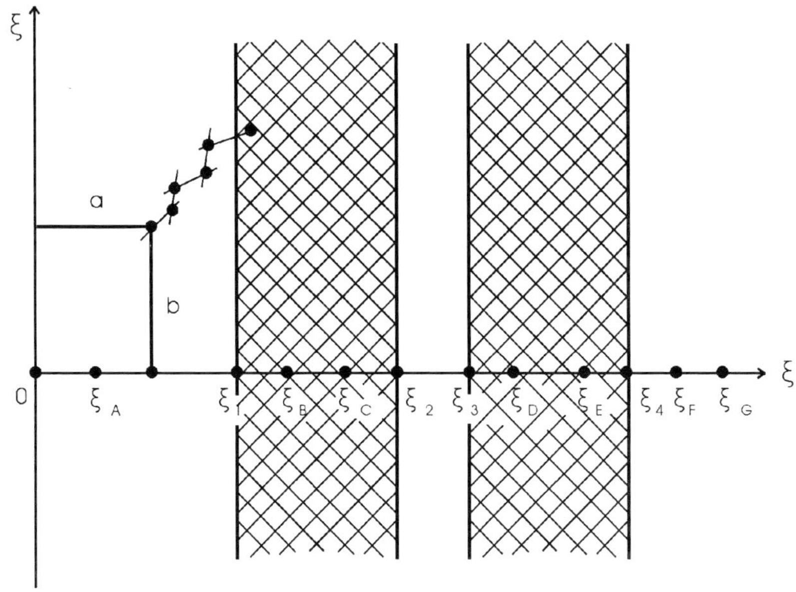

Figure 2.41 The regions of indeterminacy in the phase plane.

The end point of this small step is the initial point of the next increment, and so on, until the trajectory enters the left shaded strip of the phase plane. At the moment the solution enters this region, there are suddenly three values of the gradient, and it is no longer clear how to proceed with the solution. This solution thus ceases to be determined within the frame of the classical approach. We have come to an apparent paradox, where the usual classical approach breaks down because there is no way of knowing how to proceed with the solution for this model.

The question why this anomaly occurs and how it should be remedied is not particularly difficult to answer. The relation $P = P(\xi)$, which we used for the action of the cap as a spring, is a *projection* of the load P vs. static equilibrium solutions of the cap as a continuum, from the $1 + \infty$ dimensional space, to the *two-dimensional* space spanned by P and the apex displacement ξ. This means that, for every equilibrium state of the cap at a certain value of the load, we derive one distinct combination of P and ξ values. It must be clear, however, that it is not possible in general to find an inverse of this relationship. An arbitrary choice of the combination $(P, \xi) = [P(\xi), \xi]$ cannot represent the equilibrium state of the cap uniquely, even if we restrict the possible equilibrium states of the cap to axisymmetric deformations. Consequently, to solve the transient problem of the cap properly, we must also take into account the dynamic state of the cap as a continuum (or a proper n-dimensional discrete approximation of a continuum), and this means, in turn, that we must take into account the inertia distribution of the cap during its motion.

There are other examples of problems where it is not permitted to replace the continuous body by a system of one degree of freedom, even if the mass of the elastic body is very small compared to the mass of the weight. An interesting case is given in the book of the famous Soviet physicist L. Mandelshtam [1959]. If, as is usual, the initial conditions for the excitation of the simplest system of one degree of freedom (i.e., mass hinged to a linear spring; see Fig. 2.42) are not formulated for the mass itself, but for a point of the spring (e.g., point A in Fig. 2.42), then it is necessary to consider the wave propagation through the spring; i.e., the mass of the spring cannot be neglected and the spring must be regarded as a continuous body, at least in the initial stage of the process.

It is interesting to note that, even in the linear theory of shells, one is not always allowed to neglect the higher proper frequencies (as compared to the first frequency), because the spectrum of these frequencies is very dense at its origin (near to the area of smallest frequencies) [Benditch & Kornev, 1971; Bolotin, 1966; Kornev, 1972].

Another very instructive example is the system shown in Figure 2.43, where the beam is able to buckle, but the system as a whole is presented as a single degree of freedom model.

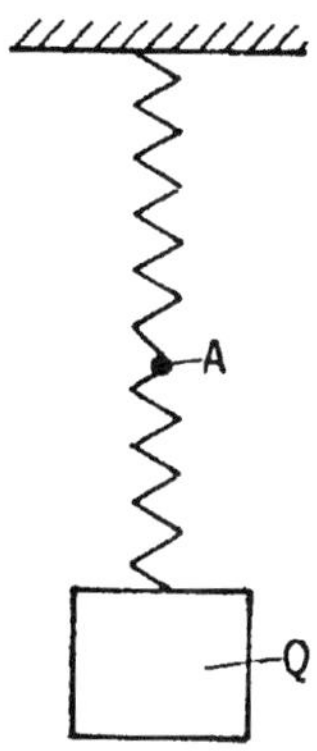

Figure 2.42 Simple mass spring system.

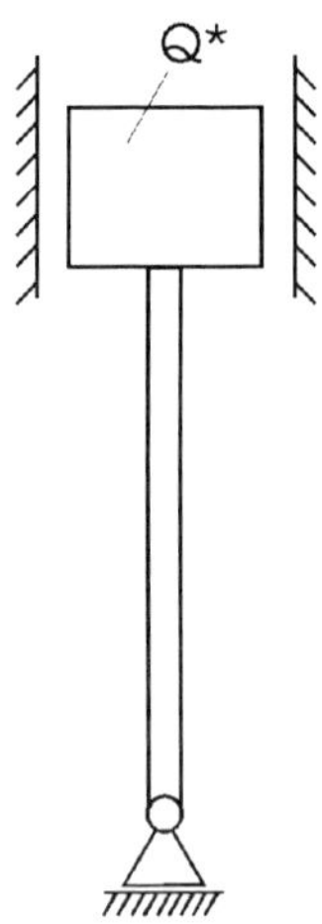

Figure 2.43 Beam column as a nonlinear spring.

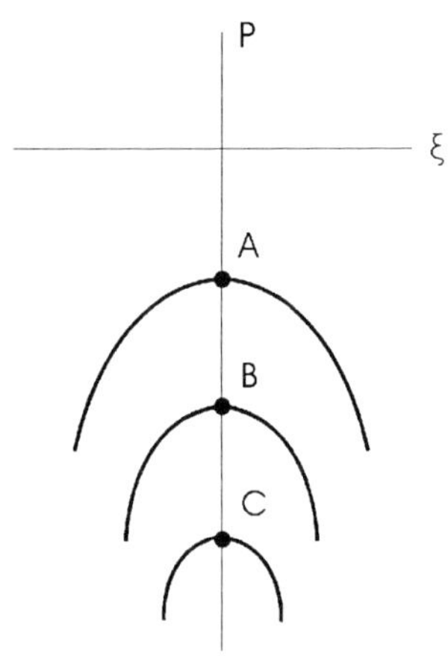

Figure 2.44 Static deformation curve; bifurcation values of P.

The static deformation curve of this structure is depicted in Figure 2.44, where A, B, C... mark the bifurcation values of P. As can be seen from this curve, an infinite multitude of values of P correspond to each value of ξ. The crucial question then arises about how one writes the equation of motion of the mass as a system, and what happens on the corresponding phase plane.

In the case of a shell shown in Figure 2.30, the complete equations of motion can directly be derived from the static equations using D'Alembert's principle. This procedure leads to an additional load term (in terms of a distributed load) given by

$$-\left[s+\frac{Q\delta(\rho)}{\rho}\right]\frac{\partial^2 w}{\partial\tau^2}+s+\frac{Q\delta(\rho)}{\rho}, \tag{2.186}$$

where $\delta(\rho)$ denotes the Dirac δ-function. The first term of (2.186) represents the inertial forces due to the distributed mass of the shell and the added mass, while the other terms stand for the forces of gravity. The parameters contained in (2.186) are all dimensionless, and conform to the conventions of Section 1.3.

2.9 Reciprocal systems of nonshallow shells.

The theory of reciprocal systems was considered earlier (Sections 2.2-2.4) for shallow shells. In the present section, we shall deal with reciprocal systems for the class of nonshallow shells of revolution that undergo static axisymmetric thermo-elastic deformations, and which are described by the basic equations and relations presented in Section 1.1.

A pair of reciprocal (symmetric) equilibrium states (denoted by additional indices 1 and 2, respectively) are those that satisfy the equations

$$\Phi_1(\xi)\equiv-\Phi_2(\xi);\;\; u_1(\xi)\equiv u_2(\xi);\;\; 0\le\xi\le\xi_K. \tag{2.187}$$

which are analogous to (2.1) for shallow states. Here we only consider isotropic shells, the rigidity of which is considered to be uniform.

It is assumed that the corresponding external excitations that generate this pair are connected by relations

$$P_{H1}\equiv P_{H2};\; P_{V1}\equiv-P_{V2};\; T_1\equiv T_2. \tag{2.188}$$

From (2.188) and (1.4)-(1.20), the following relations result:

$$V_1\equiv-V_2;\;\; r_1\equiv r_2;\;\; \varepsilon_{\xi_1}\equiv\varepsilon_{\xi_2};\;\; \varepsilon_{\theta_1}\equiv\varepsilon_{\theta_2};\;\; \alpha_1\equiv\alpha_2;$$

$$N_{\xi_1}\equiv N_{\xi_2};\;\; N_{\theta_1}\equiv N_{\theta_2};\;\; H_1\equiv H_2;\;\; Q_1\equiv-Q_2; \tag{2.189}$$

$$K_{\xi_1} + K_{\xi_2} = 2\frac{\overline{\Phi}'}{\overline{\alpha}}; \; K_{\theta_1} + K_{\theta_2} = 2\frac{\sin\overline{\Phi}}{\overline{r}}; \tag{2.190}$$

$$M_{\eta_1} + M_{\eta_2} = 2D\left(\frac{\overline{\Phi}'}{\overline{\alpha}} + \nu\frac{\sin\overline{\Phi}}{\overline{r}}\right) + 2M_T; \tag{2.191}$$

$$M_{\theta_1} + M_{\theta_2} = 2D\left(\frac{\sin\overline{\Phi}}{\overline{r}} + \nu\frac{\overline{\Phi}'}{\overline{\alpha}}\right) + 2M_T; \tag{2.192}$$

$$w_1 + w_2 = 2\int_{\xi}^{\xi_K} \overline{\alpha}\sin\overline{\Phi}d\xi, \tag{2.193}$$

where ξ_K is the value of ξ at the external contour of the shell where $w(\xi_K) = 0$. From (2.190), we get,

$$\Gamma_{\Delta 1} \equiv \Gamma_{\Delta 2}; \quad \Gamma_1 \equiv \Gamma_2. \tag{2.194}$$

which gives the connection between the Gaussian curvatures (see (1.89)) $\Gamma_{\Delta i}$ and $\Gamma_i (i = 1,2)$ of the reciprocal symmetric states "1" and "2."

If solution "1" exists, then the first two equations of equilibrium (1.1) and (1.2), and compatibility relation (1.13) are fulfilled by solution "2" connected with "1" by equalities (2.187)-(2.193). Substituting solution "2" in the third equation of equilibrium (1.3), it can be seen that this solution satisfies (1.3) if the following necessary condition holds:

$$r'D(1-\nu)\left(\frac{\overline{\Phi}'}{\overline{\alpha}} - \frac{\sin\overline{\Phi}}{\overline{r}}\right) + r\left[D\left(\frac{\overline{\Phi}'}{\overline{\alpha}} + \frac{\nu\sin\overline{\Phi}}{\overline{r}}\right) + M_T\right]' \equiv 0. \tag{2.195}$$

This relation must be an identity, that is, it is independent from r, and r' (i.e., independent of the deformed states). This can be ensured when, for example, the following two equations hold:

$$\frac{\overline{\Phi}'}{\overline{\alpha}} \equiv \frac{\sin\overline{\Phi}}{\overline{r}} \equiv \text{const.}; \quad D\left(\frac{\overline{\Phi}'}{\overline{\alpha}} + \frac{\sin\overline{\Phi}}{\overline{r}}\right) + M_T \equiv \text{const.} \tag{2.196}$$

These equations are satisfied if $M_T \equiv \text{const.}$, and the initial shape of the shell is spherical, with radius R_0, and

$$\overline{\alpha} = R_0; \; \xi = \overline{\Phi}; \; \overline{r} = R_0 \sin\overline{\Phi} = R_0 \sin\xi. \tag{2.197}$$

It is clear that, in order to ensure the reciprocicality of a spherical cap, the contour of the shell cannot be clamped, because the boundary conditions for clamping are $\Phi(\xi_K) = \overline{\Phi}(\xi_K) \neq 0$, so that the first relation (2.187) cannot be satisfied.

Let the edge be hinged. In this case, from (2.191) we have

$$M_1 + M_2 = 2D(1+\nu)/R_0 + 2M_T, \tag{2.198}$$

where the external edge moment should be interpreted as the function $M = M_r(\xi_K)$. Thus, the hinged, spherical, nonshallow shell loaded by edge moment M and $M_T =$ const. is a reciprocal system, and the properties of such systems are the same as those that were described earlier for shallow spherical caps.

It is worthwhile to emphasize that simplifying the basic equations by substituting into a number of their terms the parameters r and α of the deformed modes by their initial values (in the particular terms of these equations) [see e.g., Billington, 1967; Reissner, 1950] violates the symmetry of the system, and, in consequence, the simplified system ceases to be a reciprocal system. These simplifications sometimes facilitate the solution of the equations, but they are not necessary from the theoretical point of view.

Finally, it is interesting to note that the equation of compatibility (1.13) can be expressed in a form similar to (1.88) and (1.89) using (1.12):

$$\frac{d(\bar{r}\varepsilon_\theta)}{d\bar{s}} - \varepsilon_\xi \frac{dr}{ds} = -\int_o^s \Gamma r ds + \int_o^{\bar{s}} \bar{\Gamma}\bar{r} d\bar{s} + \cos\Phi(0) = \cos\bar{\Phi}(0), \tag{2.199}$$

where the arcs s and $\bar{s}$ have the same origin.

Equation (2.199) plays an important role in the theory of reciprocal systems, as will be shown in the next section.

2.10 About the compatibility equations

The compatibility equation (1.88) for the nonsymmetric deformation of shallow caps were given in Chapter I. This equation reveals the connection between the membrane stress function $F(r, \phi)$ and the changes of the Gaussian curvature in the initial and deformed state $\Gamma - \bar{\Gamma}$. In the case of axisymmetric deformation of shallow caps, the differential equation (1.58) for $\omega(\rho)$ can be written in a similar manner using the following transformation:

$$\theta^2 + 2\theta\bar{\theta} = \left(\theta + \bar{\theta}\right)^2 - \bar{\theta}^2 = \int_0^\rho \rho\left(\Gamma - \bar{\Gamma}\right)d\rho; \ \left(0 \le \rho \le 1\right), \tag{2.200a}$$

where

$$\rho\Gamma = \left(\theta + \bar{\theta}\right)\left(\theta + \bar{\theta}\right)'; \quad \rho\bar{\Gamma} = \bar{\theta}\bar{\theta}'. \tag{2.200b}$$

so that

$$L(\omega) = -\frac{1}{2\rho}\int_0^\rho \rho\left(\Gamma - \overline{\Gamma}\right)d\rho. \tag{2.201}$$

This shows, in the axisymmetric case, how the membrane stress function in the compatibility equation is coupled to the changes of curvature of the midsurface.

The following detail is very interesting here. The curvature change as a result of the shell's deformation in the compatibility equation is manifest in terms of the change of the Gaussian curvatures $\Gamma - \overline{\Gamma}$. Thus, if two different deformed states of the shell belonging to two different values of the external transverse forces (other things being equal) possess the same value of for the Gaussian curvature Γ, then the solutions for the membrane stresses are identical in these states. Apparently it is only the difference between the Gaussian curvature in the deformed state and undeformed state, $\Gamma - \overline{\Gamma}$, that is decisive for the membrane stress distribution.

The existence of reciprocal systems is based on this property (see corresponding formulae (2.190) and (2.194)). Another interesting property of the formulation of the shell equations is that the stress distributions belonging to a series of load steps can be added incrementally. In order to show this, the case of axisymmetric deformation in its simplest version — when, in particular, q_t, N_T are absent — will be considered. Let an arbitrary solution be given by θ(ρ) and ω(ρ), and let an incremental increase of this state be given by δθ and δω. The corresponding equation for the new membrane stress function ω + δω is then given by (2.201)

$$L(\omega + \delta\omega) = -\frac{1}{2\rho}\left[\left(\theta + \overline{\theta} + \delta\theta\right)^2 - \overline{\theta}^2\right]$$

or

$$L(\omega + \delta\omega) = -\frac{1}{\rho}\left(\Delta_1 + \Delta_2\right), \tag{2.202}$$

where

$$\Delta_1 = \int_0^\rho \rho\left(\Gamma_1 - \Gamma\right)d\rho = \left[\left(\theta + \overline{\theta} + \delta\theta\right)^2 - \left(\theta + \overline{\theta}\right)^2\right]/2,$$

$$\Delta_2 = \int_0^\rho \rho\left(\Gamma - \overline{\Gamma}\right)d\rho = \left[\left(\theta + \overline{\theta}\right)^2 - \overline{\theta}^2\right]/2. \tag{2.203}$$

Δ_1 is here the change of the Gaussian curvature during the incremental change of the shell from the state θ, ω to θ + δθ, ω + δω; Δ_2 is the change of the Gaussian curvature between the transition of the state θ, ω and the undeformed state $\theta, \omega \equiv 0$. Thus, the incremental changes are added to the previous state of stress to give the state of stress in the new state. This property remains valid in the case of nonsymmetric deformations.

When the theory of nonshallow shells is used, the compatibility equation (2.199) can be transformed in terms of the stress function S, which defines the stress resultants N_ξ and N_θ in the following way:

$$N_\xi = \frac{\cos\Phi}{r} S; \;\; N_\theta = \frac{S'}{\alpha} + \left[r p_H - \frac{1}{\alpha}\left(rVtg\Phi\right)' \right] \qquad (2.204)$$

This equation for S will contain, in the general case, the load terms p_H and p_V (via V), while, for shallow shells, the corresponding equations of compatibility do not contain any terms depending on the external loading. In the particular case when $p_H \equiv p_V \equiv 0$ (only edge loading applied), the compatibility equation of S is of the same type as the compatibility equation for the shallow shells, and it has the same properties. We finally note that the role of the membrane stresses and the Gaussian curvature in the theory of shells has been discussed in detail by Calladine [1983].

2.11 Nonregular isometric transformations of surfaces and nonregular solutions of the theory of geometrically nonlinear shells.

An isometric transformation of geometrical surfaces is, as is well-known, a change of shape that preserves the metrics and, in consequence, the Gaussian curvatures Γ. In particular, those surfaces that are capable of being developed (cylindrical, conical, and other surfaces of zero Gaussian curvatures) can be formed by regular isometric transformations from a plane. During such a transformation, no rupture or stretching of the surface result.

There are two types of isometric transformation — regular (usual) and nonregular. Regular ("smooth") isometric transformations preserve the continuity of the surface and its smoothness after the transformation. Nonregular transformations satisfy only the continuity condition, but do not satisfy the smoothness condition. It turns out that nonregular transformations are easier to find than regular transformations; for example, a regular isometric transformation of a convex surface does not exist (except for a motion of the surface as a rigid body, which, in our case, is of no interest).

On the other hand, there is an infinite multitude of nonregular isometric transformations of such a surface [see Pogorelov, 1966, 1967, 1986]. Consider the cap shown in Figure 2.45. The segment Q of the cap located above plane P is transformed into Q', which is the mirror image of Q with respect to P. This new surface, which is composed of the unchanged "lower" segment R and the reflected part Q' is obtained by a nonregular isometric transformation of the initial surface, because it satisfies all the requirements for an isometric transformation except for the loss of smoothness of the surface along the line AB in the plane *P* that connects both segments. It is noted, that it is also possible to obtain a sequence (even infinite) of such transformations by means of reflections relative to a sequence of different planes.

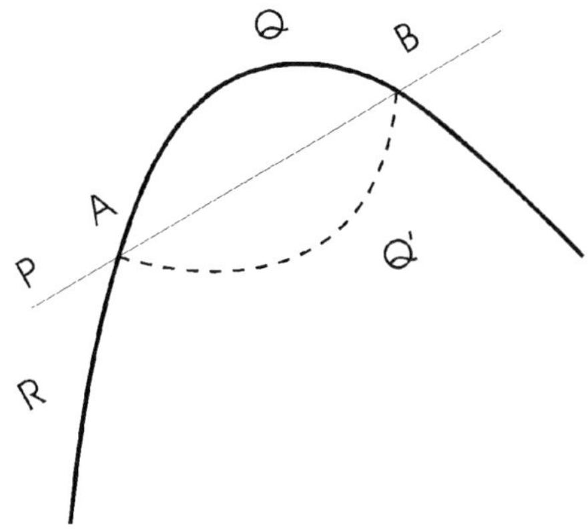

Figure 2.45 Nonregular isometric transformation by reflection.

The isometric transformation of surfaces by means of reflections of its parts is not the only possible type of nonregular transformation. For example, first separating and then connecting back a segment of a conical surface, as shown in Figures 2.46 and 2.47, is also a nonregular isometric transformation.

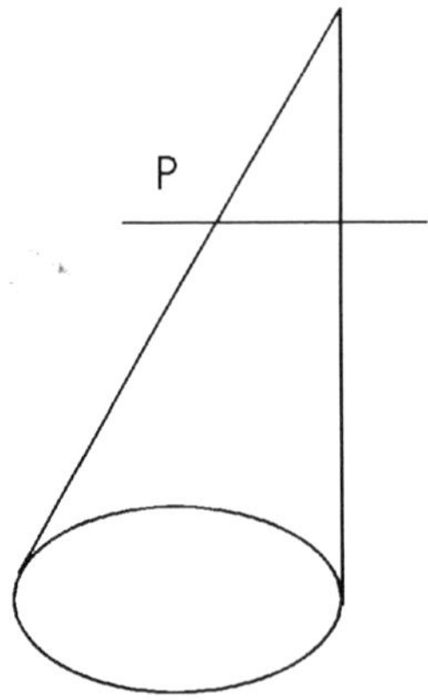

Figure 2.46 Conical surface.

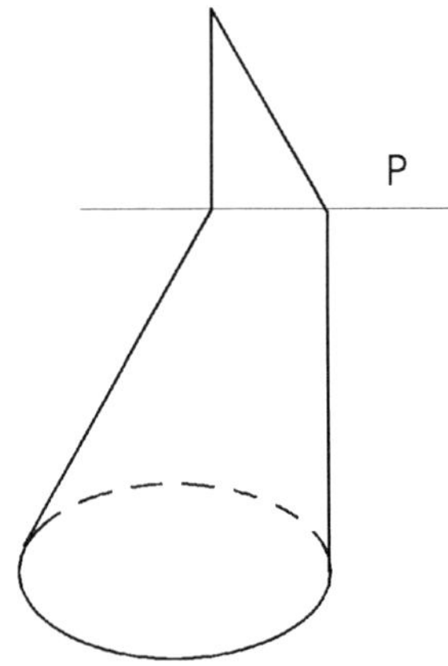

Figure 2.47 Nonregular isometric transformation of a conical surface.

Nonregular transformations of the first type have some applications to the theory of deformation of geometrically nonlinear shells. Many deformation states of shells are approximations of surfaces containing reflected parts, as is shown, for example, in Figure 2.45, where the nonregular transformation describes a solution belonging to the theory of nonlinear shells.

In order to consider some of these nonregular solutions of shell equations, we shall begin with an axisymmetric spherical cap, which undergoes a reflection of the type shown in Figure 2.45. The nonregular transformation that brings the initial symmetric cap ($R + Q$) to a new state — that is to a symmetric version of $R + Q'$ — is given by

$$F \equiv 0;\ \text{in}\ (R+Q');\quad w \equiv \begin{cases} 0 \text{ in } R \\ \neq \text{ in } Q \end{cases}, \tag{2.205}$$

which satisfy equations (1.74) and (1.88) for shallow shells when all external forces are zero. The Gaussian curvature of the reflected part is given by: $\Gamma_\Delta \equiv \overline{\Gamma}$ in $(R+Q')$. This means that the Gaussian curvature Γ remains invariant during this nonregular isometric transformation. In the unchanged part R, we have $w \equiv 0$ and membrane stress function $F \equiv 0$. In part Q', due to the reflection, $\Gamma_\Delta \equiv \overline{\Gamma}$, and, as a result of (1.88), $F \equiv 0$; but the solution for w is here $w \neq 0$. The latter, the displacement function w, can easily be obtained using the principle of a geometric reflection.

Thus, the solution (2.205) exists and satisfies the conditions for a nonregular isometric transformation; the nonregularity occurs along the line of intersection of the cap and plane P (see Fig. 2.45). Please note, that we assumed here that the cap is symmetric and spherical. It is clear, however, that similar nonregular solutions exist for nonsymmetric caps of spherical shape (for example, spherical caps with a rectangular base, as shown in Fig. 2.48).

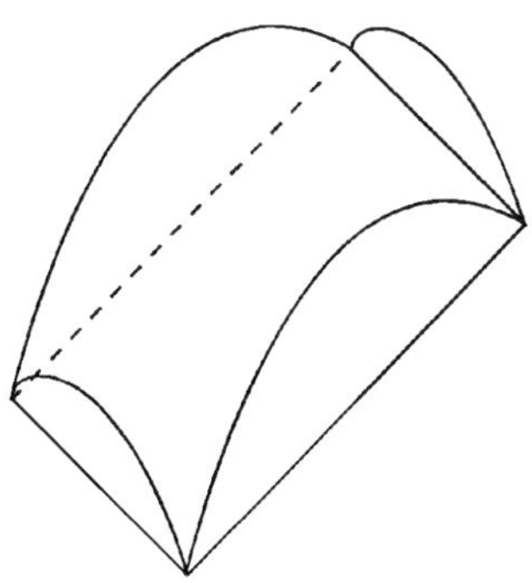

Figure 2.48 A particular example.

The possibility of applying a reflection of some part of the shell is equivalent to the possibility of creating a nonregular transformation. It follows, then, that the solution (2.205) also exists for any nonspherical cap.

In summary, it can be stated that the theory of geometrically nonlinear shells (membranes) itself describes deformation states of shells that are obtained by nonregular isometric transformations in terms of reflections of parts of the undeformed state of the shell. What we are discussing here is the class of nonregular transformations that lead to a discontinuity of the normal to the midsurface along some generator of this surface; but actual shells, which always possess some bending rigidity, are of course not able to assume a deformed shape of the type shown in Figure 2.45. The generator along which the discontinuity in slope occurs in the zero bending rigidity case is then, in reality, a narrow transition zone. But it is to be expected that deep, thin caps are able to undergo deformation modes close to the ones described by nonregular isometric transformations. The deformation of the shell is then mainly "concentrated" in this transition zone, which takes the character of a boundary layer solution. Some approximate analytical solutions have been published in the literature based on special hypotheses about the character of the deformation state in the transition zones [e.g., Aswell, 1959; Pogorelov, 1966, 1967, 1986; Yoshimura, 1951], or where asymptotic methods are used to describe the solutions in these zones [e.g., Vasyljeva & Butuzov, 1973; Vishik & Lyusternik, 1958; Vorovitch & Srubtschik, 1970].

The nonregular solutions can be used to explain why the number of equilibrium solutions of a shell for a fixed value of the load grows beyond bounds when the parameter $|\xi| \to \infty$; that is, when the thickness $h \to 0$. As can be seen from Figures 2.7-2.9, where the deformation curves of a simply supported cap are presented, the number of nontrivial states that belong to the unloaded case quickly increases with $|\xi|$. Of course, when $|\xi|$ increases beyond bounds ($h \to 0$), the shell under consideration becomes a membrane. It can then be expected that such a shell can possess nontrivial solutions at the zero load which approach the nonregular isometric changes of shape of the reference surface.

These solutions do not depend, of course, on the specific type of loading — deformation curves for any loading must cross the axis ξ (see, e.g., Figs 2.7-2.9) through the same points. Comparing the deformation curves presented in Figures 2.7 and 2.8, the influence of $|\xi|$ can clearly be seen. When $|\xi|$ = 6 (Fig. 2.7), there are 10 nontrivial solutions when loading parameter $M = 0$; but for $|\xi|$ = 15.15, there are already 28 similar nontrivial solutions. We thus come to the observation that the axis ξ (loading parameter is identically zero) is the geometric locus of nonregular solutions (nonregular isometric transformations of shells that approach the case $h = 0$). Thus, when $h \to 0$, the number of solutions goes to infinity.

The reasoning presented above is related to a simply supported cap. In the case of clamped edges, there are no nontrivial solutions at zero load for $h \neq 0$ (see, for example, the deformation curves presented in Figs. 2.1-2.6). This follows because, to the best of our knowledge, there are no numerical solutions known that point to axisymmetric nontrivial solutions of this type. It can actually be shown, that in the case of clamping, the equations of axisymmetric deformations of shallow shells of revolution do not possess any of the nontrivial solutions discussed here. The absence of these nontrivial solutions is probably a result of the existence of a thin boundary layer adjoining the edge, where large bending moments occur that are caused by the clamping condition.

In conclusion, it can be said that the notion of the nonregular isometric transformation of surfaces is a useful tool for physical explanations of some aspects of the behavior of shells. The theories of stability of geometric, nonlinear, convex shells, however, which were carried out by A. Pogorelov [1966, 1967, 1986] on the basis of this geometric transformation, unfortunately did not produce satisfactory results. For example, the theory in question does not take into account the influence of the boundary (support) conditions on critical loads, while, as is very well-known, the buckling loads of caps are extremely sensitive to boundary conditions. It is clear that the influence of support conditions decreases when the cap is deeper and thinner, but this does not apply to shallow shells, a feature which is used in this theory [Pogorelov, 1986]. According to Pogorelov's theory, the convex cap of revolution does not lose its stability by snap-through processes in the class of axisymmetric deformations when it is loaded by a concentrated force acting at the apex of the cap. This conclusion contradicts well-known theoretical results [Fitch, 1968; Fitch & Budiansky, 1970; Mescall, 1965; Shilkrut, Shevandronov, Morar & Maksimov, 1969], as well as experimental data [see Loo & Evan-Iwanowski, 1966]. The essential deficiencies (the list of which is not limited to the two cases described above) of the theory under discussion result from assumptions that were introduced by Pogorelov [1966, 1967, 1986]. In our opinion, the following is the most questionable among these assumptions.

The energy introduced by membrane stresses, σ_r, σ_φ, acting beyond the transition zone formed around the fold line AB (see Fig. 2.45), is negligible in comparison with the energy associated with the bending and circumferential stretching in the transition zone mentioned above. The neglect of the energy introduced by membrane stresses is absolutely not acceptable in the light of its decisive role in the process of loss of stability of caps, as was shown and discussed in detail in Sections 2.3 and 2.8 (see also Table 2.3). Very significant in this regard is the classical von Mises truss. The truss and its deformation curve are depicted in Figures 2.26 and 2.27, respectively. The links 1, 2 (see Fig. 2.26) can only transfer tension or compression, analogously to σ_r in the case of a cap. The existence of the stresses in the links is enough to generate a snap-through type of loss of stability, as is shown by the curve in Figure 2.27. Thus, $\sigma_r < 0$ is the main factor generating the process of loss of stability by jumping. When Pogorelov [1966, 1967, 1986] removed σ_r, the jump was removed automatically. It is interesting to remember that the maximum bending stress σ_b is proportional to h^{-2}, while the critical value of σ_r of stability loss of spherical shells, for example, is proportional to h^2. Thus, $\sigma_b >> \sigma_r$ excludes the possibility of loss of stability.

The geometrical approaches discussed above are very useful for the physical explanation of the various nonlinear phenomena tied to the loss of stability of thin, elastic, flexible (non-linear) shells, but it is unlikely that these approaches can serve as reliable tools for numerical solutions if one takes into account the present availability of very powerful numerical methods that can obtain the shell solutions in a very general context. Finally, it is noted that the famous book by A. E. H. Love [1927; see p. 553] is one of the first publications where regular isometric transformations were considered for problems of plates and shells.

Chapter 3

Numerical Investigations of Axisymmetrically Loaded Geometrically Nonlinear Shallow Spherical Caps and Circular Plates (A Survey)

3.1 Introduction and short survey of numerical methods.

There are many publications devoted to the analysis of the behavior of spherical caps, but there are only a few surveys of these results [e.g., Bushnell, 1989; Fung & Sechler, 1960; Kaplan, 1974; Shilkrut, 1980; Valishvili, 1976; Volmir, 1963; Vorovitch & Minakova, 1973] — this despite the fact that spherical shells, together with cylindrical shells, belong to the two basic models of nonlinear shell theory.

Although cylindrical shells represent the class of developable surfaces, and spherical shells represent the class of undevelopable surfaces, both of these classes are characterized by a positive Gaussian curvature. (Shells with negative Gaussian curvature must be considered separately.) The study of spherical caps makes it possible to elucidate the behavior of doubly curved, convex shells, while, at the same time, it provides a good illustration of the principal characteristics of the behavior of geometrically nonlinear shells and thin-walled structures in general.

We shall confine ourselves to closed (no central hole), axisymmetric spherical segments and circular plates of revolution with homogenous, isotropic, linear elastic material properties, which are able to undergo axisymmetric and nonsymmetric deformations while subjected to axisymmetric load systems. In our opinion, other cases that deviate from this description introduce some additional complexity or diversity in behavior, but, generally speaking, do not lead to new concepts or methodologies.

The first investigations that dealt with these types of problems were carried out using the Ritz, or Galerkin, method, or other variants of the direct method of variations, reducing the shell model to a system of one degree of freedom [see, e.g., Biezeno 1935; Panov, 1939b; von Karman & Tsien, 1939; Feodosjev, 1945]. Surveys of these results can be found in, e.g., Volmir [1956, 1963] and von Karman & Kerr [1965]. These early solutions were obtained analytically, and they played a significant role in the initial stage of the development of the theory that led to the detection of hitherto unknown phenomena, such as snap-through as a form of loss of stability. Nevertheless, because these solutions could only serve as first, let us say, rather crude approximations, they were not very reliable in either a quantitative or qualitative sense. For example, the first approximation of the problem referred to here cannot, in principle, reveal more than three solutions at one value of the loading parameter, while, in reality, the number of solutions can be very much larger (see Sections 2.1 and 2.3). Another example of the inadequacy of these first approximations is the negative sign of the lower critical pressure (lower limit load, snap-back load) of the clamped shallow spherical cap that is produced by the simplified analysis [Volmir, 1956]. In reality, the lower critical pressure (the snap-back pressure) of a clamped spherical cap is never negative (see Section 2.1), and this condition is the result of a general principle. Many other, analogous examples can be presented that show that the accuracy of the first approximation is often far from being acceptable, although its use should not be excluded in all cases.

At the end of the 1950s and the beginning of the 1960s, the rapid development of computer techniques made it possible to solve the governing equations accurately. Many different numerical methods emerged, and their development still continues. Several groups of these methods and corresponding algorithms can be classified in the following paragraphs.

a) ***Direct variational methods.***

In particular, the methods of Ritz and Galerkin* belong to this group. There is an extremely large body of literature on numerical methods that are based on variational (energy) principles. Because of the vast number of such works, we shall refer to only some of these publications, without any hope of reflecting even the bulk of them [see, e.g., Kantorovich & Krylov, 1964; Lanczos, 1949; Langhaar, 1962; Leibenzon, 1943; Mikhlin, 1964, 1971; Mushtari & Galimov, 1957; Reissner, 1958a,b; Shames & Dym, 1985; Touti, 1967; Trefftz, 1928; Vlasov, 1964; Volmir, 1963; Washizu, 1968; Svirsky, 1968].

b) ***Direct discretization methods.***

To these methods, which are now very widely used, belong, for example, the method of collocation [e.g., Kantorovich & Krylov, 1964; Kornishin, 1964; Silverman & Mays, 1973; Smirnov, Aleksandrov, Shaposhnikov & Ljastchennikov, 1964], the finite difference method [e.g., Archer, 1962; Bushnell, 1989; Kantorovich & Krylov, 1964; Shames & Dym, 1985; Thurston, 1961; Valishvili, 1976; Wilson & Spier, 1965], the finite element method, the use of which (and of a "younger" similar method of boundary elements) has increased incredibly over the last 10-20 years. The number of publications related to these methods is so large that this literature needs a series of special surveys [see the general monographs of, e.g., Banerjee & Butterfield, 1981; Adey and Brebbia, 1983; Brebbia, Telles & Wrobel, 1984; Oden, 1972; Zienkiewicz, 1977].

---

* In the Russian literature, this method has been called the "method of Bubnov-Galerkin," or just the "Bubnov method," for over seventy years. The fact is that I. G. Bubnov, in a review of a work by S. Timoshenko, presented ideas that were similar to Galerkin's two years before the latter's famous article appeared in 1915. It was, however, Galerkin who actually introduced and widely used this method in practice.

c) ***The Shooting Method, by which a boundary value problem of ordinary differential equations is transformed to a Cauchy problem (an initial value problem).***

This method has been successfully applied to many problems of nonlinear shells [see Bauer, Reiss & Keller, 1970; Keller, 1968; Keller & Wolfe, 1967; Roberts & Shipman, 1972; Shilkrut, 1992a; Shilkrut, Morar & Gerlaku, 1970; Shilkrut, Shevandronov, Morar & Maksimov, 1969]. The shooting method can be of great use when the problem at hand permits a reduction to a system of ordinary differential equations. In that case, the reduced (boundary value) problem can be reformulated in one or more (parallel shooting) initial value problems, which can then be solved numerically using a time integration method (e.g., a Runge Kutta method).

d) ***Methods of other types, as, for example, that of successive approximations.***

Budiansky [1959], was one of the first to use this method for the study of the behavior of nonlinear spherical caps, and has achieved a high degree of accuracy.

e) ***Asymptotic methods.***

The perturbation method was applied first of all for plates [see, e.g., Chien, 1947; Way, 1934], while the asymptotic series were used in particular for investigations of equilibrium states of spherical caps that are close to nonregular solutions (see Section 2.11) [e.g., Vorovitch & Srubstchik, 1970].

f) ***The transient method of analysis.***

This method starts from the equations of motion by applying very slow changes in the load with time (quasi-static loading). In combination with the Galerkin method (or other methods mentioned in a) and e), above), this approach reduces the problem to an initial value problem yielding systems of ordinary differential equations, which are relatively simple to solve numerically. At the same time, this approach enables one to determine the stable equilibrium states. It was successfully applied for a series of problems in stability [e.g., Feodosjev, 1963; Otter, 1965; Shilkrut & Vyrlan, 1977].

g) ***Combinations of methods.***

Many numerical algorithms are based on combinations of methods. For instance, in the case of nonsymmetric deformations of spherical caps, Kantorovich and Krylov [1964] solved a system of partial differential equations where each unknown function depended on two spatial arguments (r, φ). This method uses Galerkin's procedure in one direction (e.g., in the direction r), and in such a way that the original system of partial differential equations is transformed into a system of ordinary differential equations, which can then be solved analytically by any convenient approximation or by numerical means.

Very often Galerkin's method for solving differential equations of shells is implemented in the following way. The approximate expression for the unknown displacement function $w(r, \phi)$ is introduced in the right-hand side of the compatibility equations (1.76) for the stress function F. Because this equation is linear, it can be solved exactly. The solution for F (in terms of w) obtained in this way is introduced into the equilibrium equation (1.74), which is then subjected to Galerkin's procedure. As a result, a system of nonlinear algebraic equations of the third degree (in the case of shallow shells) is obtained for the unknown coefficients of the approximative expression of w. Naturally, Galerkin's method can be used directly, provided the approximate expressions of both basic unknowns F and w can be given *a priori* (for example, on the basis of physical or numerical considerations).

The nonlinear algebraic equations obtained by Galerkin's method depend on the intensity of the external loading. To solve these equations, an incremental (step-by-step) solution procedure can be used, whereby the different equilibrium states belonging to different load intensities are obtained by varying the load in small steps. Each step of such a procedure corresponds to the calculation of a sequence of linearized problems derived from the nonlinear set of equations. A solution procedure of this kind is sometimes called the "incremental-iterative procedure," because the solution along the chosen equilibrium path is carried out incrementally for the parameter that controls the step along the path, while the solution at each step itself is obtained by an iteration method (sequential linearization) [see, e.g., Bushnell, 1989; Riks, 1971, 1973; Stein, Wagner & Lambertz, 1984; Yaghmi & Popov, 1971; Zienkiewicz, 1971, 1977].

Another version of the incremental approach is based on the linearization of the basic nonlinear differential equations within each small step of the load parameter [Petrov, 1975]. It is also possible to use the solution of a series of linear differential equations of successive approximations during each increment [Shilkrut, 1965, 1967]. Many other examples of such "compound" algorithms can be enumerated, but the number of different combinations of basic methods is very large. A recent survey is given in Noor [1994]. See also the survey of Vorovitch and Minakova [1976].

Numerical methods for the solution of nonlinear ordinary systems based on the differentiation with respect to the influence parameter (e.g., the loading) have been used by Davidenko [1953], Kurdyumov [1961], Shamansky [1966], and Vorovitch and Zipalova [1965]. Transient methods for the solution of these systems were proposed and used by Feodosjev [1963], a work we have already mentioned.

An entirely different possibility is the application of analog computers to the solution of equations of nonlinear shells [e.g., Shilkrut *et al.*, 1969; Volmir, 1963]. This application is quite different from the use of digital computers, because digital computers solve the discretized models of the physical problem, while analog computers simulate the behavior of the physical problem by devices that represent the physical system itself (e.g., by means of an electronic circuit). Consequently, these devices give guaranteed qualitative results, although the solutions are not always accurate enough from the quantitative point of view.

h) *Preliminaries to the review of the behavior of circular plates and spherical caps.*

It is beyond the scope of this book to discuss the workings and intricacies of the subject of numerical solution extensively. Instead, we will devote the remaining part of this chapter to a review of solutions that have appeared in the literature in connection with the problem of shallow spherical caps loaded by axisymmetric load systems. Before we review the numerical results themselves, however, it is necessary to reiterate here the elementary geometrical relations of the initial shape of spherical caps (see Fig. 3.1, where the initial geometric parameters are presented). We have

$$|\overline{w}(0)| = R(1-\cos\alpha), \tag{3.1}$$

where $\overline{w}(0) < 0$ for the case drawn in Figure 3.1.

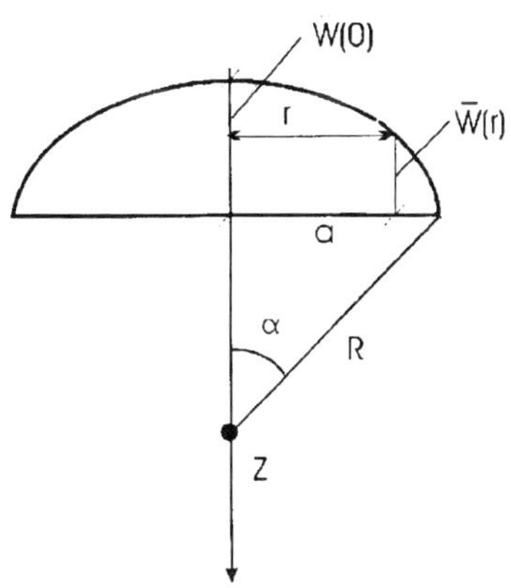

Figure 3.1. Basic conventions shallow spherical cap.

For a shallow shell, this can be written as

$$\overline{\xi} = \overline{w}(0)/h \cong R\alpha^2/(2h); \tag{3.2}$$

$$a^2 \approx 2R|\bar{\xi}|h; \quad \alpha \approx a/R. \tag{3.3}$$

Thus, the initial geometry of a shallow spherical cap can be characterized by one parameter; e.g., $\bar{\xi}$. In many cases, another parameter is used instead of $\bar{\xi}$; i.e.,

$$\lambda = 2\left[3\left(1-\nu^2\right)\right]^{1/4}|\bar{\xi}|^{1/2} = \left[12\left(1-\nu^2\right)\right]^{1/4}(R/h)^{1/2}\alpha. \tag{3.4}$$

From (3.4), we derive:

$$\lambda^2 = 6.608|\bar{\xi}| \text{ or } |\bar{\xi}| = 0.151\lambda^2 \text{ when } \nu = 0.3;$$

$$\lambda^2 = 6.532|\bar{\xi}| \text{ or } |\bar{\xi}| = 0.153\lambda^2 \text{ when } \nu = 1/3. \tag{3.5}$$

The criterion for the shell to be shallow can be taken in the form of the following inequality (see Section 1.2).

$$h/R < 1/6. \tag{3.6}$$

For convenience, the basic set of boundary conditions for the axisymmetric isotropic case without thermal stresses (see Section 1.2) are iterated here (see Table 3.1).

$$\alpha_1\omega'(1) + \beta_1\omega(1) = \gamma_1; \quad \alpha_2\theta'(1) + \beta_2\theta(1) = \gamma_2. \tag{3.7}$$

Table 3.1

Basic boundary conditions in the axisymmetric isotropic case

Type of Support	α_1	β_1	γ_1	α_2	β_2	γ_2
Fixed clamping	1	$-\nu$	0	0	1	0
Movable (sliding) clamping	0	1	N	0	1	0
Fixed hinge	1	$-\nu$	0	1	ν	M
Movable sliding hinge	0	1	N	1	ν	M

where $N = N_r(1)$; and $M = M_r(1)$ are boundary active external forces.

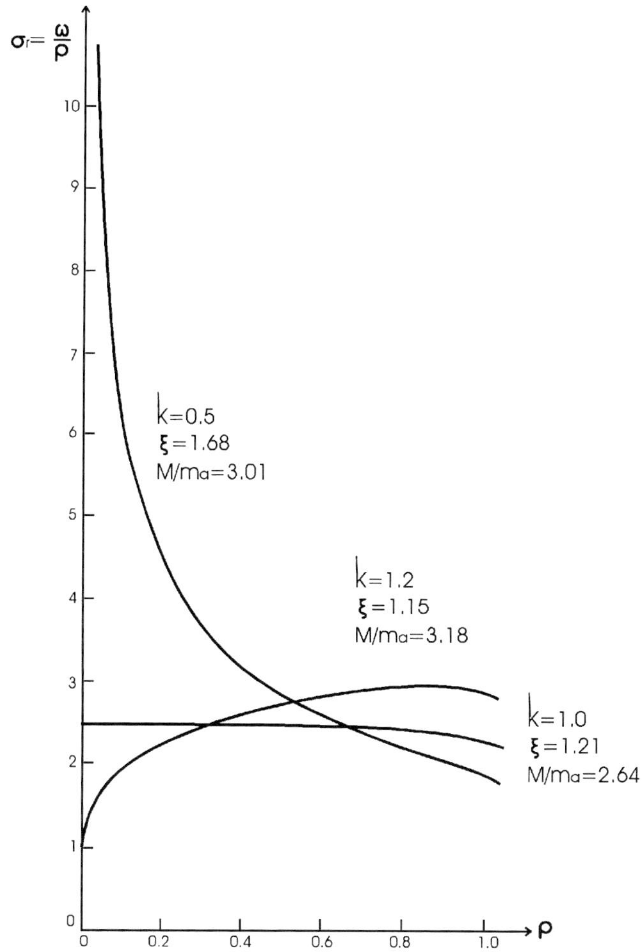

Figure 3.2 Meridional stress σ_r in the circular plate, immovable support.

3.2 Deformation of circular plates under transverse loading.*

In this section, we will only take transverse forces into account. These will include edge moments, but loads such as the edge resultant *N* and loading by thermal stresses are not considered.

When only transverse loads are acting, the equilibrium curves are of the type shown in Figures 2.1 and 2.24 (where the straight lines correspond to the linearized case) for all four

* Here, and hereafter, we will mainly focus on isotropic cases. When anisotropic cases are considered, this will be expressly stated.

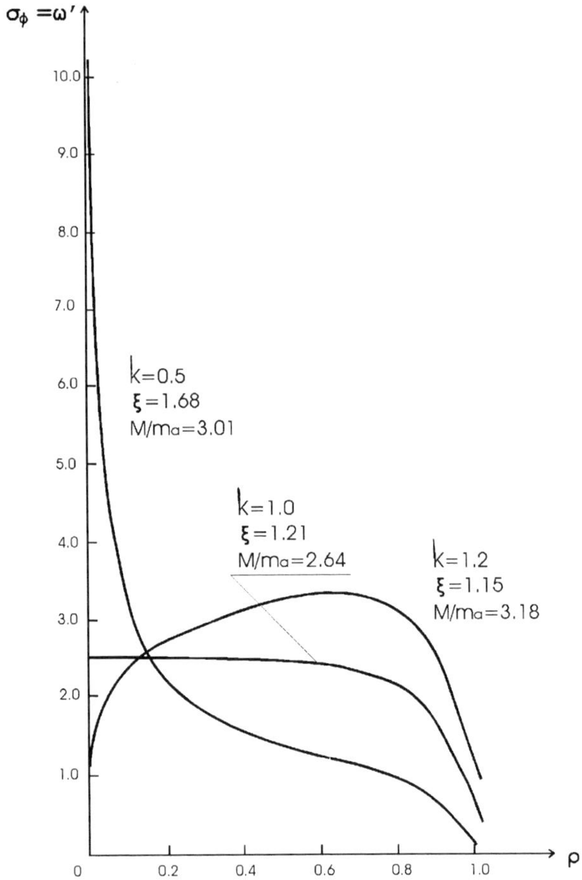

Figure 3.3 Hoop stress σ_ϕ in the circular plate, immovable support.

basic boundary conditions (see Table 3.1). At the same value of the loading parameter, the displacement of the flexible nonlinear plate is less than that of the corresponding linear problem. This effect is a result of the influence of membrane stress $\sigma_r(\rho)$, as was proven in Property 11 (Section 2.8). $\sigma_r(\rho)$ is always a non-negative monotonic decreasing function independent of the type of transverse loading, while $\sigma_\phi(\rho)$ can be non-negative anywhere (immovable support) or negative in a region near the edge (free movable support) (see, e.g., Figs. 3.2 and 3.3 for immovable support and 3.4 and 3.5 for movable support).

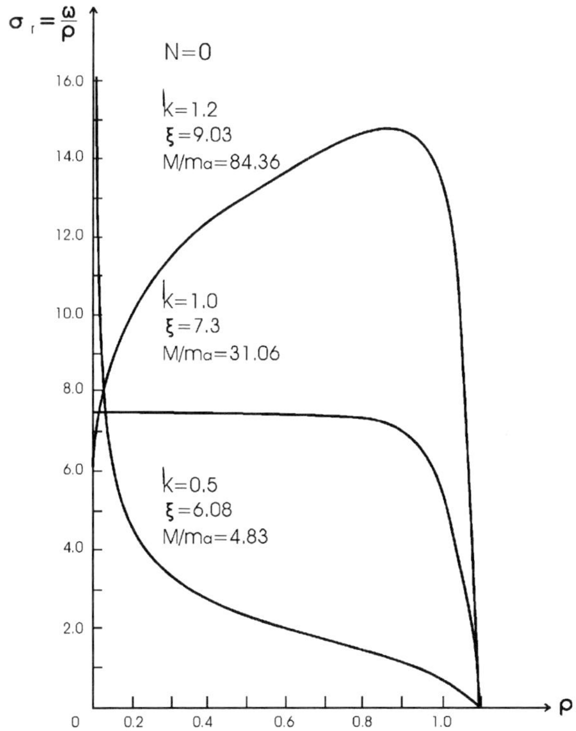

Figure 3.4 Meridional stress σ_r in the circular plate, movable support.

In these figures, results are given for "pure" bending by the edge moment M. The plates are rotationally orthotropic, where the parameter $k = \sqrt{E_\phi / E_r}$ characterizes the ratio of moduli of elasticity in the two main directions ($k = 1$ is related to the isotropic case), $m_a = 12\left(1 - k^2 \nu_r^2\right)$.

Many numerical results were obtained by Kornishin & Isanbaeva for the problem of the axisymmetric deformation of circular plates subjected to various types of transverse loading (uniform and variable external pressure, concentrated force at the apex, circularly distributed forces and several edge supports), [Kornishin & Isanbaeva, 1968].

Mescall [1968] obtained some interesting solutions for the clamped plate problem, which we present here in Fig. 3.6. Three load cases are considered: (a) a load, uniformly distributed along a circle around the center; (b) uniform pressure distributed over a circular area around the center; and, (c) a point force at the center, with all three of these cases having the same resultant P^*.

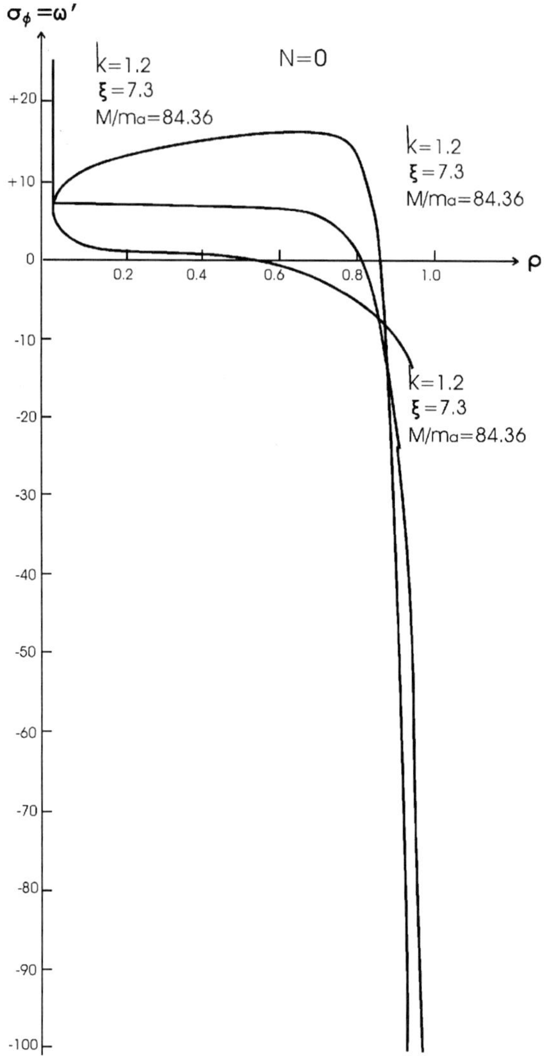

Figure 3.5 Hoop stress σ_ϕ in the circular plate, movable support.

The solutions are given in terms of the dimensionless force $P = P^*a^2/2\pi Eh^4$ versus the dimensionless apex displacement ξ for the case $a_1/a = 0.5$, where a_1 is the radius of the loading area or the radius of the circle along which the line load is applied. (Poisson's ratio is here probably $\nu = 1/3$). As can be seen, the line load (a) presents the "stiffest" case.

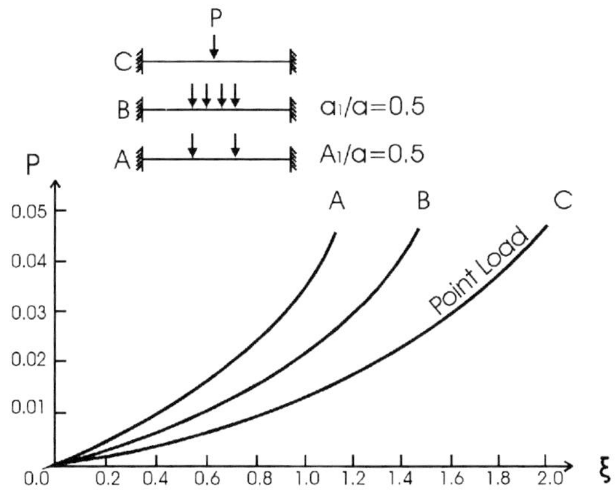

Figure 3.6 The clamped plate problem.

We shall now consider question of the stability of the solutions. When $N = 0$ and the shell material is isotropic, the membrane stress function $\omega(\rho)$ of the plate is always non-negative in the entire interval $0 \leq \rho \leq 1$. As long as we restrict the considerations to axisymmetric deformations, this property ensures the uniqueness of the solutions, and, consequently, ensures stability of these solutions (see Section 2.8). On the other hand, when the edge support is mobile, the behavior of $\omega(\rho)$ causes $\sigma_\phi(\rho) = \omega'(\rho)$ to become negative in a region close to the edge (see Fig. 3.5). This condition is the background to the occurrence of bifurcation buckling "in tension" (see Section 2.1), a process that invokes a transition of the axisymmetric state to a nonsymmetric state. The necessary conditions for this phenomenon to occur were established by the author [Shilkrut, 1983a]. Earlier, Morozov [1961] proved the existence of nonsymmetric buckled states with a potential energy that is less than the potential energy of the axisymmetric states at the same value of the load.

This particular bifurcation problem was solved for the first time by Panov and Feodosjev [1948] using Galerkin's method to a first approximation. The analysis was carried out for the case of uniform external pressure q and movable clamped edges. The problem was also investigated by Shilkrut and Gerlaku [1969], Gerlaku and Shilkrut [1970], and Cheo and Reis [1973] for the case of the combined action of external pressure q and a uniformly distributed edge force N. The results for the combined action of forces will be discussed in the next section. Here we shall only review the results for the case of external pressure.

The outcome of the analyses reported by Shilkrut and Gerlaku [1969], Gerlaku and Shilkrut [1970], and Panov and Feodosjev [1948] differ markedly from the results obtained in Cheo and Reis [1973]. The first approximation obtained by Panov and Feodosjev sets the bifurcation load at $q = 880$ (for $\nu = 0.3$), whereby the corresponding center displacement ξ (dimensionless) is given by $\xi = 9.6$. (Please note, $q = q^* a^4/(Eh^4)$, where q^* is the true (physical) pressure, E is Young's modulus, a is the base radius, and h is the thickness of the plate.) The number of waves in the circumferential direction of the buckling mode that corresponds to this load is $n = 8$.

For the same case, Shilkrut and Gerlaku obtained $q = 715$; $n = 3$; $\xi = 9.8$, while Cheo and Reis [1973] obtained $q = 1736$; $n = 14$ (no value for the apex deflection ξ is given in Cheo and Reis). Thus, the critical load q given by Cheo and Reis is a factor of 2.4 larger than the one obtained by Shilkrut and Gerlaku, and a factor of 2 larger than the value of the first Galerkin approximation produced by Panov and Feodosjev [1948]. This is an indication that the results of Cheo and Reis [1973] are grossly at fault; their numerical error probably occurred because they missed the first (smallest) bifurcation point. The value $q = 1736$ (with $n = 14$) in their paper most likely corresponds to a higher bifurcation and is thus not the critical one.

There is another essential difference between the outcome of the analysis reported in Cheo and Reis, on one hand, and that of Panov and Feodosjev, on the other. In the former, the initial part of the postbifurcation branch of the deformation curve that emerges from the bifurcation point is constructed by means of a formal asymptotic expansion, an approach which is similar to that of Koiter [1963b]. But the initial part of the bifurcation branch computed by Cheo and Reis is descending, while, in the investigations of Panov and Feodosjev, the opposite is found. This means that, according to the analysis of Cheo and Reis, loss of stability at the bifurcation load happens in the form of the snap-through process, while Panov and Feodosjev predict a smooth transition from the axisymmetric state to a nonsymmetric one.

Clearly, the results of Cheo and Reis must be in error, not only because their buckling load is far in excess of that produced by the two other analyses, but also because physical evidence obtained in experiments [see Shilkrut, 1980, 1983a], as well as mechanical insight, preclude an unstable buckling behavior in the case where the prebuckling state of the plate is in "tension." In conclusion, it is noted that, in the opinion of the author, the correct solution to this buckling problem is given in Shilkrut and Gerlaku [1969] and Gerlaku and Shilkrut [1970], where it is shown that the plate buckles at $q = 715$ with a mode of $n = 3$ circumferential waves.

The solution of the bifurcation problem when the plate is subjected to a uniformly distributed edge bending moment M_K (pure bending) is presented in Shilkrut and Gerlaku [1969] and Gerlaku and Shilkrut [1970] for the case of a mobile hinged edge and $\nu = 0.3$. The value of the bifurcation load is $M_K = 318.07$ with a center displacement $\xi = 7.4$. (M_K is defined as: $M_K = 12\left(1-\nu^2\right)M_K^* a^2/\left(Eh^4\right)$, with M_K^* the true (physical) moment.)

Although it seems evident that this type of buckling must also occur for other types of loading and edge supports (as long as the latter are movable (see Property 11 in Section 2.8)), to the best of our knowledge, solutions other than we have described here have not yet appeared in the literature.

Higher order approximate solutions are described by Valishvili [1976; Valishvili & Sylkin, 1975] for some axisymmetric problems of circular plates with small corrugations subjected to uniform pressure. The case of annular plates loaded by uniform pressure is dealt with by Valishvili [1976] and Kayuk [1974]. An interesting problem is considered by Klemm and Wozniak [1967], who investigated the bending of a perforated flexible plate with a regular set of holes periodically distributed along the meridians and circumferential circles.

Axisymmetric deformations of plates supported by intermediate supports in addition to the more common edge supports were studied by Aleksandrov and colleagues [Aleksandrov, Ljastchenikov, Shaposhnikov & Smirnov, 1976]. One of the main peculiarities of this problem is that both the bending moments, M_r and M_ϕ, have peaks at the intermediate supports, where the shear forces are discontinuous. These effects are not caused by the nonlinearity of the problem, because they are already present in the linear case. Here, the variation of the stress function $\omega(\rho) \geq 0$ is similar to the variation of $\omega(\rho)$ in the original case without additional support, but the variation of the rotation $\theta(\rho)$ is now different; i.e., it is a function of variable sign.

It is interesting to see how the transverse shear effect influences the deformation of the plate [see Pian & Tong, 1971]. This study was based on Reissner's equations for nonshallow (shear deformable) shells, which we presented in Section 1.1. Figure 3.7 presents the case $G/E = 0.02$, where G is the shear modulus, $a/h = 0.1$ and $\nu = 1/3$. The dashed curves in this figure are the solutions with transverse shear effect, while the solid curves correspond to the solutions without the shear effect ($G/E \rightarrow \infty$). According to these results, the transverse shear effect is only substantial when the shear modulus is relatively small. We note here that the effect of transverse shear is rarely investigated in the theory of nonlinear plates and shells.

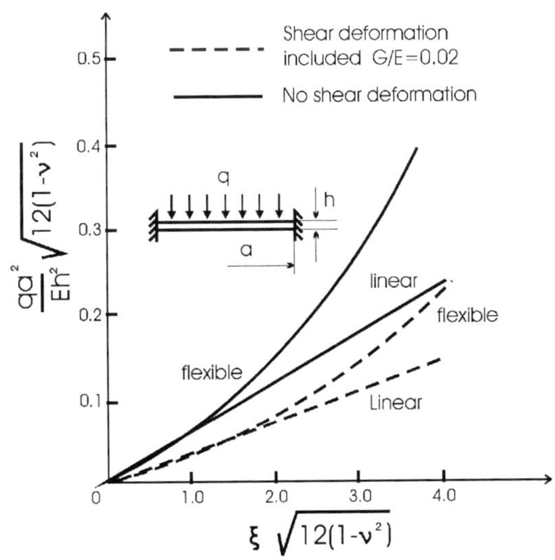

Figure 3.7 A study of the transverse shear effect [Pian & Tong, 1971].

3.3 Buckling and postbuckling behavior of circular plates under compression or under a combination of compression and bending.

The simplest buckling problem (bifurcation) for circular plates is the case of radial compression induced by a uniformly distributed edge load $N = -N_r^*(1)a^2 / D$. This is a problem that has been considered frequently [e.g., Timoshenko & Woinowsky-Krieger, 1959] since it appeared in the pioneering work by G. H. Bryan [1891] on the stability of plates. The bifurcation phenomenon consists here of a transition of the flat fundamental state of the plate to a buckled state exhibiting bending deformations. The buckled state can be either axisymmetric or nonsymmetric. The determination of the buckling load and buckling mode is carried out by the solution of an eigenvalue problem similar to the classical problem of the buckling of bars (Euler's problem). As an example, the first five entries of the spectrum of bifurcation loads N_b for axisymmetric bifurcation modes are given in Table 3.2 for two types of supports [see Vainberg, 1973]. (The supports are movable in the plane of the plate; the plate is isotropic.)

Table 3.2

$\nu = 0.3$

4.20	29.04	73.48	137.6	221.4	Hinged support
14.68	49.22	103.5	177.5	271.3	Clamping

The axisymmetric bifurcation modes are of the form :

$$w(\rho) = C[J_0(\rho\alpha a) - J_0(\alpha a)]; \quad (0 \le \rho \le 1), \tag{3.8}$$

where $\alpha^2 = -N_r^*(1)/D$ and J_0 is the Bessel function of the zero order [Volmir, 1963]. The infinite discrete spectrum of bifurcation loads N_b is determined by the roots of a transcendental equation that enforces the boundary conditions. The first buckling mode, which corresponds to the smallest buckling load N_1, does not contain nodal circles — that is, the radial shape of the mode is a monotonic curve — but the axisymmetric buckling mode determined by N_n contains $n - 1$ nodal circles.

The nonsymmetric modes are described by the following expression:

$$w(\rho,\phi) = C\left[J_m(\alpha a\rho) - \rho^m J_m(\alpha a)\right]\sin m\phi; \quad (0 \le \rho \le 1); \quad (0 \le \phi \le 2\pi); \quad (m = 1,2,...). \tag{3.9}$$

J_m is the Bessel function of the order m [Volmir, 1963]. Such a buckling mode is of a complex shape, which, in addition to the undulations in radial direction, also displays m half-waves in the circumferential direction.

Thus, each value of parameter m generates an infinite spectrum of bifurcation loads N_{mn}. In the case of a clamped edge, for example, the first term of the spectrum of bifurcation loads, determined by the first root of $J_{m+1} = J_2$ for $m = 1$, is given by $N_{11} = 26.38$. The second term for $m = 1$ is $N_{12} = 70.85$, and so on. Thus, the bifurcation to a nonsymmetric mode takes place at a load which is higher than the one that belongs to a bifurcation into a symmetric mode. This, means that the flat plate will buckle into the symmetric mode.

The postbifurcation behavior of the plate has been studied using rigorous mathematical tools of analysis [see, e.g., Friedrichs & Stoker, 1941, 1942; Keller, Keller & Reiss, 1962; Morozov, 1963; Yanowitch, 1956], or numerical methods [e.g., Bodner, 1955; Friedrichs & Stoker, 1942; Volmir, 1963], or perturbation methods (asymptotic series) and computer oriented algorithms [e.g., Keller & Reiss, 1958; Valishvili, 1976; Volmir, 1963]. Global pictures of the deformation and stability of plates subjected to single and multi-parametric load systems have been obtained by means of the so-called "deformation map" developed by the author [Shilkrut, 1974, 1982, 1983b, 1992a]. These results will be considered in detail in Sections 4.3 and 4.4.

The main character of the behavior of an ideal (no imperfections) plate loaded in compression can be described as follows. At the first bifurcation point, with corresponding load N_b, the plate bifurcates into an axisymmetric form with one semi half-wave in the radial direction. The post bifurcation state is stable. When the load is further increased beyond N_b, a zone develops near the edge, where both σ_r and σ_ϕ are negative. (The central part of the plate can be stretched or compressed — both possibilities can occur.) The existence of this zone of compression close to the edges presents a condition whereby a secondary bifurcation can occur with a transition to a new one that has multiple lobes in the circumferential direction (see Section 3.3, and Shilkrut [1983a]). Upon further loading, the development of this state can lead to yet another bifurcation with yet another transition to another mode (with another number of circumferential waves) and so on. These successive bifurcations along the loading path have not been analyzed as yet, but the evidence for their existence has been found in experiments.

The description given above is a qualitative picture of the "dynamics" of the post-bifurcation behavior of the plate. We observe that the post-bifurcation behavior of plates and caps is very "rich" owing to a great number of possible equilibrium states. It is clear, however, that many of these states are unstable, so that the actual postbuckling behavior of a cap or plate can only be established after an inspection of the stability of these states. The various aspects of this question are considered in Chapter 6 [see also Shilkrut, 1980; Shilkrut & Vyrlan, 1977].

We shall now consider in more detail the phenomenon of postbifurcation behavior in the range of axisymmetric deformations. As was established in Section 2.8 (see Property 11), when the plate is loaded in compression ($N < 0$), the stress function $\omega(\rho)$ can only behave in two ways: (a) $\omega(r)$ is non-positive, monotonically decreasing in the interval [0,1]; or (b) $\omega(r)$

is monotonically decreasing in the interval [0,1], of positive sign in a sub-interval $(0 \le \rho < \bar{\rho})$, and of negative sign in the remaining part $(\bar{\rho} < \rho \le 1)$. Consequently, the radial stress distribution $\sigma_r(\rho) = \omega(\rho)/\rho$ is of the same form and has the same sign as $\omega(\rho)$. This behavior is characteristic for the axisymmetric postbifurcation range and prevails even for large values of $\Lambda = -N / N_b$ (N is, as usual, the current value of the edge load).

Consider, for example, Figure 3.8, which displays results that are taken from Valishvili [1976]. In this figure, the (nondimensional) membrane stress resultant N_r is plotted as a function of ρ for different values of the ratio of the load parameter Λ. N_r is related to the physical stress resultant by $N_r(\rho) = 12(1-\nu^2)N_r^*(\rho)a^2/(Eh^3)$. The case considered refers to clamped but movable edges. As can be seen from these results, when $\Lambda < 2$, N_r is negative everywhere; but when $\Lambda > 2$, the force N_r becomes of variable sign.

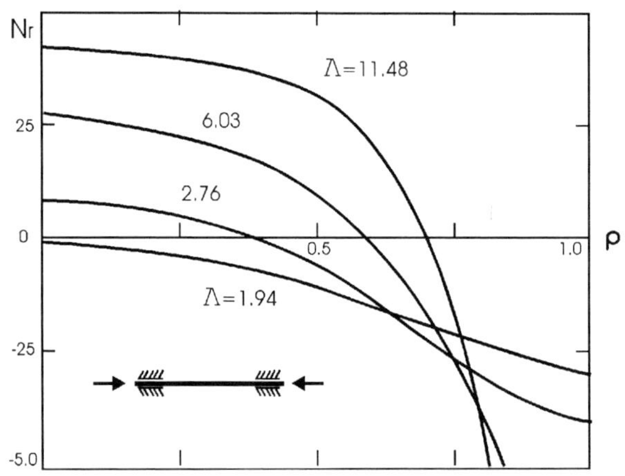

Figure 3.8 Distribution stress resultant N_r at various intensities of the external load.

The corresponding deflections are depicted in Figure 3.9, where the nondimensional displacement $w(\rho)$ is defined as: $w(\rho) = 12(1\text{-}\nu^2)\, w^*(\rho)/h$ [Valishvili, 1976; Volmir, 1963]. These results indicate that the shape of the plate when $|N|$ is increasing attains a shape where the central part of the deformed plate is practically a stretched flat membrane, while bending occurs near the edges. The variation of the bending moment resultant M_r illustrates this edge effect, as is shown in Figure 3.10 [Friedrichs & Stoker, 1942; Valishvili, 1976; Volmir, 1963]. M_r is here defined as $M_r(\rho) = \left[12(1-\nu^2)\right]^{3/2} M_r^*(\rho)a^2 / Eh^4$.

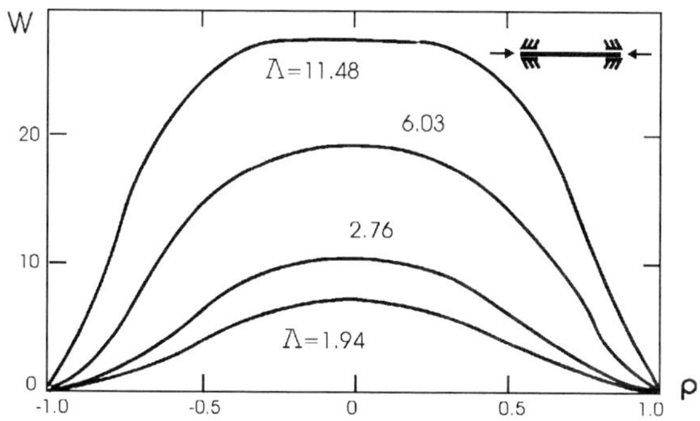

Figure 3.9 Deflections of the plate at various intensities of the external load.

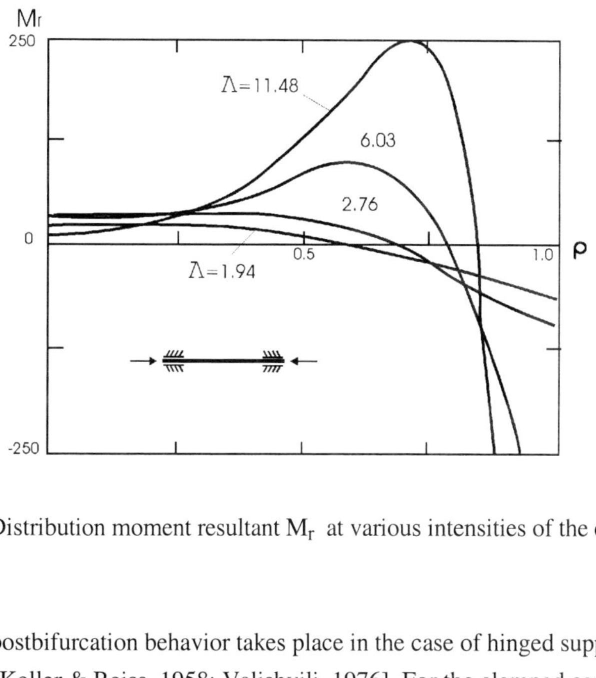

Figure 3.10 Distribution moment resultant M_r at various intensities of the external load.

A similar postbifurcation behavior takes place in the case of hinged supports [Friedrichs & Stoker, 1942; Keller & Reiss, 1958; Valishvili, 1976]. For the clamped case, it is expected that, when Λ reaches the magnitude of about 15 $(N \approx -63)$, a secondary bifurcation occurs with a transition to a nonsymmetric shape [Volmir, 1963]. In the case of a hinged plate, this event occurs at $\Lambda = 7.49$ $(N = -110)$ [Cheo & Reis, 1973].

Until now, it was thought that buckling only occurred under axisymmetric compression induced by edge forces alone. We shall now look at buckling phenomena that occur when two different independent loading systems are simultaneously applied; i.e., an edge force N and normal uniform pressure q. Each of these load cases by themselves are capable of generating bifurcation buckling phenomena in the plate. The buckling phenomenon that is caused by N alone has just been described. The pressure load q can cause bifurcation in tension, a mechanism we considered in the previous section. But when both loads are acting together, they can cause a complex nonlinear interaction between these two bifurcation mechanisms.

The pioneering article by Cheo and Reis [1973] is the first work where this problem was considered for a clamped plate. These authors used the following approach. As a first step, the axisymmetric buckled state generated by load N (N is taken positive in compression) is obtained numerically, and this buckled state is considered as the fundamental state I. After this, the load q is applied and the bifurcation load q_b is determined by solving the corresponding linear eigenvalue problem for bifurcation from state I into a new nonsymmetric state II (with m waves in the circumferential direction). The buckling load q_b is the smallest eigenvalue of the bifurcation equations to be found at a fixed value of N. This is a result of a search procedure whereby N is chosen and fixed, the wave number m is changed, and the corresponding load $q_b(m)$ is examined in order to find its minimum (for $m = 1, 2,...$). The procedure is repeated for several values of N. We note here that, for this problem, the load q can be seen as internal pressure for the plate that is deformed into a cap by N. (The case that the pressure q can be viewed as external pressure, $q \to -q$, is not considered by these authors.)

But there is a serious flaw in the work of Cheo and Reis [1973]. This has something to do with the fact that the authors seemed to have selected an axisymmetric state that was not appropriate as a fundamental state in cases when there are more than one axisymmetric equilibrium states at the considered value of the load N. We clarify this point by investigating their results, which are summarized in Table 3.3 for $\nu = 0.3$ and the case of clamped edges.

It can be seen from Table 3.3 that, for $N = -5$, 0, and 10, there is pure bifurcation in tension where a transfer occurs to a buckled state from the axisymmetric fundamental (prebifurcation) state. This follows because, when N is acting alone, the first buckling load is given by $N_b = 14.68$ (see Table 3.2). At the value $N = 20$, the fundamental state is the axisymmetric postbuckling solution for $q = 0$. In this case, the buckling load q_b is the load that belongs to the secondary bifurcation.

Table 3.3

$\nu = 0.3$

N	-5	0	10	20	40	60	80	100	110	120	130
m	14	14	13	13	13	12	11	10	6	6	10
$12(1-\nu^2)q_b$	6496	6260	5775	5290	4365	3410	2450	1450	0	750	1000
	0	0	0	0	0	5	0	0			

N	131	133	134	135	140	150	152.5	160	180	200
m	10	10	6	6	11	12	12	13	15	17
$12(1-\nu^2)q_b$	600	0	1500	1000	800	600	0	600	1000	1600

It is important to pay attention to the case $N = 110$ ($m = 6$ and $q_b = 0$). Because $q_b = 0$, buckling occurs by N alone, and the corresponding mode has here m circumferential waves. But it is known from other sources that $N = 110$ is not the smallest buckling load at which nonsymmetric buckling occurs. Between $N = 0$ and $N = 110$ there are also other bifurcations [see, e.g., Timoshenko & Woinovsky-Krieger, 1959], which are apparently lost in Cheo and Reis [1973]. For $m = 1$, for example, the smallest corresponding buckling load is, as is noted above, equal to 26.38, and the following is 70.85, and so on. The data from Table 3.3 are also presented graphically by Cheo and Reis [1973], and, in these graphs, the buckling loads 26.30, 70.85,..., are absent. This means that many bifurcation loads are overlooked by these authors. (In the previous section, it was shown that the second column in Table 3.3. does not present the smallest bifurcation loads.)

These observations, and the erroneous initial directions of the postbifurcation branches that we discussed earlier in Section 3.2, show that the data in Cheo and Reis's paper are far from complete or perfect. Despite these deficiencies, we want to stress that the work of Cheo and Reis presented an important attempt to investigate the complicated phenomenon of the deformation of plates by a two-parameter load system.

As was mentioned, Cheo and Reis did not consider the case where the pressure q can be viewed as an external pressure that is applied to the bulging plate; i.e., when external pressure is applied to the plate after it is loaded into the axisymmetric postbuckling state by the edge load N. In this case, loss of stability can take place by snapping. However, we will encounter and examine the whole set of solutions that belongs to this case in Section 4.3 — including the solutions determined by the pressure q applied in both directions — to illustrate the use of the deformation map for the examination of the action of multi-parametric loads on axisymmetrically deformed, geometrically nonlinear structures. The question of the influence of the loading path on the establishment of the final equilibrium state, which is a very important question in the case of multi-parameter loaded systems, will be investigated in Section 4.5.

3.4 Clamped spherical caps subjected to uniform external pressure.

One of the most popular problems in the theory of geometrically nonlinear shells is the case of clamped spherical caps. The study of this problem was preceded by the study of the stability of the complete spherical shell subjected to external pressure, which served as one of the first problems of the theory of geometrically nonlinear shells in general. All numerical solutions given below are valid for isotropic shells with $\nu = 0.3$. When deviations from this standard are encountered, this will expressly be mentioned.

a) Axisymmetric deformations under uniform external pressure.

The conditions for clamping in the isotropic case are:

$$\omega(0) = \theta(0) = 0; \;\; \omega'(1) - \nu\omega(1) = 0; \;\; \theta(1) = 0. \qquad (3.10)$$

small, the limit loads are not yet influenced by the geometry of the cap. On the other hand, λ also increases when the height of the cap increases, while the thickness h remains unchanged. Thus, also, for deep caps, the limit loads η^+ and η^- reach a stationary value independently of the magnitude of λ. Experimental confirmation of this feature of the critical values $\eta^+(\lambda)$ are mentioned in a review article by Wang [1967]. Another detail one should pay attention to is that energy-based numerical methods lead to an overestimation or (underestimation) of η^+ or (η^-).

The curves $w_0(\rho) = \sqrt{12(1-\nu^2)}w^*(\rho)/h$ (w^* is the true (physical) deflection) for different λ at $q = q^+$ are presented in Figure 3.12 [Valishvili, 1976]. These results show essential changes of the waviness of the deformation mode with an increasing magnitude of λ (or, what amounts to the same, $|\xi|$). At the value q^-, the function $w_0(\rho)$ is much simpler, being similar to the one depicted in Figure 3.12 for $\lambda = 5$.

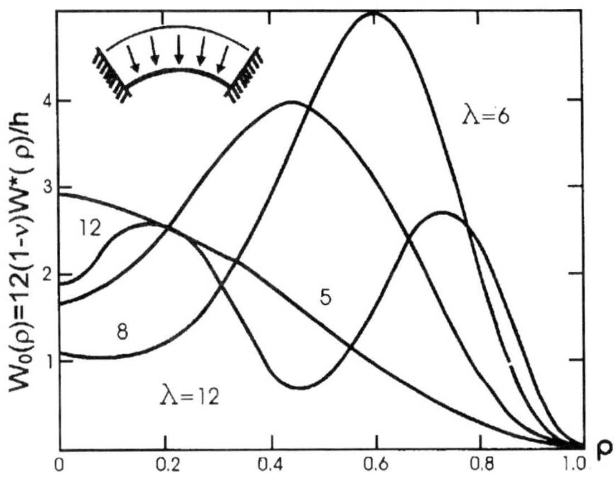

Figure 3.12 Displacement solutions for the clamped cap at $q = q^+$.

This is confirmed by many numerical results [e.g., Keller & Wolfe, 1967; Shilkrut *et al.*, 1969; Afanasjeva & Kantor, 1969]. Figures 3.13-3.16 show the distribution of M_1, M_2, and N_r, N_ϕ at q^+ for different values of λ [Valishvili, 1976]. These nondimensional quantities are defined by $M_{1,2}(\rho) = M^*_{r,\phi}(\rho)a^2/(2Dh|\xi|)$ and $N_{1,2}(\rho) = N^*_{r,\phi}(\rho)\sqrt{12(1-\nu^2)}a^2/(2Eh^3|\xi|)$ (the components marked by an asterisk * are the corresponding physical quantities).

It transpires that N_1 is negative everywhere at $q = q^+$ [e.g., Afanasjeva & Kantor, 1969; Kornishin, 1964; Kornishin & Isanbaeva, 1968; Shilkrut *et al.*, 1969; Valishvili, 1976]. The negative sign of N_2 in regions close to the edge indicates the possibility of a bifurcation into a buckling pattern with multiple waves in the circumferential direction. This phenomenon is recognized as bifurcation in compression (see Section 2.1, and part *c* of the present section).

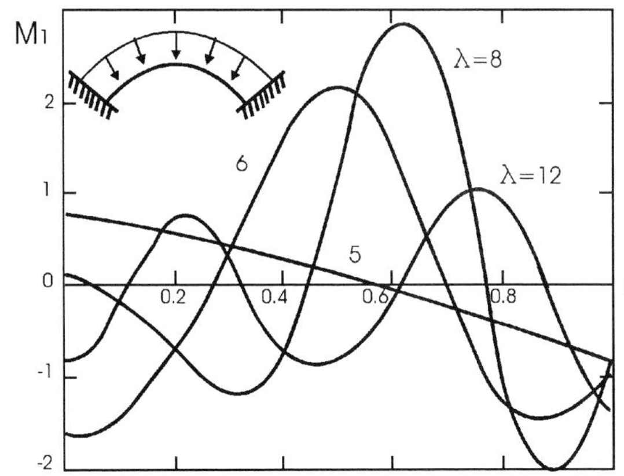

Figure 3.13 Moment resultant (M_1) distributions for the clamped cap; $M_1 = M_r^*(\rho)\dfrac{a^2}{(2D|\xi|h)}$.

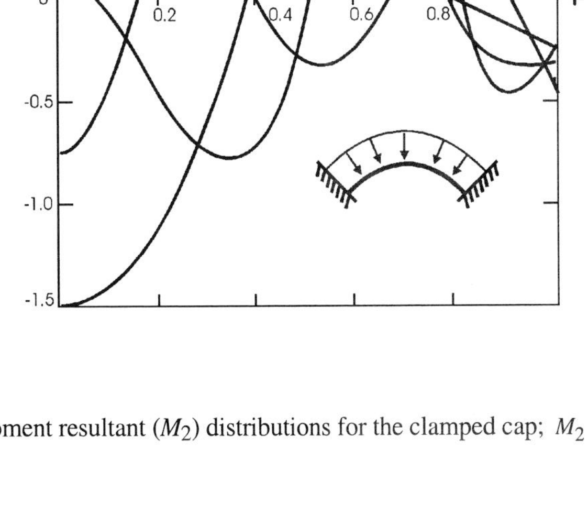

Figure 3.14 Moment resultant (M_2) distributions for the clamped cap; $M_2 = M_\phi^*(\rho)\dfrac{a^2}{(2D|\xi|h)}$.

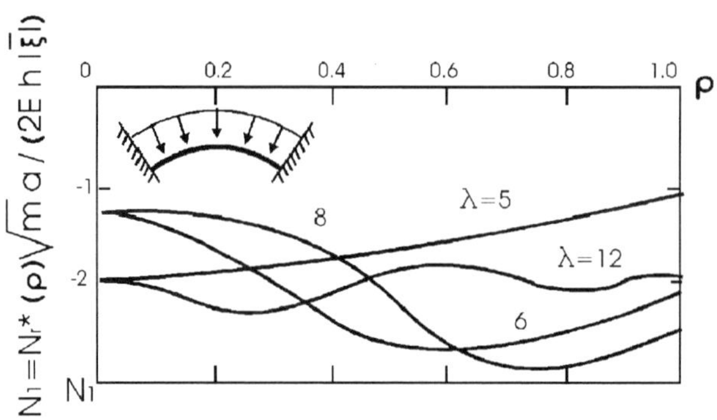

Figure 3.15 Meridional stress distributions for the clamped cap.

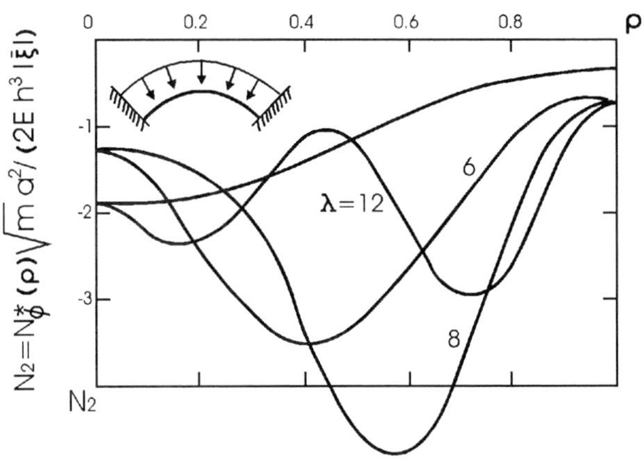

Figure 3.16 Hoop stress distributions for the clamped cap.

Results obtained by Huang [1964] and Uchiyama and Yamada [1974] show a variation in the magnitude of η^+ by 2.3% if ν is changed from 0.3 up to 1/3. To the best knowledge of the author, there are no other results available where the influence of the magnitude of ν on q^+ or η^+ is checked. In our opinion, this influence may be important for very small or large values of ν (by large, we mean close to 0.5).

In Pian and Tong [1971], a specific case is examined by means of the finite element method taking into account the transverse shear effect (the cases that are considered correspond to $G/E = 0.02$; 0.1 and ∞, with $h/a = 0.0934$ and the central angle of the cap equal to $2\alpha = 70^o$, Poisson ratio $\nu = 1/3$). These results show a noticeable increase of the upper limit pressure when the shear rigidity G is small. In this publication, the solution curve is not continued up to the lower limit point and consequently the influence of the transverse shear on the lower limit load remains unknown. Again, it is necessary to emphasize that the results obtained for one specific case do not give sufficient ground for a definite conclusion about the influence of a physical parameter on the process of loss of stability.

b) Axisymmetric deformation of spherical caps subjected to pressure distributed on a part of its external surface around the apex.

This is a case that is less well investigated. We shall base our survey on the results produced by Kornishin and Isanbaeva [1968]. The loading area is here characterized by the dimensionless parameter $\bar{\rho} = a_1 / a \leq 1$, where a_1 is the radius of the loaded part of the cap, and a is, as before, the radius of the cap's base.

It could have been expected that the behavior of the cap in this case would lie somewhere in between the case of a concentrated force acting at the cap's apex $(\bar{\rho} = 0)$ and the previous load case $(\bar{\rho} = 1)$. This preliminary assumption is confirmed by the data presented in the following table, where $q_1^+ \div q_5^+$ represent the values of the limit loads for $\bar{\rho}$ = 0.3125, 0.4375, 0.5625, 0.6875, and 0.8125, respectively.

Table 3.4

$\lvert\xi\rvert$	q_1^+	q_2^+	q_3^+	q_4^+	q_5^+
0.5	-	-	-	-	-
1	-	-	-	-	-
1.5	-	-	-	-	-
2	-	-	-	11.6	10.7
2.5	-	-	18.4	16.0	16.4
3	-	-	24.8	22.5	22.8
3.5	-	39.7 ÷ 41.5	31.3	28.9	31.3
4	-	46.9 ÷ 48.7	38.5	38.5	42.7
4.5	-	54.1 ÷ 55.9	44.9	48.1	57.0
5	-	61.3 ÷ 64.9	52.9	57.7	No data

In Table 3.4, no entry means that, under the given conditions, snapping does not occur. From the data in this table, it can be concluded that, for $\lvert\xi\rvert \leq 5$, there is no loss of stability for

$\overline{\rho} = 0.3125$. This behavior can be understood if one considers that, when a concentrated force is acting ($\overline{\rho} = 0$), snap-through occurs for $|\overline{\xi}| \geq 9.316$ [Fitch, 1968], while, in the other extreme case (when $\overline{\rho} = 1$), loss of stability by snapping takes place for caps with $|\overline{\xi}| \geq 1.677$. It is clear that, for $\overline{\rho} = 0.3125$, which is close to the case of a concentrated load, snap-through does not occur at $|\overline{\xi}| = 5$. When $\overline{\rho} = 0.8125$, the jumping process starts to occur for the apex rise $|\overline{\xi}| = 2$, close to case of $\overline{\rho} = 1$. The data in Table 3.4 also show that, from the point of view of loss of stability, it is expedient to concentrate the external loading in the area defined by $0 \leq \rho \leq 0.3125$.

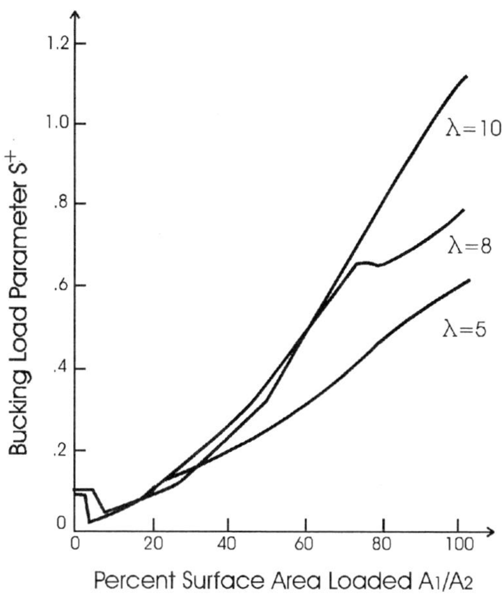

Figure 3.17 Results for the cap when the loading is applied to a restricted area of the cap.

Results similar to those just presented were obtained by Stephens and Fulton [1969] and Chia [1976]. In Figure 3.17, for example, limit loads are presented for the geometries $\lambda = 5$, 8 and 10 expressed in terms of $S^+ = \eta^+ A_1/A_2$,where η^+ is defined by (3.11), and A_1, A_2 are the areas of the loaded part and the entire surface of the cap, respectively. (This corresponds to the cases $|\overline{\xi}| = 3.78$, 9.66, and 15.1, respectively; see (3.4)).

For the geometry $\lambda = 5$, and $A_1/A_2 \leq 0.2$, no snapping takes place. Note that, at $A_1/A_2 \approx 0.05$, the limit load S^+ drops drastically for the geometries $\lambda = 8$ and 10. These curves are irregular because the basic state changes in shape when the magnitude of A_1/A_2 is varied. Thus, in general, S^+ increases with A_1/A_2, except in the special case mentioned above.

Some numerical results pertaining to the case of uniform pressure acting on a zone near the edge are presented in Kornishin and Isanbaeva [1968]. In Chia [1976], results are also presented for η^+ when the pressure load acts on a belt-like zone $0.2 \le \rho \le 0.3$ (see Table 3.5 for the case $\nu = 1/3$).

Table 3.5

$\nu = 1/3$

$\lambda/\lvert\xi\rvert$	5/3.83	8/9.80	10/15.31
η^+	2.22	1.24	1.34

It seems that for the geometries $\lambda = 8$ and 10, the data presented in Chia [1976] present the actual critical loads. But the value of η^+ given for $\lambda = 5$ does not seem to be correct, because all available results show that the limit load increases with the increase of λ (see, e.g., Table 3.4 and Fig. 3.17).

A very interesting solution (see Fig. 3.18) for a particular case is presented by Makowski and Nolte [1987], who suggest the possibility of the occurrence of two successive snapping processes. The same type of solutions were obtained by Mescall [1968] (see Figs. 3.19-3.22). Figures 3.19-3.21 show deformation curves $P = P(\xi)$, where $P = RP^*/(2\pi Eh^3)$; here, R is the radius of the spherical cap, and P^* is the physical value of the resultant of the uniformly distributed pressure, which is acting on the area of the conical section around the apex with radius a_1). In Figures 3.19 and 3.20, the equilibrium curves are presented for the geometries $\lambda = 8$ and 12, respectively (see (3.5)). The solutions are presented for several ratios $\overline{R} = a_1/a$, where a is the radius of the cap's base). In Figure 3.21, the deformation curves are given for different values of λ when $\overline{R}$ is kept fixed at $\overline{R} = 0.4$. As can be seen from these results, when λ and $\overline{R}$ increase, the value of the first limit point along the deformation curve also increases. This feature is in complete agreement with the data from Table 3.4.

When $\overline{R} = 0$ (see Fig. 3.19), we have the concentrated force P acting at the pole. In Figures 3.20 and 3.21, we can see that the equilibrium curves for $\lambda = 12$ and $\overline{R} = 0.3, 0.4$ are of the same type as the one given in Figure 3.18, except for the case of $\lambda = 8$. The solution for $\lambda = 12$ and $\overline{R} = 0.3$ (Fig. 3.20) leads to a very important conclusion. There must be two successive jumps: the first occurs at the first maximum, and the second at the second maximum. The same conclusion must follow from the result given in Figure 3.18. Graphs of P_{cr} vs. λ (not entirely complete) are given in Figure 3.22 for different values of $\overline{R}$.

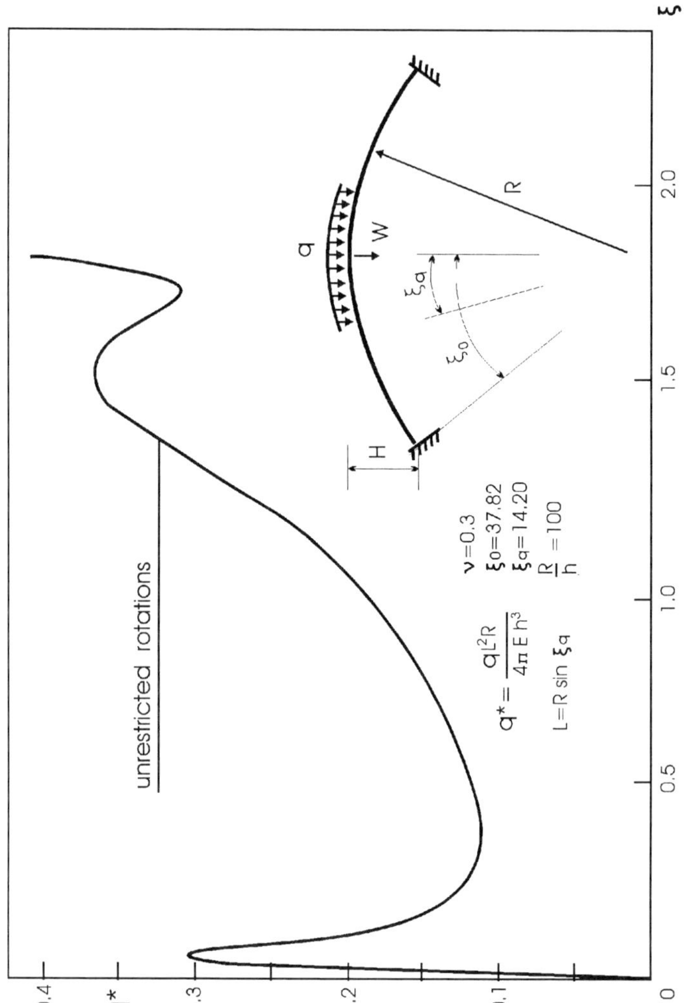

Figure 3.18 Results obtained by Makowski and Nolte [1987]

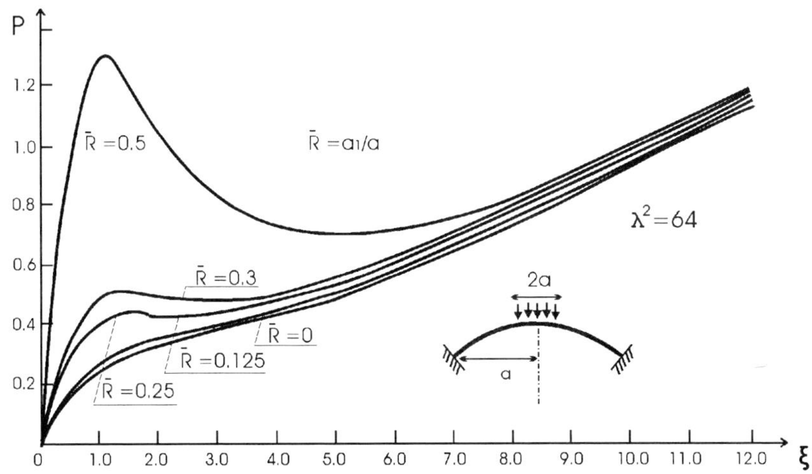

Figure 3.19 Mescall's results for the cap [1968], part 1.

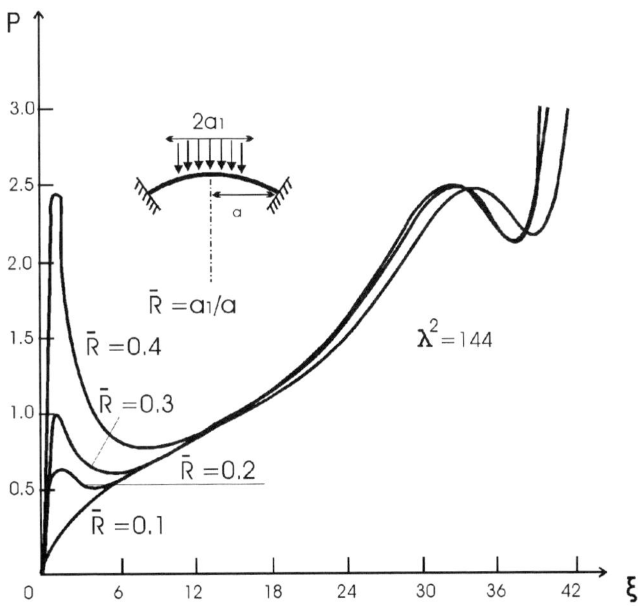

Figure 3.20 Mescall's results for the cap [1968], part 2.

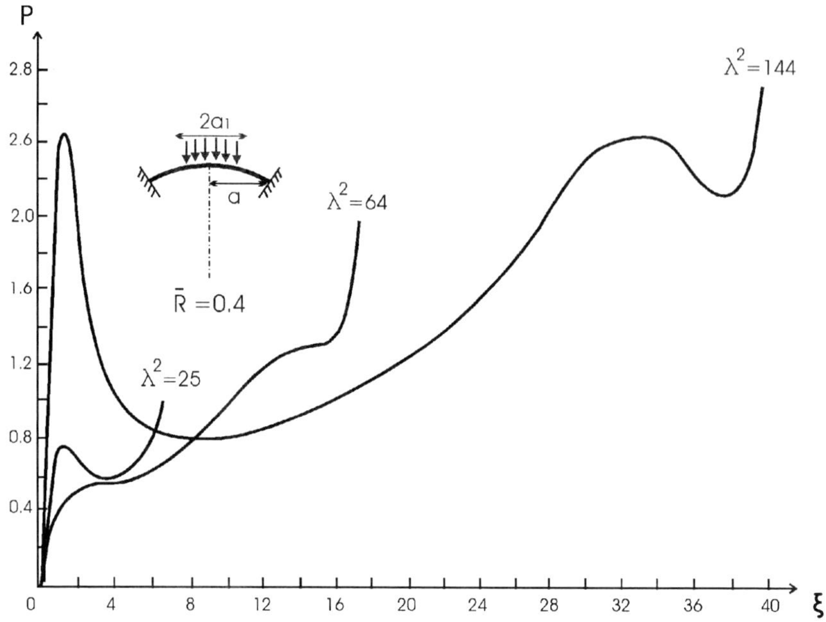

Figure 3.21 Mescall's results for the cap [1968], part 3.

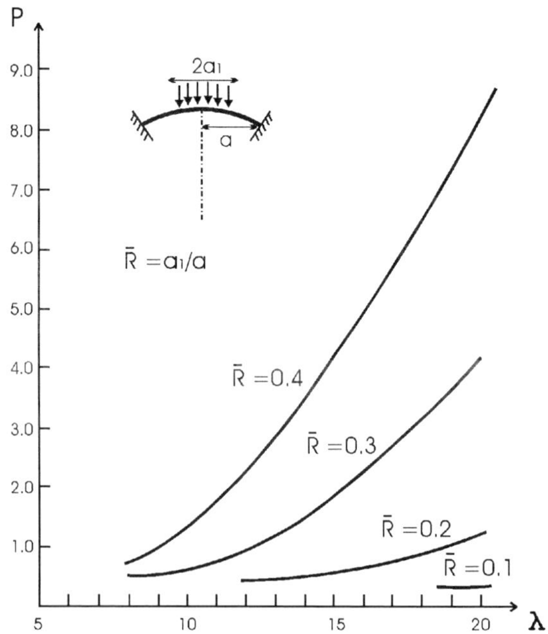

Figure 3.22 Mescall's results for the cap [1968], limit loads.

c) ***Bifurcation and the initial postbifurcation behavior of caps subjected to pressure load distributed over a part or over the whole of its surface.***

The first attempts to solve this bifurcation problem concerned caps loaded over their entire surface. These solutions (first approximations) were obtained using Galerkin or other direct methods of the calculation of variations [Gjelsvik & Bodner, 1962; Grigolyuk, 1960; Krivosheev, 1959]. A reliable numerical solution of this problem was obtained by Huang [1964], and this was followed by a series of other contributions [e.g., Bushnell, 1989; Archer & Famili, 1965; Famili & Archer, 1965; Fitch & Budiansky, 1970; Gerlaku & Shilkrut, 1970; Shilkrut & Gerlaku, 1969; Stephens & Fulton, 1969; Valishvili, 1976; Weinitschke, 1965], some of which present rather complete solutions that cover many different aspects of the problem.

As far as the postbuckling behavior is concerned, the results are scarce [see Archer & Famili, 1965; Fitch, 1968; Fitch & Budiansky, 1970; Valishvili, 1976] despite the availability of a general asymptotic postbuckling theory originated by Koiter that points the way toward the construction of these solutions [see Koiter, 1945, 1963b; Budiansky, 1974; Koiter, 1969b, 1982].

Let us present the solution of the bifurcation problem and the initial postbifurcation branch for clamped spherical caps subjected to uniform pressure on a part of their surface, as worked out by Fitch and Budiansky [1970]. The bifurcation loads for the cap in question were obtained by a classical (standard) approach which is described in detail in several papers [e.g., Huang, 1964; Shilkrut & Gerlaku, 1969; Valishvili, 1976]. After the fundamental state is determined, the initial postbuckling path is developed in accordance with Koiter's asymptotic approach taking into account all the relevant nonlinear effects. The essence of this approach is the following.

In the neighborhood of the bifurcation point, which is assumed to be a simple, the unknown functions (displacement w and stress function F) are presented in the asymptotic series defined for $0 \le \varepsilon << 1$:

$$\begin{bmatrix} w \\ F \end{bmatrix} = \begin{bmatrix} w_0 \\ F_0 \end{bmatrix} + \sum_{i=1}^{N} \begin{bmatrix} w_i \\ F_i \end{bmatrix} \varepsilon^i, \tag{3.14}$$

where w_0 and F_0 describe the fundamental axisymmetric state of equilibrium; w_1 and F_1 represent the buckling mode (the eigenmode resulting from the first bifurcation equation); and the remaining sum, the terms $i \ge 2$, represent the modifications that are necessary to describe the initial postbifurcation path properly. The small parameter ε, which controls the distance of the postbuckling solution to the bifurcation point ($\varepsilon = 0$ at the bifurcation point) can be introduced in different ways.

In Fitch and Budiansky [1970], ε represents the amplitude of the difference beween the postbuckling solution and the prebuckling state. This is the distance between these states projected onto the "direction" provided by the buckling mode (w_1, F_1). The relation between the load intensity q and ε along the postbifurcation path can also be written in form of a power series:

$$q/q_b = 1 + \sum_{i=1}^{N} a_i \varepsilon^i, \tag{3.15}$$

where q_b is the bifurcation load, and the coefficients a_i, which do not depend on q, are expressions involving terms containing the solutions w_i,F_i(i = 0,1,2,...). In practice, the determination of (3.15) beyond $n = 2$ is seldom executed because the corresponding equations are very difficult to solve, and, also, because the solutions are sought only for a small range of ε.

In the case considered here, $a_1 = 0$, which refers to a symmetric bifurcation point (i.e., a pitchfork bifurcation). The curvature of the bifurcation branch is determined by a_2. If $a_2 > 0$, the postbifurcation path is rising and the transition from the axisymmetric deformation state of the cap to the buckled state then takes place in a smooth fashion. This means that, in this case, the load carrying capacity of the cap is not exhausted. When $a_2 < 0$, however, the bifurcation branch is descending, and loss of stability occurs when the load reaches the critical value q_b, which is followed by a transient transition to a new stable state. This process is called snap-buckling. The new state that is reached after the snap can either be axisymmetric or nonsymmetric. Koiter's theory is restricted to solutions of equations of equilibrium in a small neighborhood of the bifurcation point. It is therefore not able to predict what particular final state will be reached.

It turns out, that when one or more bifurcation branches are descending, the buckling load of the cap is very sensitive to geometric imperfections, at least when these imperfections contain a component that corresponds to the buckling mode that is associated with the bifurcation point. For caps with such imperfections, the bifurcation point and its branches disappear in the load displacement diagram. They are replaced by smooth curves, one of which (the one connected with the undeformed state) contains a limit point with a corresponding load that is smaller than the bifurcation load [see Koiter, 1945,1963, Budiansky, 1974].

The buckling loads of the perfect cap (when no imperfections are present) are given by the top diagram in Figure 3.23 for the geometry $\lambda = 12$, $\left(|\xi| \cong 22\right)$, and Poisson's ratio, ν = 1/3. Here the dimensionless critical load $\Pi_c = 3\left(1-\nu^2\right)\bar{\rho}^2 q_c / |\xi|$ is plotted against the parameter $\bar{\lambda} = \bar{\rho}\lambda$ which determines the area of the loaded part of the cap's surface; q_c is the critical value defined by $q_c = q_c^* a^4/Eh^4$, $\bar{\rho} = \bar{r}/a$, where $\bar{r}$ is the radius of the loaded area of the cap. The case of a concentrated force is presented by $\overline{\rho}=0$, while $\overline{\rho}=1$ refers to uniform pressure. For convenience, in the right side of this diagram, the vertical axis is also presented in terms of η_c (see (3.11)).

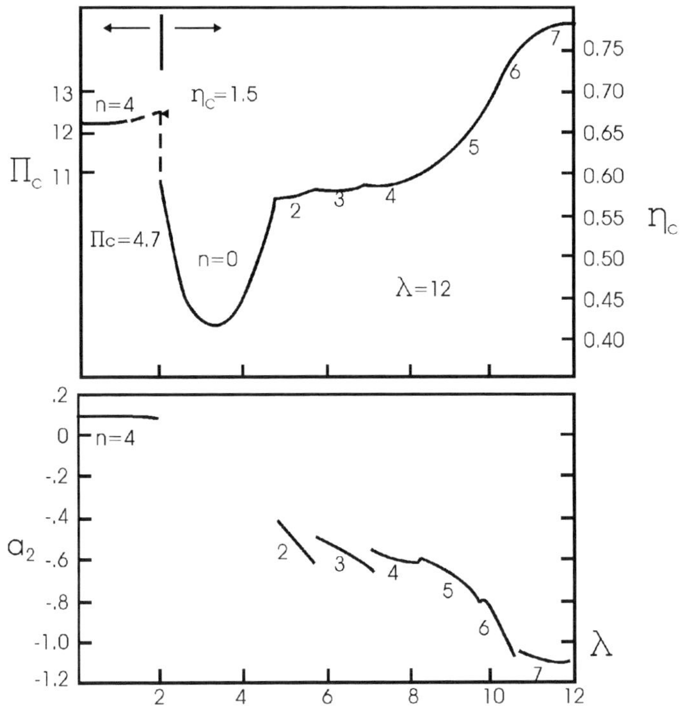

Figure 3.23 Buckling and postbuckling coefficients for the cap.

The results for the postbuckling coefficient $a_2(\overline{\lambda})$ are shown at the bottom part of Figure 3.23. The physical significance of these values is the following. For $0 \le \overline{\lambda} < 2$, the bifurcation takes place with a buckling mode with $n = 4$ circumferential waves. $a_2 > 0$ in this interval of $\overline{\lambda}$, and the buckling process is stable; i.e., the transfer of the deformation at the buckling point happens smoothly with the variation of the load. In the interval $2 < \overline{\lambda} \le 4.8$, there are no bifurcation points. In this range of $\overline{\lambda}$, loss of stability occurs at a limit point of the axisymmetric basic state before a bifurcation point is encountered. The possibility cannot be excluded, however, that, for $\overline{\lambda} < 2$, unstable bifurcation points exist beyond the first limit point along the basic state. This circumstance requires still further investigation.

In the range $4.8 < \overline{\lambda} \le 12$, bifurcation takes place (before the limit point) into nonsymmetric states with n circumferential waves that increase with $\overline{\lambda}$. The postbuckling coefficient is negative here, and it decreases with increasing $\bar{\lambda}$. In Fitch and Budiansky [1970, Fig. 5], computational results were also presented for very large values of λ; that is, $\lambda \to \infty$

These results, as far as the bifurcation analysis is concerned, permit the following conclusions. In the case of a clamped spherical cap subjected to a concentrated force applied at the apex, the initial postbifurcation path is rising, while, when the external load corresponds to uniform pressure applied to the entire external surface, the situation is the opposite — the initial postbifurcation branch is descending, with all the consequences that result from this condition.

These conclusions hold for all values of λ for which bifurcation occurs. We note here that the asymptotic values of η_c and a_2 (when $\lambda \to \infty$) presented in Figure 5 of Fitch and Budiansky [1970] are probably not correct. This follows, because, according to their article, the asymptotic value of η_c is equal to 0.636, while η_c for an entirely loaded surface is equal to about 0.76 for $\lambda = 12$. It must be clear that the asymptotic value of η_c for $\lambda \to \infty$ cannot be smaller than 0.76, because η_c, for the case of a pressure distribution over the entire surface, must increase with λ [see Huang, 1964; Valishvili, 1976]. The same discrepancy is also noted for a_2. For $\lambda = 12$, the minimum value of a_2 is given by $|a_2| = 1.1$, while the asymptotic value is $|a_2| = 0.64$ [Fitch & Budiansky, 1970, Fig. 5]. In reality, the last number must be greater than 0.64, as is shown in Figure 8 of Fitch and Budiansky [1970], where the asymptotic value is given by $|a_2| = 1.19$.

In the same study, the initial postbifurcation path for the case $\bar{\lambda} = \lambda(\bar{\rho} = 1)$ was also investigated, and it was demonstrated that a_2 is negative for all λ. Moreover, the authors found that the postbifurcation path descends backwards in the load vs. generalized displacement diagram. These facts indicate the seriousness of the imperfection sensitivity of the behavior of the cap from the point of view of loss of stability. (The principal question of the meaning of the imperfection sensitivity mentioned above is discussed in point d of the present section.)

The general picture of the cap's behavior in the context of the notion of loss of stability can be given as follows. In the case of $\bar{\rho} = 0$ (concentrated force) when $\lambda < 7.8$, there is no loss of stability in general. In the range $7.8 < \lambda < 9.2$, there is a loss of stability in terms of axisymmetric snap-through, while, for $\lambda > 9.2$, the bifurcation buckling occurs whereby the axisymmetric prebuckling state is changed into a buckled nonsymmetric state without loss of the load-carrying capacity, and, practically, without sensitivity to small geometric imperfections.

In the second extreme case ($\bar{\rho} = 1$), there is generally no loss of stability when $\lambda < 3.5$. From $\lambda = 3.5$ to $\lambda \approx 5.5$, axisymmetric snapping occurs. When $\lambda > 5.5$, the behavior is dominated by a bifurcation point with descending branches. This means that snap-buckling occurs at the bifurcation point. The state at which the snap will end is not revealed. It is possible that the post-snapping state is again an axisymmetric state, despite the fact that the trigger for the jump is provided by bifurcation into a nonsymmetric shape.

Similar results (falling initial postbifurcation path) for $\overline{\rho} = 1$ have been obtained for the cases $\lambda = 6$ and 10 ($\nu = 1/3$) [Famili & Archer [1965]; for the cases $\lambda = 6$ and 15 ($\nu = 1/3$) [Batoz, Dhatt & Prost, 1975]; for the cases $\lambda = 6$ and 7 ($\nu = 0.3$) in [Valishvili, 1976]; and for $|\xi| = 5$ and 7 ($\nu = 0.3$) [Uchiyama & Yamada, 1975]. But contradictory post-bifurcation results were obtained in other studies [e.g., Famili & Archer, 1965; Fitch & Budiansky, 1970; see also Tillman, 1970; Uchiyama & Yamada, 1975; Valishvili, 1976]. These conflicting results indicate that further study is necessary.

The dependence of bifurcation load η_c on λ in the case that $\overline{\rho} = 1$ is presented by the dashed curve 4 in Figure 3.11. As can be seen, the magnitude of η_c very slowly increases with λ (being of a weak oscillating character). For example, $\eta_c = 0.764$ ($n = 2$) for $\lambda = 6$, and 0.798 ($n = 31$, 32, and 33) when $\lambda = 42$; while for $\lambda = 30$, we have $\eta_c = 0.793$ ($n = 22$ and 23). With regard to the wave pattern in the circumferential direction, the situation is the following: as λ increases, the number n of the circumferential waves of the critical mode increases continuously, as can be observed from the data given above (see, e.g., the cases $\lambda = 30$ and 42). At $\lambda = 18$, the bifurcation loads that belong to the modes with $n = 12$ and 13 are very close. The existence of two modes at the same value of the critical load is very important. Due to the essential nonlinearity of the problem, it is natural to expect that the nonlinear interaction of these modes can generate new, very complicated phenomena. In this regard, there are the findings of the investigations of Koiter [1969b, 1982], Hutchinson [1967], Gräff, Scheidl, Tröger and Weinmüller [1985], and others, concerning the interaction of coupled buckling modes in connection with the problem of a complete (closed) spherical shell subjected to external pressure. In this case, a strong imperfection sensitivity of the initial postbifurcation path was discovered. Unfortunately, to the best of the writer's knowledge, we do not have similar analyses of compound buckling phenomena in spherical caps, despite the fact that this issue is of great interest from the practical, as well as theoretical, point of view.

The curves in Figure 3.11 and the corresponding numerical results are taken from Valishvili [1976], where $\nu = 0.3$. These results deviate about 6% from those given by Huang [1964], who used $\nu = 1/3$. Our calculations [Shilkrut & Gerlaku, 1969; Gerlaku & Shilkrut, 1970] for $\nu = 0.3$, produced critical loads that are also lower than those obtained by Huang by a margin of 3%. At the same time, it should be remembered that the asymptotic value of η_c (for $\lambda \to \infty$) obtained in Fitch and Budiansky [1970] is 7% below the corresponding value of η_c in Huang's study. We are thus inclined to believe that the numerical algorithms used by Huang overestimate the actual values of the buckling loads. The data from Archer and Famili [1965], where the bifurcation values of η_c are obtained by means of small vibrations (at the bifurcation point the frequency of proper small nonsymmetric vibrations must be equal to zero), are practically equal to those given in Fitch and Budiansky [1970] and Valishvili [1976]. There are some results for bifurcation loads η_c in Bushnell [1967], but these results do not appear to be correct.; e.g., for the case $\lambda \approx 41$, Bushnell found a critical buckling load of $\eta_c =$

$$c_i = \alpha_i^{(1-i)}\left(1-\alpha_i\right)^{-2};\;\; \alpha_i = (i-1)/(i+1);$$

$$c_i^* = \alpha_i^{*-[m+2(i-1)]/2}\left(1-\alpha_i^*\right)^{-2};\;\; \alpha_i^* = [m+2(i-1)]/[m+2(i+1)]. \tag{3.20}$$

ϕ is the coordinate (angle) in the circumferential direction.

Numerical solutions were obtained for $N = 6$; $N^* = 3$; $\nu = 0.3$, and $m = 2$, 3, and 4. Figures 3.25 and 3.26, show the relations $q^+(A_1)$ (for $A_2 = 0$) and $q^+(A_2)$ (for $A_1 = 0$), respectively for different values of $\left|\bar{\xi}\right|$.

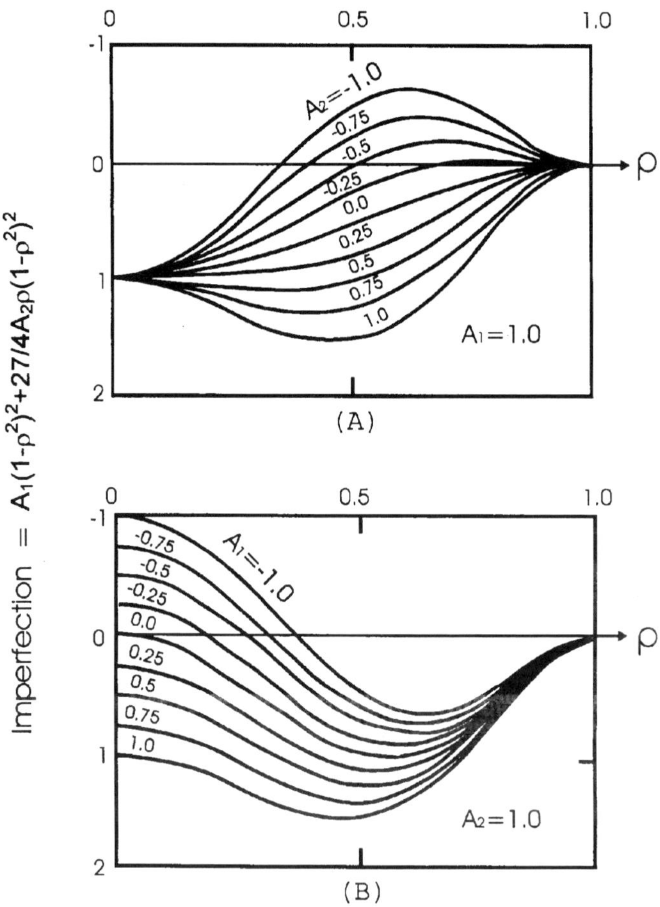

Figures 3.24a,b Axisymmetric imperfections used in the study of Uchiyama and Yamada [1974].

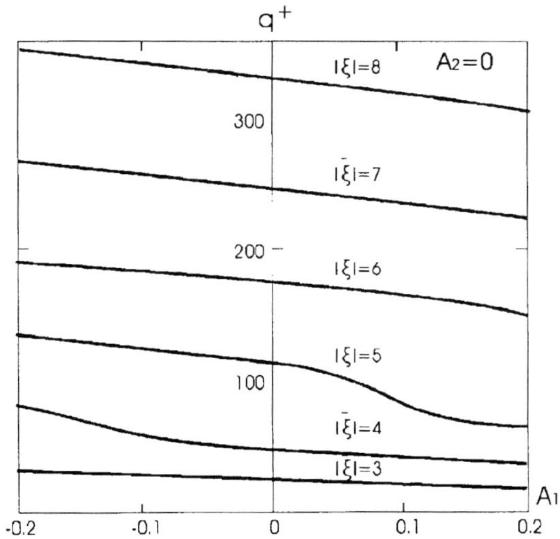

Figure 3.25 The imperfection study of Uchiyama and Yamada [1974], the limit load $q^{+}(A_1)$.

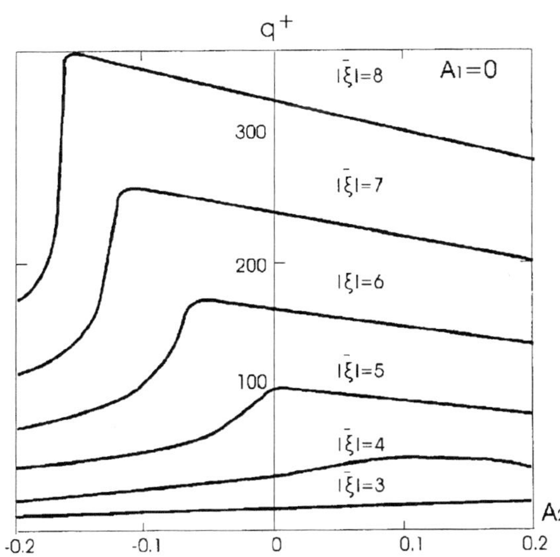

Figure 3.26 The imperfection study of Uchiyama and Yamada [1974], the limit load $q^{+}(A_2)$.

Figure 3.27 shows $q^{+}(A_2)$ and $q_b(A_2)$ for different values of A_1 and $|\bar{\xi}| = 6$. The analysis of these figures reveals that, when $A_2 = 0$, the limit load q^{+} and the buckling load q_b decrease if $A_1 > 0$ (imperfection represents a central dimple); but when $A_1 < 0$, the effect is the opposite. This case was also investigated in Budiansky [1960], and the same qualitative results were obtained. When $A_1 = 0$ (Fig. 3.26), the investigated phenomenon is more complicated. There is a range of negative values of A_2 for which q^{+} decreases substantially. This is not surprising, because, in this case, the imperfection is similar to the deformation mode of the ideal cap; thus, q^{+} is bound to decrease.

The results of this analysis leads to the conclusion that any imperfection that facilitates the development of the buckling mode of the cap diminishes the limit load q^{+} and the buckling load q_b (see Fig. 3.27). On the other hand, any stiffener (the role of which is to introduce a useful geometrical imperfection of the initial shape) will offer resistance to the development of the buckling mode, leading to an increase of the critical load. It is interesting to note that (as shown in Figs. 3.24a,b), the parameter A_2 plays the principal role in the shape of the imperfection given by (3.17). The results presented in Figures 3.26 and 3.27 show that some types of axisymmetric imperfections are able to essentially diminish the limit load q^{+}.

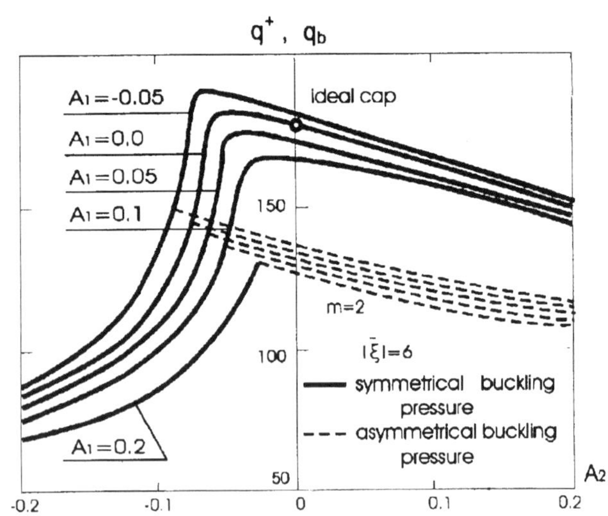

Figure 3.27 Results of Uchiyama and Yamada [1974], the limit and buckling load, $q^{+}(A_2)$, $q_b(A_2)$.

Goncalves and Croll [1992] presented an interesting study about the influence of large axisymmetric imperfections on pressure-loaded clamped shallow caps. Part of these results of this study are reproduced in Figure 3.28, which show that large imperfections are able to eliminate the danger of buckling (see the deformation curve G in Fig. 3.28).

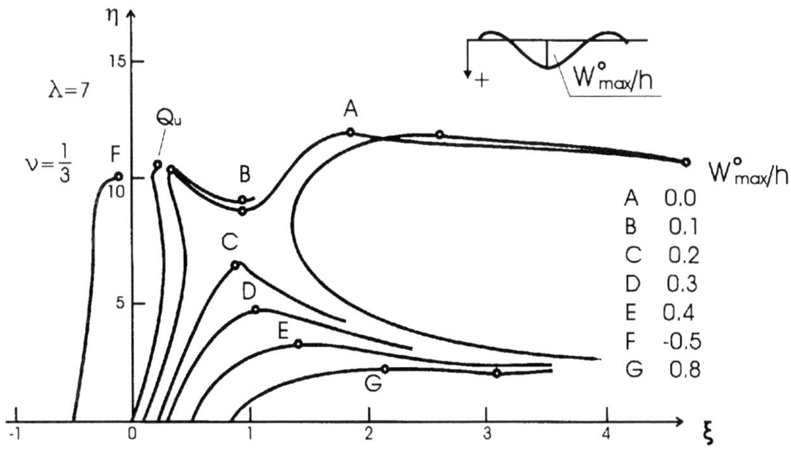

Figure 3.28 The analysis of Goncalves and Croll [1992] for large imperfections.

The shape of the imperfection (imperfection mode) with amplitude $w^0_{\max}/h$ is presented in the upper right corner of Figure 3.28, while the numerical values of the imperfection amplitudes that were used to construct the deformation curves are also listed. These results were obtained for the case $\nu = 1/3$, using Galerkin's method with a large number of terms .

Some other types of axisymmetric imperfections were considered by Freskasis and Morris [1972] (a local axisymmetric imperfection close to the edge), and by Bushnell [1967b, 1989] (a local decrease in the cap's thickness, and a local increase of the cap's radius near the apex).

An interesting analysis whereby the imperfections of laboratory specimens were measured was conducted by Thurston and Penning [1966]. On the basis of these measurements, the imperfections (the axisymmetric part) were reproduced in analytical form and introduced in the simplified version of Reissner's equations (see Section 1.1), which were then solved numerically. The test results and numerical solutions turned out to be close. This study showed that axisymmetric imperfections are able to reduce the value of q_{cr} with a factor of two to three. Finally, there exist a series of studies that deal with stiffened spherical shells, which we will not discuss further here [e.g., Bushnell, 1967a,b,c, 1989; Afanasyeva, 1975; Simitses & Cole, 1970; Simitses & Blackmon, 1975].

e) ***The influence of nonsymmetric geometric imperfections on the behavior of clamped spherical caps subjected to external pressure.***

This problem is considered by Uchiyama and Yamada [1975], who described the initial imperfect (nonsymmetric) shape $\overline{w}$ of the cap by the following expression.

$$\overline{w}(\rho,\phi) = \xi\left(1-\rho^2\right) + \overline{w}_b(\rho,\phi), \tag{3.21}$$

where $\overline{w}_b(\rho,\phi)$ is a nonsymmetric part of the imperfection which is taken in the same form as w_b in (3.19), and (3.20), but here it is with known (given) constants. The analysis is based on shallow shell equations, which are solved using Galerkin's method. The approximate numerical solution w is presented in the form of (3.18), (3.19), and (3.20). Specifically considered was the case of $N^* = 3$ in (3.19) and $m = m_{cr}$ (the number of circumferential waves of the bifurcation mode of the ideal cap). The expression of Galerkin's approximation contains five symmetric and three nonsymmetric terms.

$$w = \sum_{i=1}^{5} \alpha_i c_i \rho^{2(i-1)}\left(1-\rho^2\right)^2 + \left[\sum_{i=1}^{3} \alpha_{i1} c_{i1} \rho^{2(i-1)}\left(1-\rho^2\right)^2\right] \rho^{m_{cr}} \cos m_{cr}\phi$$

$$+\left[\sum_{i=1}^{2} \alpha_{i2} c_{i2} \rho^{2(i-1)}\left(1-\rho^2\right)^2\right] \rho^{2m_{cr}} \cos 2m_{cr}\phi. \tag{3.22}$$

Some of the results of this analysis are given in Figure 3.29 for various values of the rise parameter $|\xi|$. The abscissa ζ is here the amplitude of imperfection $\overline{w}_b$ divided by the shell thickness. At $\zeta = 0$, the value of the limit load q^+ coincides with bifurcation load q_b in the ideal case. One can observe the decrease of the collapse load q^+ with the increase of ζ, and the transitions that occur when the wave pattern of the buckling mode change from one value of m to another when ζ increases. When $|\xi| = 7$, for example, a transition from $m = 3$ to $m = 2$ takes place. This is because $m_{cr} = 3$ gives the smallest value of $q+$ when $0 < \zeta < \approx 0.1$, but for $\zeta > 0.1$, $m_{cr} = 2$ gives the smallest value of $q+$.

These investigations could have been carried out by an asymptotic approach for the post-bifurcation paths mentioned above. But the asymptotic expansions have only a limited range of validity. It is doubtful, therefore, that this method can be used to reveal the very interesting changes in wave numbers that occur in the Uchiyama and Yamada study when the amplitude ζ increases — changes that show that, for a certain amplitude ζ, one particular imperfection with wave number n has a dominating influence on the lowering of the buckling load with respect to imperfections with a different wave number.

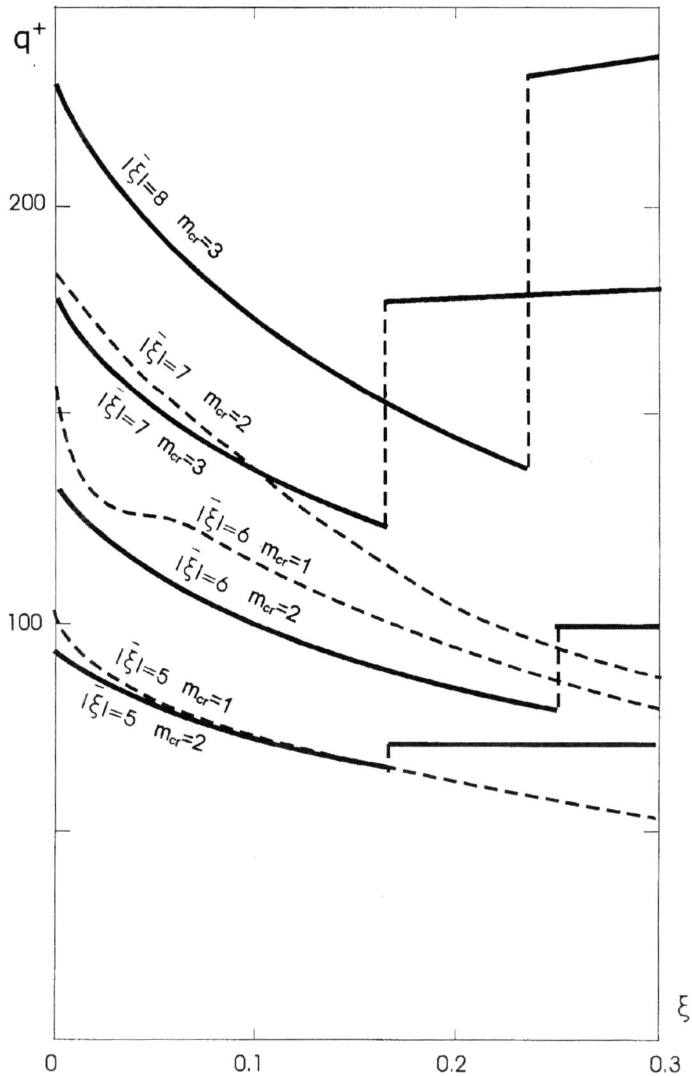

Figure 3.29 The influence of nonsymmetric imperfections; results of Uchiyama and Yamada [1975].

A significant drop in the load q^+ due to nonsymmetric imperfections was observed by Freskasis and Morris [1972], who considered caps for values of λ up to $\lambda = 28$. A particularly strong effect was observed of an imperfection of the meridian in the form of a dimple at $\rho \approx 0.5$, which is distributed periodically in the circumferential direction with the number of half waves given by m_{cr} (the wave number that corresponds to the bifurcation mode). For each

known about the snap-back load P^- [see Valishvili, 1976; Stegniy, 1969; Zipalova & Nenastjeva, 1966], because, as far as this author knows, that particular part of the solution has never been investigated systematically.

Table 3.6

$\nu = 0.3$

$\bar{\rho}$ \ $\lvert\bar{\xi}\rvert$	2	2.5	3	3.5	4	4.5	5
0.1875	-	-	-	-	-	-	-
0.3125	-	4.87	6.56	8.06	9.75	11.4	13.5
0.4375	-	7.69	11.1	15.0	18.0	24.7	30.0
0.5625	8.25	15.0	22.1	28.5	37.5	45.7	*

$\bar{\rho}$ \ $\lvert\bar{\xi}\rvert$	5.44	9.66	15.1
0.125	-	-	-
0.25	10.7	27.3	71.4
0.5	52.7	123.7	260.5
0.75	141.7	335.3	733.8

* No data

From the results presented above, it can be concluded that the increase of $\bar{\rho}$ leads to an increase of P^+ (when the value of $\lvert\bar{\xi}\rvert$ is fixed). This effect is fairly obvious. When $\bar{\rho}$ is increased, the load moves toward the clamped edge, so that the shell is able to resist the action of loading more effectively. Thus, from the point of view of stability, it is much more effective to spread the force out along a circle than concentrate it at the apex.

Results, qualitatively similar to the ones discussed above, were obtained by Mescall [1968], who used the simplified equations of Reissner for nonshallow shells (see Section 1.1). These results are represented in Figures 3.31-3.34 (no indications about the value of ν appear in Mescall's original work, but we believe that ν was set at $\nu = 1/3$).

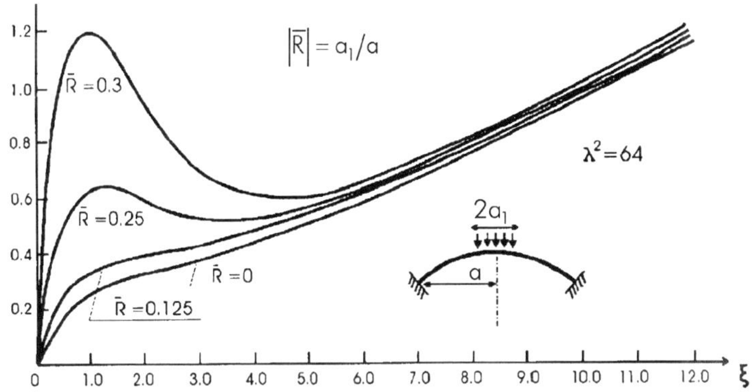

Figure 3.31 Equilibrium curves obtained by Mescall [1965], P versus ξ, $\lambda^2 = 64$.

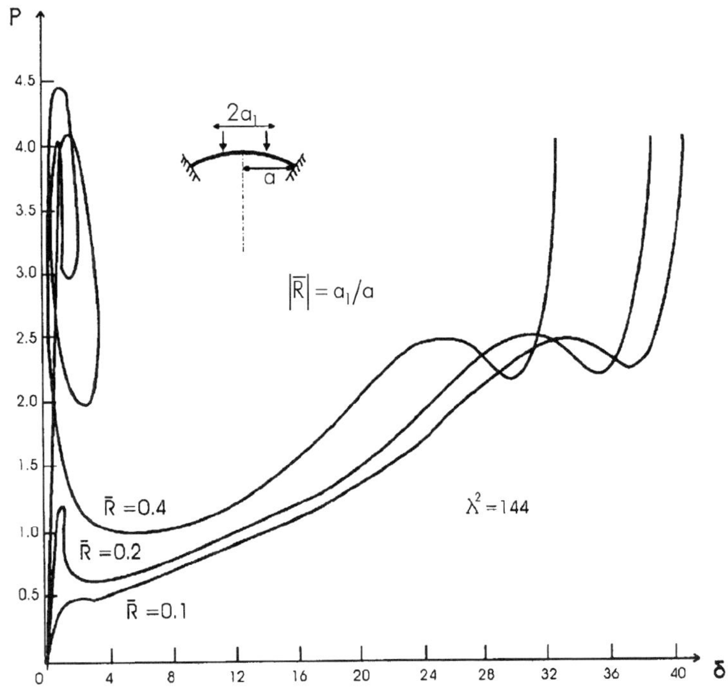

Figure 3.32 Equilibrium curves obtained by Mescall,[1965], P versus δ; $\lambda^2 = 144$.

In all these figures, the dimensionless load parameter is $P = RP^*/(2\pi Eh^3$, where P^* is the physical value of the resultant of the distributed load). Figures 3.31 and 3.32 show the equilibrium curves $P(\xi)$ and $P(\delta)$ for $\lambda = 8$ and 12, respectively. As is usual, ξ is the dimensionless displacement of the cap's apex, $\delta = w_p/h$ is the dimensionless displacement of the points where the forces are applied, and R is the radius of the spherical surface. The various branches corresponding to different ratios a_1/a are presented in Figure 3.34 (where a_1 is the radius of the load application and a is the radius of the cap's base).

The results presented in Figure 3.31, show clearly that increasing the radius a_1 — i.e., distributing the load further away from the apex — has a beneficial influence on the load-bearing capacity of the cap, as we mentioned earlier. In Figure 3.33, deformation curves $P(\delta)$ are drawn for different values of λ at constant ratio $a_1/a = 0.4$. Graphs of P_{cr} versus λ for different values of the ratio a_1/a are shown in Figure 3.34.

It turns out that the numerical results displayed in Figure 3.34 are very close to those given in Table 3.6. It can be observed in Figure 3.34 that the larger a_1/a is, the smaller λ needs to be to create a limit point in the axisymmetric response curve of the cap. When $a_1/a = 0.1$, for example, the smallest value of λ at which a limit point is created is $\lambda \approx 12$, while the corresponding value of λ is about 8 for $a_1/a = 0.2$, and is less than 5 for $a_1/a \geq 0.3$. This conclusion can also be deduced from Table 3.6. When $\overline{\rho} = 0.5625$, the basic state of the shell with $|\overline{\xi}| = 2$ contains a limit point, while in the case of $\overline{\rho} = 0.25$, a limit point occurs for the first time when $|\overline{\xi}|$ reaches $|\overline{\xi}| = 5.44$.

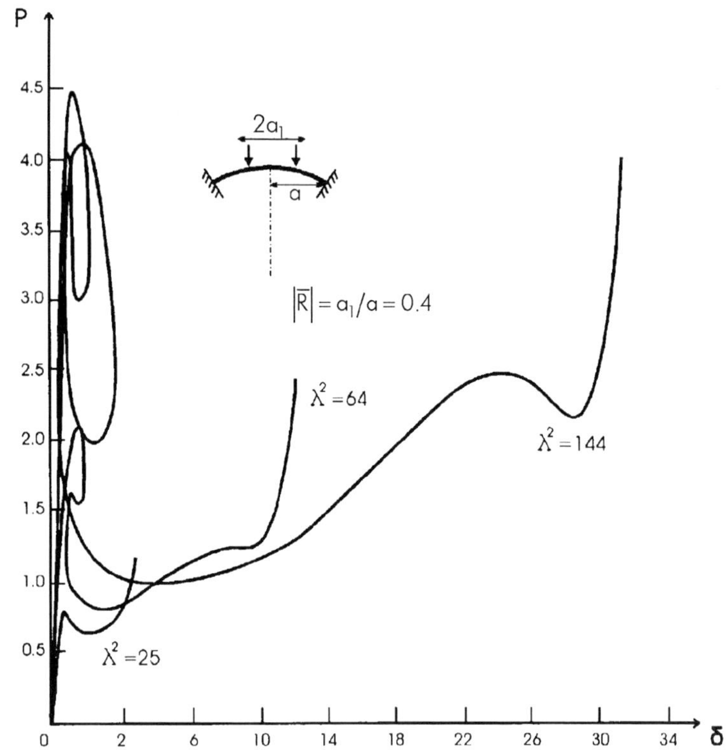

Figure 3.33 Equilibrium curves obtained by Mescall [1965], P versus δ; $a_1/a = 0.4$.

Another important conclusion is that, with the increase of λ and a_1/a, the solution curves become more and more complex (see Figs. 3.32 and 3.33) and the limit load P_{cr} increases. Another interesting characteristic is the emergence of multiple limit points in the solution curves (Figs. 3.32 and 3.33) of a type that suggest that successive jumps will occur when the load is increased (see also Figures 3.18, 3.20 and 3.21, which display similar behavior). It seems that such jumping phenomena are characteristic for partly loaded caps defined by relatively large values of λ.

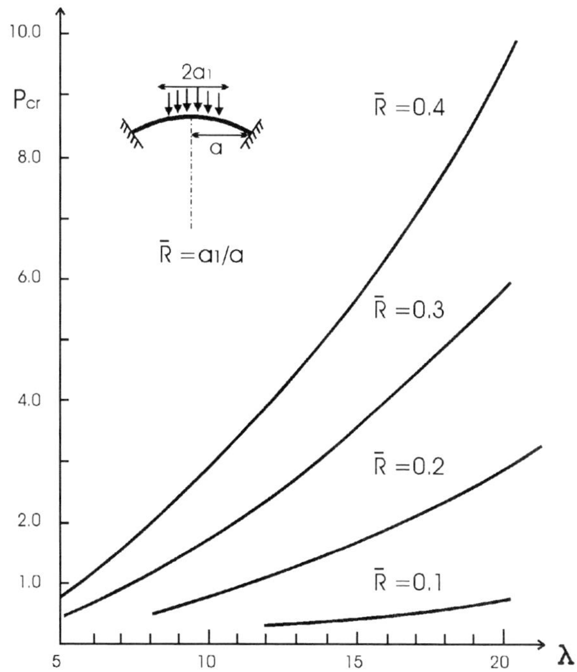

Figure 3.34 Critical loads obtained by Mescall [1965], P_{cr} versus λ.

c) Bifurcation and initial postbifurcation behavior of clamped caps subjected to a concentrated apex load or an axisymmetric line load.

These problems were considered by Fitch [1968] (apex load) and Akkas and Bauld [1971] (line load). Interesting results were also found by Bushnell [1967d]. Fitch's [1968] main results are presented in Figure 3.35a and 3.35b. Figure 3.35a shows the bifurcation load P_c as a function of λ. Here P_c is defined as $P_c = P_c^* R/(2\pi D)$; $D = Eh^3/\left(1-\nu^2\right)$, with P_c^* standing for the true (physical) load. Figure 3.35b shows the post-buckling coefficient a_2 as a function of the shell parameter λ. (Refer to Section 3.3c and the asymptotic expression (3.15) for the significance of a_2).

The heavy lines in Figure 3.35a give the solution to the problem. The number n near each curve segment denotes the number of waves in circumferential direction of the corresponding bifurcation mode. Analysis of these results shows that, for $\lambda < 7.8$ ($\left|\xi\right| \approx 3.31$), there is no loss of stability in general. In that case, the corresponding deformation curves are monotonically rising. In the interval $7.8 \le \lambda \le 9.2\left(\left|\xi\right| \approx 13\right)$, there are no bifurcation phenomena — the cap

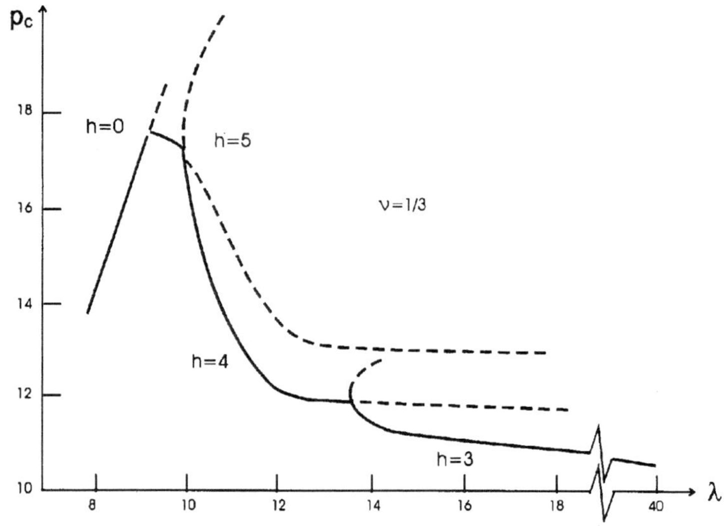

Figure 3.35a Results obtained by Fitch [1968], P_c vs. λ.

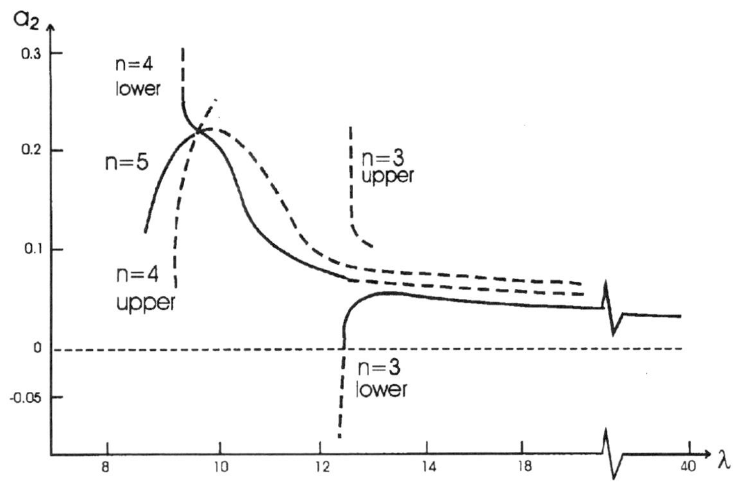

Figure 3.35b Results obtained by Fitch [1968], P_c vs. λ.

loses its stability at a limit point ($n = 0$) with limit load P^+. Beyond $\lambda = 9.2$, bifurcation points occur before the limit points are reached; i.e., the behavior of the cap is dominated by a bifurcation point that is located along the basic state at $P_c < P^+$. The postbuckling coefficient

a_2 turns out to be positive for all cases considered here, which means that buckling at P_c occurs smoothly, without loss of the load bearing capacity of the cap. It is of interest to note that, because $a_2 > 0$, the curvature term $\frac{d^2\lambda}{d\xi^2}$ is positive, and thus the postbuckling branch of the (pitch fork) bifurcation point is rising. It is of interest to note that, because $a_2 > 0$, the curvature term $\frac{d^2\lambda}{d\xi^2}$ is positive, and thus the postbuckling branch of the (pitch fork) bifurcation point is rising. A rising equilibrium branch is by itself not a sufficient condition for stability (as shown in Section 6.4), but, in the case analyzed here, the rising branch is connected with the first bifurcation point along the initially stable basic state. In that particular case, a rising secondary branch implies a stable branch [Koiter, 1963b, 1982; Budiansky, 1974].

It is also interesting to note that, when λ increases (the shell is more flexible), the wave number n decreases. As shown by Bushnell [1967d], for large values of λ, $\lambda \geq 30$, and thin shells R/h = 3040, the number n = 3. This value of n = 3 is related to the first bifurcation point, but secondary, higher bifurcations are not excluded. Thus, if one takes into account the complete picture of all possible solutions, the behavior of the caps in the postbifurcation range, beyond the interval of validity of the initial first postbifurcation solution, remains unknown. This limitation can be the cause (beside imperfections) of the discrepancy between the results produced by the mathematical model and experimental data. Figure 3.35a shows that when $\lambda \to \infty$, the load P_c approaches the asymptotic value P_c = 11. At the same time, for large λ and large R/h (a very thin shell), the influence of the boundary conditions is diminished, in which case the behavior of the cap and a complete spherical shell become practically identical. (In the latter case, the complete sphere is "compressed" by two diametrically opposing concentrated forces.) In other words, any thin-walled, deep spherical cap behaves as an inseparable part of a complete spherical shell. This fact is very well demonstrated by Figure 3.36 [Bushnell, 1967d], where the deformation curves for increasing λ (up to λ = 20) and for increasing values of the ratio R/h approach the deformation curve of a complete shell asymptotically. The same tendency can also be observed for the bifurcation points. The first three bifurcation loadings are presented in Bushnell [1967d]. It is further noted that these results agree very well with those given by Fitch [1968].

The case of axisymmetric, uniformly distributed line loads was examined in Akkas and Bauld [1971]. In this work, the snap-through, buckling, and initial postbuckling behavior is investigated by the same method that was used in Fitch [1968] and Fitch and Budiansky [1970]. This behavior turns out to be very similar to the phenomena described in Fitch and Budiansky [1970] for a cap loaded by external pressure over a part of its external surface.

qualitative aspects of the problem, but which, in terms of a quantitative solution, cannot be considered adequate.

Figure 3.38 presents a family of curves that are graphs of the critical values $P^+ = P_+^*\left[12\left(1-\nu^2\right)\right]^{3/2} a^2/\left(8\pi Eh^4\right)$ of the concentrated force P^* applied at the apex as a function of the external pressure $q = q^*\left[12\left(1-\nu^2\right)\right]^{3/2} a^2/\left(4Eh^4\right)$, for different values of λ (P^* and q^* are the corresponding real physical magnitudes). These results show that, when the pressure q increases, the critical value of the point force P^+ decreases. This is obvious, because the basic equilibrium curve $P = P(\xi)$ (when q = const.), which represents the response of the cap, must "straighten up" when q is decreasing; i.e., the difference between P^+ and P^- becomes smaller and smaller. In fact (see Section 3.4, point a), when $q = 0$, the snap-through phenomenon occurs "late," starting at the geometry given by $\lambda \approx 7.8$, while, when q is applied, the snapping due to the load P already occurs at $\lambda < 7.8$.

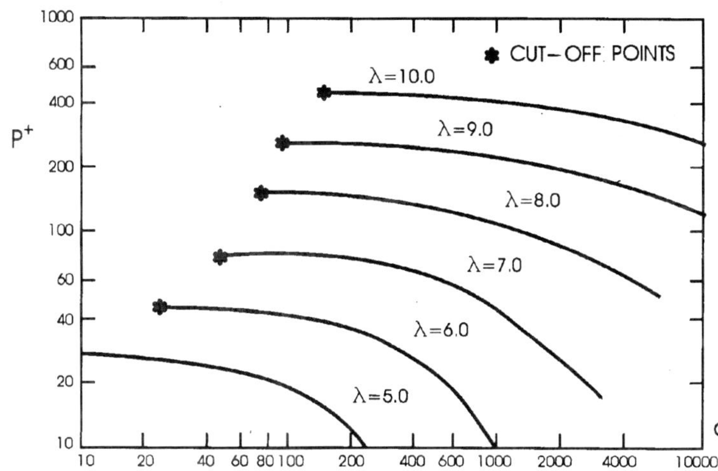

Figure 3.38 The case of a point force and external pressure, from Loo and Evan-Iwanowski [1966].

The paper by Loo and Evan-Iwanowski, however, also contains some errors. In Figure 5 of their work, for example, the case $\lambda = 9.1$ at the pressure $q = 1978.3$ sets the limit load for P at $P^+ = 43$, which should roughly correspond to our results presented in Figure 3.38, $\lambda = 9$, $q \approx 2000$ yielding the critical value $P^+ \approx 200$. This discrepancy leads us to believe that the numerical values of the critical load P^+ given by Loo and Evan-Iwanowski are grossly in error, although the qualitative picture these solutions portray is probably correct.

Another qualitative result that these authors produced concerns the application of $q < 0$ (i.e., internal pressure instead of external pressure). This has the effect that the limit load P^+ increases, but that the difference between P^+ and the lower limit load P^- decreases, so that eventually snap-through, as a possible way of stability loss, is eliminated. The role of internal pressure on the branching diagram for bifurcation was not investigated by Loo and Evan-Iwanowski [1966]. Also not investigated is the opposite case — the influence of P (inward or outward directed) on the maximum pressure q^+.

In view of the above, it is natural to ask what the influence is of the loading path on the equilibrium state that is finally reached in the case of a two-parameter load system with which we are dealing here. An equilibrium state at a certain value of the load combination P_1 and q_1 can be reached by specifying the (quasi-static) variation of P and q from the zero state to P_1, q_1 in an infinite number of possible ways; for example, let $P = 0$ and q be applied from zero to q_1. After this, P is applied until the value P_1, keeping $q = q_1$ constant. Of course, the opposite path of loading could also have been chosen; i.e., P is applied first until the value P_1 is reached and after that q is added. Another way is to let P and q be applied simultaneously in such a way that they reach their final values P_1 and q_1 at the same time, and so on.

At first glance, the possibility of finding an answer to the question of the influence of the loading path on the final result seems to be hindered by contradictions. On one hand, our problems are nonlinear; consequently, it seems reasonable that the loading paths have an influence on the final outcome when the loading process is completed. On the other hand, our systems are conservative, and, it seems that we must therefore conclude that the loading paths cannot have an influence on the final states, for such an influence contradicts the definition of conservatism. We can already mention here that this (apparent) paradox will be discussed and explained in detail in Section 4.5, where it will be shown that the question regarding the influence of the loading path cannot be answered by studying the solutions of the equations of statics alone. It can only be answered adequately if the problem under consideration is conceived as a transient problem; i.e., when the equations of motion of the system are taken into account.

It is worth noting that the experiments conducted by Kaplan [1974] and Chia [1967] showed that the order of the load application (first P and then q, as opposed to first q and then P), did not influence the final result. This observation, however, concerns a particular set of cases and is not a sufficient reason to conclude that the influence of the loading path should be ruled out. On the contrary, we will show in Section 4.5 that, in general, the loading sequence can have a decisive influence on the final equilibrium state of the system.

Chia [1967, 1976] presented results about the connection between upper limit loads of distributed forces and the external pressure in the case $\lambda = 5$ ($\nu = 1/3$). These results are similar to those described above. The axisymmetric deformation of a cap under a pressure load distribution $q = q_0(1-\rho^k)$ (where q_0 = const., and k = 1 or 2) was considered in Valishvili [1976], while Chia [1976] examined axisymmetric deformations for a load $q = q_0\cos(\pi\rho/2)$ [see also Perrone & Kao, 1970].

A series of investigations of the axisymmetric deformation of stiffened spherical caps, with different types of stiffeners and loaded by external pressure, are presented in, for example, Bushnell [1989, 1967a,b,c], Simitses and Cole [1970], and Simitses and Blackmon [1975]. Some interesting comparisons between the effect of nonsymmetric loading acting on a part of the cap's surface and that of the axisymmetric case of a fully loaded surface is given in Kao and Perrone [1971], Kao [1972], and Perrone and Kao [1970].

3.6 Caps with a movable clamped edge subjected to different types of loading.

When there are no edge forces applied, the boundary conditions have the following form

$$\omega(0) = \theta(0) = 0; \quad \omega(1) = \theta(1) = 0. \tag{3.23}$$

This case has not received so much attention as the previous case of an immovable (fixed), clamped edge. Results for the axisymmetric deformation states and the bifurcation buckling loads along these states are presented in Valishvili [1976], Kornishin and Isanbaeva [1968], and [Valishvili, 1976] for the load case of external pressure acting on the entire surface of the cap. These results correspond to the curves marked by the index 3 in Figure 3.11. The heavy curve 3 presents the limit loads of the axisymmetric basic state, while the dashed curve 3 gives the bifurcation loads. (All the calculations were carried out for $\nu = 0.3$.) Snap-through starts to occur at a value of $|\xi|$ between 3 and 3.5 ($4.46 < \lambda < 4.8$) in the class of axisymmetric states. At $|\xi| = 3.5$, the limit load is given by $\eta^+ = 0.253$ (see (3.11)). Bifurcations start to appear at λ = 8 ($|\xi| = 9.66$), with the corresponding bifurcation load at $\eta_c = 0.348$ and a buckling wave pattern with $n = 2$. When λ increases, η^+ also increases, but η_c decreases, the wave number n increases steadily with the increase of λ. At $\lambda = 42$ ($|\xi| = 266.4$), we have $\eta^+ = 0.43$; $\eta_c = 0.241$, with n = 17, 18, 19, and 20. It is important to pay attention to the appearance of compound bifurcation points, or, to phrase it with more precision, the appearance of clustered bifurcations, where a number of bifurcation points are grouped together with very little space between them. This occurs for some specific cases. For example, when $\lambda = 18$ ($|\xi| = 48.92$), we have, at $\eta_c = 0.301$, two-buckling modes with n = 8 and 9. Compound bifurcation points are an additional complication in the asymptotic analysis of the initial postbifurcation behavior.

It is interesting to note that, even for large values of λ (up to $\lambda \approx 42$, $|\xi| \approx 266.4$), there are large differences between the critical loads (limit and buckling loads) of caps with a movable clamped edge and those of caps with an immovable clamped edge (see Fig. 3.11). This difference does not occur when the loading consists of a concentrated force at the apex. In that case, even at $\lambda \approx 30$, the boundary conditions do not yet have an influence, and the critical values for P for a movable and immovable clamped edge practically coincide [see Bushnell, 1967d].

Results for the lower limit loads are almost nonexistent. (The author knows of only one case — that of $q^- \approx 20$ ($\eta^- \approx 0.258$) for $|\xi| = 4$ and $\nu = 0.3$ [Shilkrut *et al.*, 1969; Zipalova and Nenastjeva, 1966].) Also, no results are available for the postbuckling behavior of a cap with a mobile clamped edge.

The values of the limit load q^+ for a distributed load acting over an area, with the apex as center and the radius a_1 determining the perimeter (in nondimensional form $\bar{\rho} = a_1/a$), are presented in Table 3.7, where the pressure is given in terms of the intensity parameter $q = q^* a^4/(Eh^4)$ [Kornishin & Isanbaeva, 1968]. Table 3.8 presents the values of a distributed load $P = 2P^* \bar{\rho} a^3/(Eh^4)$ acting at radius a_1. The left part of this table is compiled on the basis of Kornishin and Isanbaeva [1968], while the right side corresponds to results compiled by Valishvili [1976]. No entry means that there is no result available, while a dash means that, for the corresponding geometrical parameters, no limit points occur.

Only a few results are available for the case of a concentrated force $P = P^* R/(2\pi E h^3)$ at the apex [e.g., Valishvili, 1976; Kornishin & Isanbaeva, 1968; Zipalova & Nenastjeva, 1966]. These lead to the following conclusion. For the case $\nu = 0.3$, no limit points occur until the rise parameter $|\xi|$ reaches the value $|\xi| \cong 5$. At $\lambda = 8$ and 10 ($|\xi| = 9.66$ and 15.1), the limit loads for P are equal to 0.6 and 0.907, respectively. In the case of a movable clamped edge, the limit loads P^+ are much smaller than the corresponding values P^+ in the case of rigid clam-ping (see Section 3.4, point a). The same trend can be observed in the case of a cap that is partly loaded by external pressure (compare Tables 3.7 and 3.4).

Compound loading, consisting of a point force P acting at the apex and a pressure distribution q over the entire surface, is briefly discussed in Chia [1976]. Kornishin and Isanbaeva [1968] consider the case of uniform pressure distributed in a belt-like area adjacent to the cap's edge.

There is a general characteristic that spherical shells with clamped (movable or immovable) edges possess. The equilibrium branches of these systems never cross the horizontal axis ξ; in other words, there are no nontrivial solutions at the zero load. This can be proven mathematically using the properties of the governing equations. In contrast, as will be shown below (Section 3.8), in the case of simple support (a hinged edge), this characteristic of the solutions no longer exists.

Table 3.7

$\nu = 0.3$

	$\|\bar{\xi}\|$	2.5	3	3.5	4	4.5	5
$\bar{\rho}=$	0.1875	-	-	-	-	-	-
	0.3125	-	-				73-75.7
	0.4375	-	-		32.5	36.1	46.9
	0.5625	-	-		24.4	30.5	33.7
	0.6875	-	-		20.8	28.1	32.1
	0.8125	-		15.7	19.9	26.4	33.5
	1	-	-	15.0	21.0	25.5	33.7

Table 3.8

$\nu = 0.3$

	$\|\bar{\xi}\|$	2.5	3	3.5	4	4.5	5		$\|\bar{\xi}\|$	5.44	9.66	15.1
$\bar{\rho}=$	0.1875	-	-	-	-		6.75	$\bar{\rho}=$	0.1	-	23.02	-
	0.3125	-	-	5.44	6.37	7.50	7.87		0.3		44.07	177.66
	0.4375	-	-	7.50	9.56	12.2	15.6		0.5	25.28	82.24	227.18
	0.5625	-	10.4	12.6	17.2	21.7	26.6		0.7	60.59	144.49	
	0.6875		21.4	35.6	33.7	43.5	48					

3.7 The hinged cap in pure bending. Axisymmetric deformation states and the phenomena of bifurcation in compression and tension.

a) Axisymmetric deformations.

The term "pure bending" is used for the state of deformation of a cap that is loaded by a uniformly distributed, edge bending moment M^*, represented by the dimensionless intensity parameter $M = M_r^*(1)12\left(1-\nu^2\right)a^2/\left(Eh^4\right)$. For the isotropic case, the boundary conditions are given by:

$$\alpha_1\omega'(1) + \beta_1\omega(1) = \gamma_1;\ \theta'(1) + \nu\theta(1) = \gamma_2, \tag{3.21}$$

The coefficients α, β, γ are specified as $\alpha_1 = 1$, $\beta_1 = -\nu$, $\gamma_1 = 0$, and $\gamma_2 = M$ for the immovable case, and $\alpha_1 = 0$, $\beta_1 = 1$, $\gamma_1 = N$, and $\gamma_2 = M$ for the mobile case. $N = N_r(1)$ is the horizontal component of the edge membrane stress resultant. The parameters M and N can also have different physical meaning; i.e., they can be related to a thermo-elastic problem (see Section 2.7). Let us consider the case of a linear elastic, isotropic cap with uniform rigidity, for example. Let us also assume that the temperature field $T(\rho,\zeta)$ (where ζ is the coordinate of a point of the cap's body along the normal measured from the middle surface) does not depend on ρ. Then $M_T \equiv$ const.; $N_T \equiv$ const. (see (1.61)), and the basic equations (1.58)-(1.59) do not contain M_T and N_T explicitly. The temperature load factors M_T and N_T then enter the formulation via the boundary conditions (see (1.66)) in the following way: $\gamma_1 = -N_T(1)$, (instead of zero, in the case of immovable support), and $\gamma_2 = M+(1+\nu)M_T(1)$ (see Table 1.1). If T, being independent from ρ, is an odd function of ζ, then $N_T(1) = 0$, and, in this case, there is no difference between bending induced by the edge load M or bending induced by the combination of M and the thermo-elastic effect given by $M_T(1)$. In this section, we shall exclusively deal with generalized pure bending.

The pure bending problem was investigated analytically with the help of the theory of reciprocal systems [see Shilkrut, Morar & Gerlaku, 1970; Shilkrut & Vyrlan, 1977; Shilkrut, 1967, 1969a, 1974], but also numerically [Shilkrut, 1980, 1982, 1983; Shilkrut *et al.*, 1969; Shilkrut, Morar & Gerlaku, 1970; Shilkrut & Vyrlan, 1977; Shilkrut & Gerlaku, 1969; Gerlaku & Shilkrut, 1970; Chia, 1971]. (In all these publications just enumerated (except the last), "reciprocal systems" were referred to as "symmetrical systems.") The theory of reciprocal systems is discussed in Sections 2.2-2.6. Here, we shall mainly follow the work of Shilkrut [1980, 1982, 1983], Shilkrut *et al.* [1969], and Shilkrut, Morar and Gerlaku [1970]. Typical results for the case of immovable edge support are shown in Figures 2.7, 2.8 and 3.39. For the case of mobile edge support the results are given in Figure 2.9.

One of the characteristic features of these curves is that, in the case of mobile hinges, the branching diagrams possess self-intersections, while, when the supports are fixed, the self-intersections do not exist. It must be emphasized, however, that the self-intersections displayed in these diagrams have nothing to do with bifurcations. A point where the curves inter-sect in the diagram marks only the existence of two equilibrium states that share the same deflection of the apex. In general, when we perceive the cap as a continuum, the solutions trace a spatial "curve" in an $(\infty + 1)$ dimensional space, and this means that, at a point of the so-called self-intersection, only "one component" of the infinite dimensional vector of displacements is shared by the two separate solutions (at the same value of the load). Of course, all of this has something to do with the circumstance that the plots M versus ξ are merely projections of the solutions onto the plane spanned by the load factor M, and the apex displacement ξ. We will return to this subject in Chapter 6.

As mentioned earlier, the spherical shell, undergoing pure bending, is a reciprocal system (see also Chapter 2). Its response curve is skew-symmetric, where the center of symmetry C has the coordinates $\xi = |\bar{\xi}|$; $M = 2|\bar{\xi}|(1+\nu)$. Any equilibrium state of the shell that relates to this point C has a flat shape (is straightened) and is unstable. When $M = 4|\bar{\xi}|(1+\nu)$ and

The bifurcation in tension can take place when the deformed meridian in the prebuckling state is stretched everywhere (take, for example, plates subjected to any type of transverse forces and caps loaded by forces that act outwards). This condition can be fulfilled for any value of $|\bar{\xi}|$ but only when the edge is moveable. The bifurcation points in tension are located along the branches of the response curves for pure bending; that is, they are present along the descending branch for $M \leq 0$ and along the rising branch for $M \geq 4|\bar{\xi}|(1+\nu)$ (the latter case corresponds to the state whereby the cap is turned inside out). These branches are symmetric relative to the center of symmetry of the deformed curve in conformity with the theory of reciprocal systems. As a consequence, the bifurcation points will be distributed along both these branches by symmetric pairs, and the corresponding bifurcation loads M_i are connected by means of the same relation that exists for M^+ and M^- (3.25). For the other parameters of any pair of symmetric bifurcation states, the following relations hold:

$$\xi_1 + \xi_2 = -2\bar{\xi};\;\; \omega_1(\rho) \equiv \omega_2(\rho);\;\; \theta_1(\rho) + \theta_2(\rho) \equiv -4\bar{\xi}\rho;$$

$$F_1(\rho,\phi) \equiv F_2(\rho,\phi);\;\; w_1(\rho,\phi) + w_2(\rho,\phi) = -2\bar{w}(\rho), \tag{3.26}$$

where F_i, w_i are the solutions for the buckling modes. Examples of bifurcation in tension for a plate and very shallow caps $\bar{\xi} = \pm 1$ are presented in Figure 2.13. The plate's bifurcation loads are $M_{1,2} = \pm 318$, with the corresponding deflections $\xi_{1,2} = \pm 7.4$. For $\bar{\xi} = 1$, $M_1 = 366$ and $\xi_1 = 6.84$, and for $\bar{\xi} = -1$, $M_1 = -372$ with $\xi_1 = 8.9$.

Bifurcation points in "compression," when they occur, are also distributed by symmetric pairs, and their corresponding parameters are also coupled by (3.25) and (3.26). The "dynamics" of the appearance and distribution of these points is as follows. For very small values of $|\bar{\xi}|$, bifurcation in compression does not exist in general. At some threshold value $|\bar{\xi}| = |\bar{\xi}_0|$, bifurcation points start to appear at the center of symmetry C, where the cap is turned into a flat plate representing an equilibrium state that is unstable. As $|\bar{\xi}|$ increases beyond $|\bar{\xi}_0|$, bifurcation points "move" away from the center C in opposite directions along the branches of the unstable states that pass through C. Further, at some other particular value of $|\bar{\xi}| = |\bar{\xi}_\beta| > |\bar{\xi}_0|$, bifurcation points cross the limit points of the axisymmetric equilibrium curves. The first of these points has then become the critical equilibrium state of the prebuckling equilibrium branch, a role that it has taken over from the limit point.

Thus, the picture is the following. When $|\bar{\xi}| < |\bar{\xi}_{00}|$, there is no loss of stability in general ($|\bar{\xi}_{00}|$ is the value of $|\bar{\xi}|$, at which jumping in the class of axisymmetric modes starts to occur). In the range $|\bar{\xi}_{00}| < |\bar{\xi}| < |\bar{\xi}_\beta|$, the loss of stability can happen only by axisymmetric snap-through. In the range $\left|\xi_{00}\right| \leq \left|\xi\right| \leq \left|\xi_{\beta}\right|$, bifurcation points in compression appear, but they are located along the unstable branches of the deformation curve. But as soon as $\left|\xi\right| > \left|\xi_{\beta}\right|$, the loss of stability of the basic axisymmetric prebuckling state takes place at a bifurcation point before a limit point is reached. It is worth noting here that the knowledge of the bifurcation load and the corresponding buckled mode(s) is not sufficient to assess what will happen when the structure is loaded up to and beyond this point. To be able to make this assessment, it is necessary to compute the postbuckling branch of the bifurcation point.

Generally speaking, the bifurcation buckling behavior of the cap can be divided into two separate cases. In the first case, (1), the cap passes smoothly to a new path undergoing non-symmetric deformations when the load is increased. This means that the postbuckling solution is stable, although, it can become critical again in a secondary bifurcation point somewhere along the bifurcation branch (where a change in wave pattern takes place, and so on). In the second case, (2), the structure will snap; i.e., move in a transient manner to either an axi-symmetric or a nonsymmetric postbuckling state, because the postbuckling branch is partly or completely unstable. Where the transition, which starts at the unstable bifurcation point, will take us is dependent on the particulars of the problem at hand. Finally, if bifurcation is not an option, we deal with a scenario, (3), where loss of stability occurs at a limit point of the axisymmetric fundamental state. This stability loss is followed by snap-through, which will end at an axisymmetric (stable) equilibrium state.

Values of $\left|\xi_{0}\right|$ for different support conditions and wave numbers n are presented in Table 3.9 [see Shilkrut, 1974; Shilkrut & Gerlaku, 1969; Gerlaku & Shilkrut, 1970; Shilkrut & Vyrlan, 1977]. The case of $n = 0$ is only relevant to the range $\left|\xi_{00}\right| < \left|\xi\right| < \left|\xi_{\beta}\right|$.

Table 3.9

$\nu = 0.3$

Types of Supports	n = 0	1	2	3	4
Mobile hinge	1.92	3.38-3.40	5.40-5.50	7.60-7.70	9.60-9.70
Immovable hinge	0.69	1.20-1.22	1.70-1.80	2.20-2.30	2.70-2.80

The results presented in the table were actually obtained analytically, making use of the circumstance that, at the center C, the fundamental axisymmetric solution is known, so that the corresponding linear boundary value problem for the equations of type (2.153) and (2.154) can be solved almost exactly.

It can thus be stated that, when $|\bar{\xi}| > |\bar{\xi}_\beta|$, bifurcation points appear along the certain stretches of the axisymmetric solution diagram (i.e., along those parts defined by $-\infty < M < M^+$ and $Mw^- < M < \infty$) that bring about a change in the behavior of the cap. (Here, M^+ and M^- are the first maximum and the last minimum of the deformation curves when we move along the curve in the positive direction of axis ξ.) The threshold values obtained by Shilkrut and Gerlaku [1969] and Gerlaku and Shilkrut [1970] are: $|\bar{\xi}_\beta| = 7.8$ for mobile support and $|\bar{\xi}_\beta| = 6 \div 7.5$ for immovable support. In the latter case, a more accurate number is not available, although we can be certain that, for immovable support, $|\bar{\xi}_\beta|$ must be smaller than for movable support.

The dynamics of the "motion" of bifurcation points when $|\bar{\xi}|$ increases is very well illustrated by the curves in Figure 2.14 for mobile supports for wave number $n \geq 2$. Similar curves are given in Figure 2.15 for immovable supports. Some numerical data based on results given in Shilkrut and Gerlaku [1969] and Gerlaku and Shilkrut [1970] are presented in Table 2.1 for mobile support and in Table 3.10 for immovable support.

Table 3.10

$\nu = 0.3$

$\lvert\bar{\xi}\rvert$	n	M_{cr}	ξ_{cr}	Notes
6	0	73.85	0.174	Snapping
6	9	70.98	0.440	Bifurcation along the intermediate
7.5	0	102.30	0.32	unstable branch
7.5	13	84.39	-0.06	Bifurcation along the first stable branch

M_{cr} in this table means M^+ (snap-load in the axisymmetric case) or M_b (the bifurcation load). The results related to the bifurcation phenomena appear in the first four columns. It is interesting to note that, in the case of pure bending, bifurcation points with multiple buckling modes were not discovered. Nevertheless, there are cases when there are bifurcation points that are very closely spaced and each is associated with buckling modes with different wave number n.

Studies of the influence of axisymmetric imperfections on the limit and bifurcation loads have been given in Shilkrut [1980] and Shilkrut and Gerlaku [1969]. The initial imperfect shape is taken here as:

$$\overline{w}(\rho) = \bar{\xi}\left(1-\rho^2\right) + 6.75\zeta\,\rho^2\left(1-\rho^2\right)^2. \tag{3.27}$$

The first term in (3.27) describes the reference geometry of the cap, while the second term is the (axisymmetric) imperfection with an amplitude ζ, which is measured at $\rho = 3^{-1/2}$. A positive ζ means that the imperfection is directed inward at $\rho = 3^{-1/2}$. According to the definition (3.27), the rise of the cap $|\bar{\xi}|$ is independent of the imperfection (3.27). The picture of the influence of imperfections on the behavior of the cap is very clearly illustrated in Figures 3.41 and 3.42. which refer to the case of mobile support, for both positive as well as negative amplitudes ζ.

These results admit the following conclusions. The limit loads M^+ and buckling loads M_b belonging to the first part of the solution branch decrease when the amplitude ζ of the imperfection is increased from zero, while the opposite occurs when the sign of the imperfection is reversed. This is not surprising, however, because, when $\zeta < 0$, the presence of the imperfection can be compared to the addition of a stiffener to the cap. A particular case is encountered for the geometry $|\bar{\xi}| = 14$ where M^+ and M_b increase for positive as well as negative values of ζ.

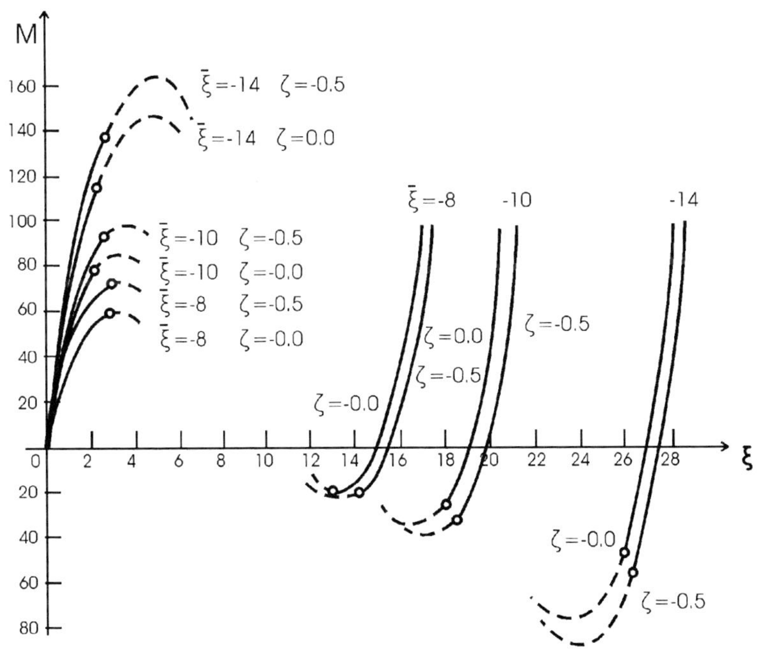

Figure 3.41 The influence of axisymmetric imperfections, from [Shilkrut & Gerlaku, 1969; Shilkrut, 1980]; $\zeta \leq 0$.

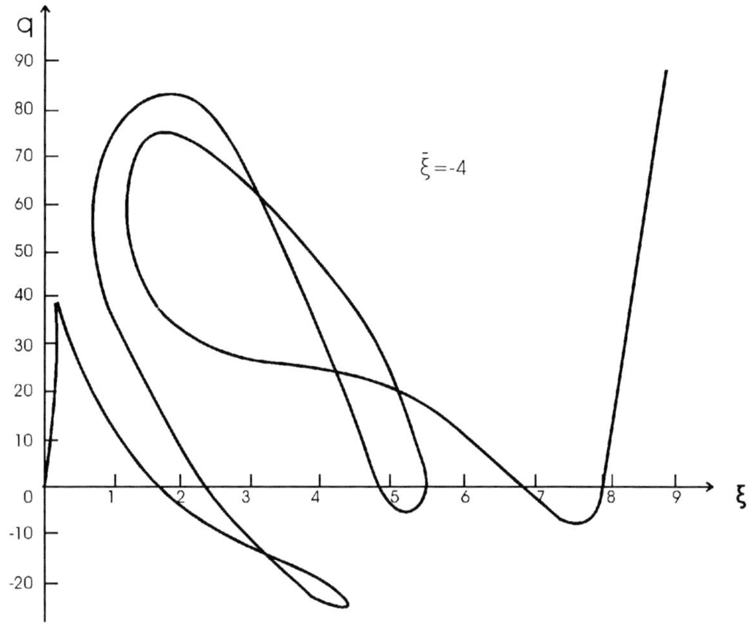

Figure 3.44 Uniform pressure in the case of immovable supports; $\bar{\xi} = 4$ [Shilkrut *et al.*, 1969; Shilkrut, 1980].

Not many results for the limit load q^- are available. Its values for mobile supports when $|\bar{\xi}| \leq 5$ are presented in Shilkrut *et al.* [1969] and Zipalova [1967]. It turns out that, for $|\bar{\xi}| > 4$, q^- is negative. A comparison of the solutions in Figure 3.11 reveals that the behavior of clamped and hinged caps is qualitatively very similar. More detailed comparisons will be discussed in Section 3.9.

b) A hinged cap partly loaded by external pressure in the area around the apex.

To the best knowledge of the present author, this case was only considered by Kornishin and Isanbaeva [1968] who restricted their investigation to axisymmetric deformations. Table 3.11 presents critical load values q^+, where q^+ is defined as $q^+ = \overset{*}{q}_+ a^4 / Eh^4$ [*ibid.*] ($\bar{\rho}$ is the dimensionless radius of the boundary of the loading area, and $\overset{*}{q}_+$ is the actual physical value of the critical pressure.

The analysis of these data shows that, when $\bar{\rho}$ increases, the values of q^+ decrease. This tendency does not imply that it is more convenient, from the point of view of stability, to concentrate the external loading in the neighborhood of the apex. On the contrary, the resultant of the external loading, being proportional to the square of $\bar{\rho}$, becomes larger together with the

Table 3.11

$\nu = 0.3$

Mobile Support							Immovable Support				
$\lvert\bar{\xi}\rvert$	2.5	3	3.5	4	4.5	5	1	1.5	2	2.5	3
$\bar{\rho}$ =0.3125	16.2	24.3	32.5	43.3	58.2	71.7	7.16	18.6	30.8	44.3	59.6
=0.4375	0.02	12.6	18.0	27.1	36.1	46.9	4.73	10.8	18.0	25.2	32.5
=0.5625	6.16	9.62	13.6	19.2	25.6	34.5	3.21	6.41	12.8	18.4	24.0
=0.6875	5.11	7.21	10.4	15.2	20.0	25.6	2.40	4.81	11.2	16.0	22.4
=0.8125	4.26	5.70	8.91	11.4	16.4	20.7	2.00	4.99	9.97	17.1	22.8
1	3.75	5.25	7.50	10.5	13.5	18.0	1.78	4.69	9.00	18.7	28.1

increase of $\bar{\rho}$. Thus, from the stability point of view, it is better to spread the distributed load (with a given resultant) over as large an area as possible. The same characteristic is observed in all previous cases for other boundary conditions.

It is necessary to emphasize that the given values for q^+ are related to the first maximum (limit point) of the deformation curve. Results for the lower limit load q^- or buckling loads q_b are not available.

c) *Hinged caps under the influence of a concentrated force at the apex or axisymmetric line loads.*

1. *The case of a concentrated force* $P = P^*R/(2\pi Eh^3)$. This problem is considered in a number of publications [e.g., Shilkrut *et al.*, 1969; Kornishin & Isanbaeva, 1968; Mescall, 1965, 1967; Valishvili, 1976; Valishvili & Stegniy, 1970; Zipalova, 1967]. From these investigations it follows that, for both types of support (movable or immovable), P^+ increases monotonically with $\lvert\bar{\xi}\rvert$. For a movable edge ($\nu = 0.3$), there is no snap-through in the range $\lvert\bar{\xi}\rvert \approx < 2$. At $\lvert\bar{\xi}\rvert = 2$, snap-through occurs at $P^+ = 0.148$, while at $\lvert\bar{\xi}\rvert = 15.1$, $P^+ = 0.876$. In the case of immovable supports, at $\lvert\bar{\xi}\rvert = 1$, $P^+ = 0.199$, while $P^+ = 2.175$ at $\lvert\bar{\xi}\rvert = 15.1$. The results in the range $1 \leq \lvert\bar{\xi}\rvert \leq 15.1$ were determined by Kornishin and Isanbaeva [1968] and Valishvili [1976], and they are in good agreement with other studies [see, e.g., Mescall, 1965, 1967; Shilkrut *et al.*, 1969; Zipalova, 1967]. The following empirical formula,

$$P^+ \cong \lambda^2/110, \tag{3.28}$$

which was proposed by Mescall [1965], can be used in the interval ($4 \leq \lambda \leq 10$) (the case of $\nu = 1/3$) for immovable supports.

The available results for the critical load P^- are scattered over various publications [see Mescall, 1965, 1967; Shilkrut, 1980; Shilkrut *et al.*, 1969; Valishvili, 1976; Zipalova, 1967]. What can be deduced from these results is that the load P^- is still positive for small values of $|\xi|$, but because it decreases with the increase of $|\xi|$, it will eventually become negative (at $|\xi| \approx 1$ for an immovable hinge, and, at $|\xi| \approx 2$ for a movable hinge). As soon as P^- changes sign, the equilibrium curves intersect the ξ-axis at the same point as in the case of pure bending, or in any other loading case.

The equilibrium curves become very contorted when $|\xi|$ becomes larger, in particular when the support is hinged (see, e.g., the deformation curves for immovable support in Figs. 2.34 and 2.35). In Figure 2.36, the deformation curve is given for mobile support for the geometry $|\xi| \approx 12.3$. We should note here that this curve was probably first obtained by Mescall [1967] using the simplified Reissner equations for nonshallow shells. The same solutions were obtained from the unabridged equations of Reissner for nonshallow shells (see Section 1.1) using the shooting method [described in Shilkrut *et al.*, 1969], and also from the von Karman-Marguerre type equations for shallow shells. Thus, the type of equations are of no consequence in this case, but the type of support is probably decisive.

It is necessary to mention, that the postbuckling branches of deformation curves for $|\xi| =$ 8 and 10 in the case of mobile hinged edge, as given in Zipalova [1967], cannot be considered to be correct. This is because their results violate the above-mentioned feature (see Section 3.7, point a); i.e., the independence of the location of the point of intersection of the equilibrium curves with the ξ-axis for any type of loading. The points of intersection for a concentrated force P and uniform pressure q do not coincide in Zipalova.

In Mescall [1965], it is stated that experimental evidence does not indicate that snapping takes place when the theoretically determined limit load P^+ is reached (the experiments themselves are not described in this work.). A possible explanation for this observation is that it was not the force, but the central deflection, which was used as a loading device. In that case, the experiment and the mathematical model do not correspond. We can finally note that, as far as is known, the buckling analysis for this problem has not yet been carried out.

2. *Axisymmetric uniformly distributed line loads.* The most important results that are available for this case are given by Valishvili [1976] and Kornishin and Isanbaeva [1968], although some others can also be found in Stegniy [1969] and Zipalova and Nenastjeva [1966]. Limit loads P^+ (where $P = 2P^*\bar{\rho}a^3 / Eh^4$; $0 < \bar{\rho} \leq 1$ is the dimensionless radius of the loading ring) are presented in Tables 3.12 and 3.13. These results are based on Kornishin and Isanbaeva [1968] (the left-hand side of the tables) and Valishvili [1976] (the right-hand side). We are not aware of any other results for this particular problem.

Table 3.12

Mobile hinges. $\nu = 0.3$

$\lvert\xi\rvert$	2	2.5	3	3.5	4	4.5	5	$\lvert\xi\rvert$	5.44	9.66	15.1
$\bar{\rho}$ =0.1875	1.23	1.50	2.29	3.00	3.75	5.18	6.37	$\bar{\rho}$=0.1	7.14	22.3	No data
=0.3125	1.36	1.87	2.62	3.75	5.25	6.00	9.00	=0.3	10.8	46.6	108.9
=0.4375	1.58	2.25	3.19	4.50	6.37	8.55	10.5	=0.5	15.1	56.3	163.2
=0.5625	2.02	2.62	4.12	5.15	7.50	9.00	12.0	=0.7	20.5	60.7	No data
=0.6875	3.75	4.50	7.50	10.5	13.5	16.5	21.0				

Table 3.13

Immovable hinges. $\nu = 0.3$

$\lvert\xi\rvert$	1	1.5	2	2.5	3	$\lvert\xi\rvert$	5.44	9.66	15.1
$\bar{\rho}$ =0.1875	0.750	1.69	2.86	4.12	5.62	$\bar{\rho}$=0.3	27.1	113.5	103.8
=0.3125	0.937	2.06	3.37	4.97	6.00	=0.3	27.1	113.5	103.8
=0.4375	0.937	2.62	4.87	7.50	9.00	=0.5	46.1	246.4	249.5
=0.5625	1.31	3.37	7.50	12.9	19.5	=0.7	67.2	246.4	595.8
=0.6875	1.87	5.25	11.2	18.0	24.0				

3.9 A comparison of the behavior of hinged and clamped caps: Returning to the role of membrane stresses in the buckling process.

Figure 3.11 shows that the limit and buckling loads depend, first of all, on the boundary conditions for the stress function ω (movable or immovable support). Indeed, the curves 1 (hinged) and 3 (clamped), which are both related to movable supports in this figure, are close to one another. The same can be observed about the curves 2 and 4. This circumstance shows once more the crucial role the membrane stresses play in the process of loss of stability. Of course, the bending stresses also affect the snapping and buckling phenomena, as can be deduced from the fact that curve 3 is higher than curve 1. A cap with a clamped edge behaves as if it is more rigid than a cap with a simply supported edge. The differences between the results for hinged and clamped supports, however, are markedly smaller than those between movable and immovable supports.

3.10 Shallow spherical caps with flanges. A model for imperfect support conditions.

In practice, the actual support of a cap is realized by means of a ring stiffener or flange. This circumstance is usually not taken into account when the buckling behavior is analyzed. However, such a support can be compared to an imperfection, and it can have a serious influence on the value of critical loads, so that its effect must be understood and quantified. As far as we know, this aspect has only been studied by Kochavi and Shilkrut [1984]. We shall consider the case of mobile and immovable clamping of shallow spherical caps with a support flange connected to the edge which is subjected to uniform external pressure q (see Fig. 3.45).

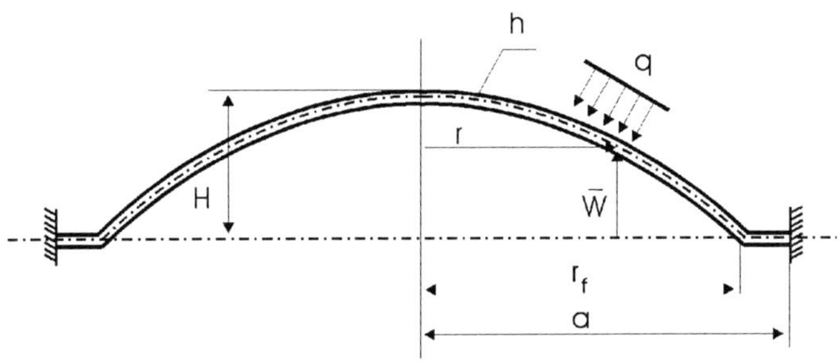

Figure 3.45 Cap with flange under uniform external pressure, from Kochavi and Shilkrut [1984].

The notations and equations remain unchanged for the present problem, except that definition of the initial shape is now given by:

$$\overline{w}(\rho) = \begin{cases} \overline{\xi}\left[1-\left(\dfrac{\rho}{\rho_f}\right)^2\right]; & 0 \le \rho < \rho_f; \\ 0; & \rho_f < \rho \le 1; \end{cases} \tag{3.29}$$

where $\rho_f = r_f/a$ (see Fig. 3.45); (3.30)

$$\overline{\theta}(\rho) = -\frac{d\overline{w}}{d\rho} = \begin{cases} 2\overline{\xi}\dfrac{\rho}{\rho_f}; & 0 \le \rho < \rho_f; \\ 0; & \rho_f < \rho \le 1. \end{cases} \tag{3.31}$$

There are numerical results for three types of shells; a shallow cap with $\bar{\xi} = -1.5$; a not-so-shallow cap with $\bar{\xi} = -7$; and a deep shell with $\bar{\xi} = -15$. Figures 3.46 and 3.47 show a series of deformation curves for different values of the flanges width: $(1 - \rho_f)$ (for $\rho_f = 1$, the flange is absent). These results are obtained for the intermediate case $\bar{\xi} = -7$ and two types of boundary conditions: immovable and movable clamped edges.

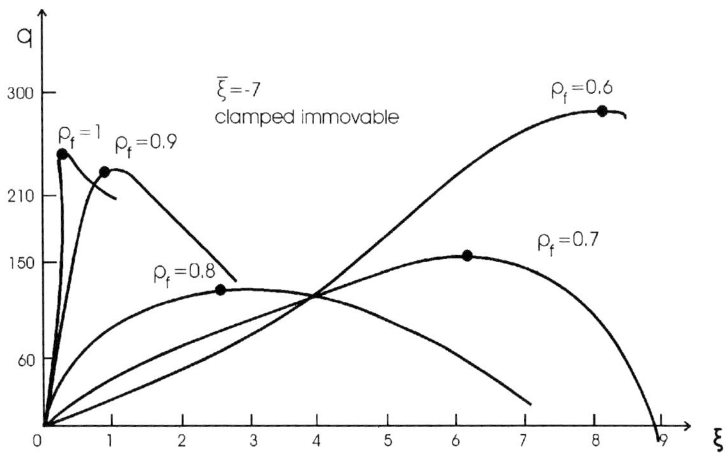

Figure 3.46 Basic results for cap with flange, $\bar{\xi} = -7$, immovable, from Kochavi and Shilkrut [1984].

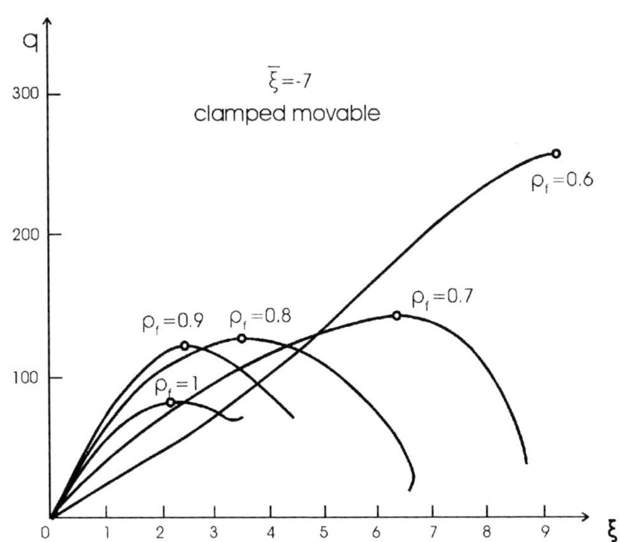

Figure 3.47 Basic results for cap with flange, $\bar{\xi} = -7$, movable, from Kochavi and Shilkrut [1984].

The critical loads (for $\bar{\xi} = -7$) as a function of the width of the flange are given in Figure 3.48. The data in Figures 3.49 (only immovable clamping) and 3.50 are related to $\bar{\xi} = -15$. The case of a shallow cap, $\bar{\xi} = -1.5$, is presented in Figures 3.51 and 3.52.

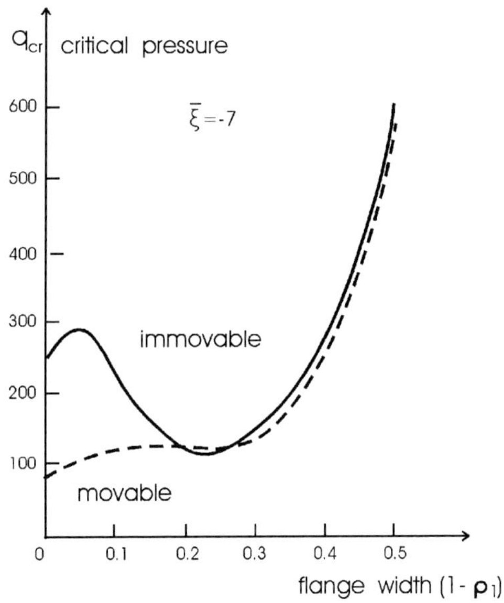

Figure 3.48 Critical loads for cap, $\bar{\xi} = -7$, from Kochavi and Shilkrut [1984].

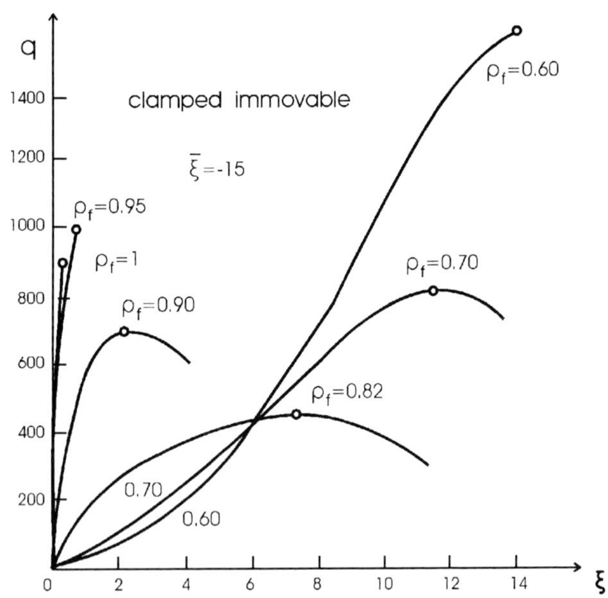

Figure 3.49 Basic results for cap with flange, $\bar{\xi} = -15$, immovable, from Kochavi and Shilkrut [1984].

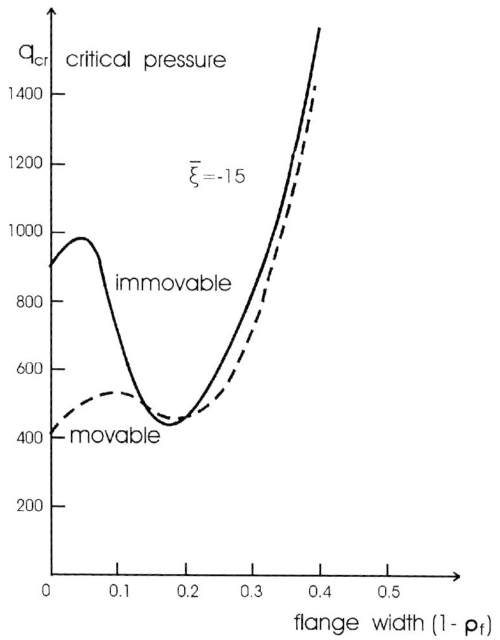

Figure 3.50 Critical loads for cap, $\bar{\xi} = -15$, from Kochavi and Shilkrut [1984].

The analysis of these results leads us to the following conclusions. From Figure 3.48 ($\bar{\xi} = -7$), we can observe the strong effect the flange width has on the maximum pressure loads of both movable and immovable clamped caps. The limit load increases initially until it reaches a maximum (at $1-\rho_f = 0.05$ for the immovable flange, and at $1-\rho_f = 0.17$ for the movable flange). After that, it decreases until it reaches a minimum (at $1-\rho_f = 0.22$ for both cases), after which it rises back rather steeply. This behavior can be understood with the help of the results presented in Figures 3.46-3.50 for the cap with $\left|\bar{\xi}\right| = 7$ and 15, respectively. When the width of the flange is increased in our model (see equations (3.29-3.31)) two opposing effects try to influence the critical load. The first is due to the circumstance that, according to the way our model is defined, the rise parameter H/r_f becomes larger with the increase of the width of the flange. This change in geometry stiffens the response of the cap and raises the critical load. On the other hand, the more material is transferred to the flange the more flexible the support between flange and cap will become. This reduction in the degree of clamping is responsible for a reduction of the critical load of the cap.

For the two cases presented in Figures 3.46-3.50, the effect of the deepening of the cap is initially stronger than the effect of the relaxation of the edge conditions, with the result that at first the critical load rises with the increase of the flange's width $(1-\rho_f)$ until a maximum is reached, after which the boundary relaxation effect takes over and the critical load starts to decrease. When the width of the flange becomes larger and larger, the structure will gradually start to look like a circular plate and behave likewise. This means that, after a certain value of $(1-\rho_f)$, the critical load of the system starts to rise again until it reaches infinity (which marks the point at which buckling disappears).

For shallow caps that are almost flat ($\bar{\xi} = -1.5$; Figs. 3.51-3.52), the behavior is somewhat different. For very small flanges $(1-\rho_f < 0.1)$, there is no snapping. For flanges in the interval $0.1 < 1-\rho_f < 0.3$, there are limit points; but when the width $1-\rho_f$ exceeds .3 the limit points disappear again, because, in this range, the system becomes effectively a flat plate.

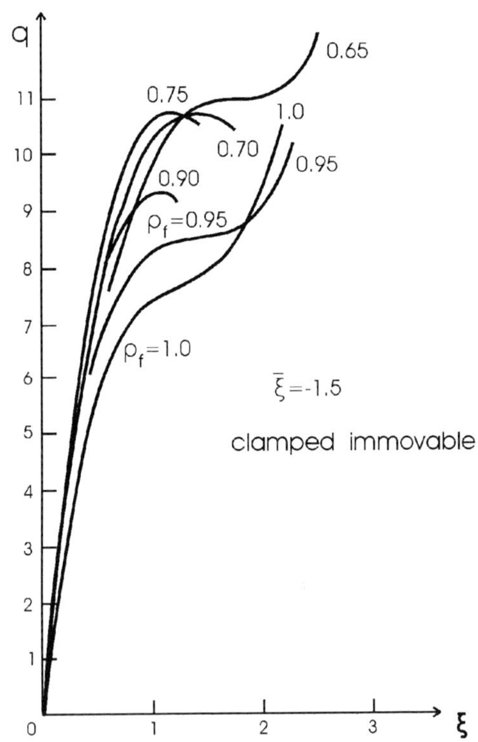

Figure 3.51 Basic results for cap with flange, $\bar{\xi} = -1.5$, immovable, from Kochavi and Shilkrut [1984].

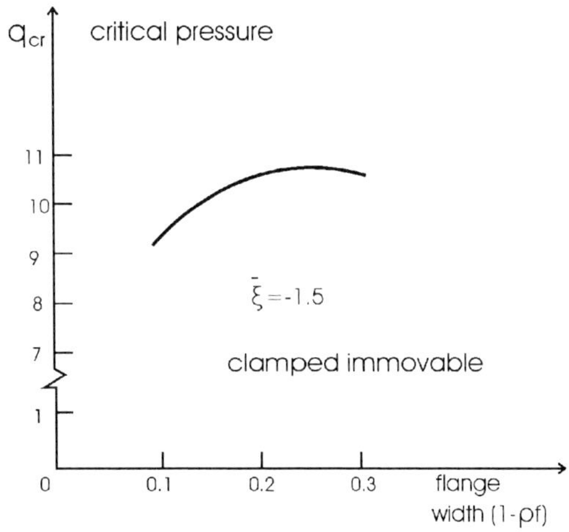

Figure 3.52 Critical loads for cap, $\bar{\xi} = -1.5$, from Kochavi and Shilkrut [1984].

In order to compare both cases of supports (mobile and immovable), let us return to Figures. 3.48 and 3.50. The differences in the limit loads of these cases are large for narrow flanges ($1-\rho_f < 0.15$) and small for wide ones. This effect can be explained when it is realized that wide flanges reduce the mobility of movable edges by stiffening the cap in the base plane, thus reducing the movable case to that of the immovable one. It can be illustrated by comparing the load-deflection curves of both cases (Figs. 3.47 and 3.50) for $\rho_f = 0.6$ and 0.7. Another observation is that, for both $\bar{\xi} = -7$ and $\bar{\xi} = -15$ (Figs. 3.48 and 3.50), there is a small range of widths (near $1-\rho_f = 0.2$) where the critical pressure for the movable edge is higher than for the immovable edge. The phenomenon just indicated is interesting, but the differences in critical pressures remain small and it occurs only in a very narrow band of ρ_f values.

The findings discussed above can be summarized as follows. The buckling load is strongly influenced by the presence of a flange at the edge of the cap. But it is very difficult to decide *a priori* what the effect will turn out to be in an actual situation that is not covered by the results of [Kochavi & Shilkrut]. Moreover, the analysis in that case should be conducted in a more general context and should also cover the phenomenon of bifurcation into non-symmetric states (which, for as far as is known, has never been considered for this case). We can therefore conclude that a problem where flanges are used to connect the shell to its surroundings should be treated as a fresh new case, unless other information is available.

deformation curves is depicted in Fig. 2.10).

As can be seen, the load deformation curves $M(\xi)$ become monotonic at $q \approx 40$. This effect can also be observed in Figs. 2.11 and 2.21 (the case of the joint action of M and a concentrated force P at the apex). A similar trend is revealed in the case of three-parameter loading, shown in Figure 2.12, for the case $\left|\bar{\xi}\right| = 4$, where the support is movably hinged, and, in addition to the loads M and q, the edge force $N = -5$ is added.

Part of the family of curves of Figure 2.12, mapped onto the plane (ξ, q), is shown in Figure 4.1. The latter, together with the figures mentioned above, show very clearly the process of straightening, as well as the decrease of the degree of non uniqueness of the solutions at fixed values of the loading parameter. All the results described in the present section for $\nu = 0.3$ are taken from Shilkrut, Shevandronov, Morar and Maksimov [1969] and Shilkrut [1980]. The features of symmetry of these results completely agree with the theory of reciprocal systems given in Sections 2.3-2.5.

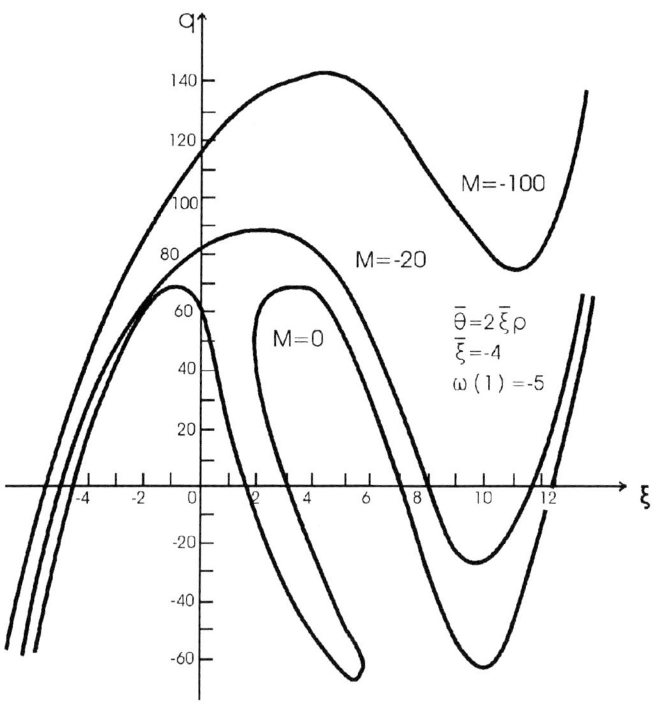

Figure 4.1 Detail of the solution shown in Figure 2.12, mapped in the q, ξ plane.

The global picture of the behavior of plates and caps, which undergo axisymmetric deformations under multi-parameter loading, can be given in a very general way by means of the so-called "deformation map." Its concept and construction are described in Section 4.3.

4.2 Nonsymmetric bifurcation phenomena in the case of a two-parameter loading.

This section is based on results obtained by Shilkrut and Kochavi [1983, 1986], a study that was mainly focused on the bifurcation phenomena that are generated by the action uniform pressure q and the edge force $N = N_r(1)$. Unless stated otherwise, the results hold for $\nu = 0.3$.

a) Internal pressure only.

We will first present the case of internal pressure, with $N = 0$. This case is characterized by negative values of η (see (3.11)). Snap-buckling is not expected in this case, but buckling into nonsymmetric wave patterns may occur (this is known as "bifurcation in tension"; see Section 2.1). The bifurcation pressures for the case of a movable clamped cap under internal pressure are presented in Figure 4.2 for the parameter values $\lambda = 4 - 8$ (see (3.4), (3,5)).

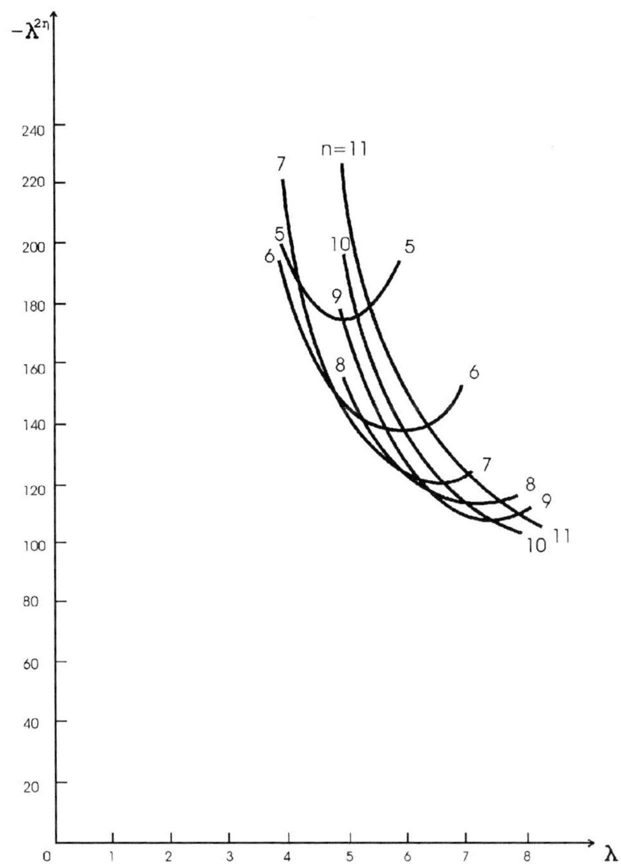

Figure 4.2 Bifurcation pressures for the case of a cap with a movable clamped edge.

Each curve in the diagram represents the variation of the bifurcation load as a function of λ (in terms of $-\lambda^2\eta$ vs. λ) that belongs to a wave pattern with a fixed number n of waves in the circumferential direction. The critical bifurcation load, and the particular wave number that corresponds to it, is determined by the lowest intersection of the vertical λ = const. with any of the given curves. Figure 4.3 shows the distribution of the axisymmetric circumferential stresses for different boundary conditions and values of λ. In this figure, x is defined by $x = \lambda\rho$, so that $0 \leq x \leq \lambda$.

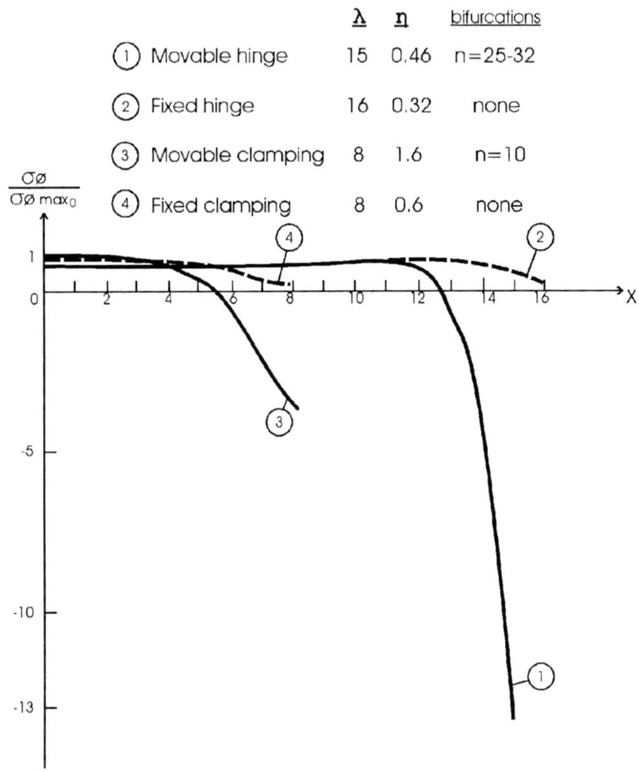

Figure 4.3 Hoop stress solutions for various edge conditions.

The main conclusions of this investigation are:

1. Buckling into a nonsymmetric mode was not found for fixed edges. This is not surprising because there are no circumferential negative (compressive) stresses, and, thus, the necessary condition for bifurcation is not satisfied in these cases (Fig. 4.3, cases 2 and 4).

Buckling into a nonsymmetric pattern can only occur when the edge is movable (Fig. 4.3, cases 1 and 3). A theoretical substantiation of this fact is given in Shilkrut [1983a], but it was also demonstrated experimentally [see Shilkrut, 1980, 1983a; Shilkrut & Kochavi, 1983, 1986]. We should also refer to Section 2.8 (property 10 and others).

Figure 4.4 shows photographs of two test articles that served in an experiment whereby hydraulic pressure was applied [Shilkrut & Kochavi, 1983]. The samples, were manufactured by forging thin steel sheet into the form of a spherical cap with flanges. The edge of the left sample on Figure 4.4 was immovably clamped and bifurcation buckling did not occur during the test. The edge of the right sample was clamped, but movable, and in this case a buckling pattern developed after the internal pressure was applied. Although the post-buckling state shown in the photograph already belongs to the plastic range, buckling started before yielding took place.

Figure 4.4 The outcome of two experiments (immovably clamped, movably clamped).

2. In the case of movable clamped caps under internal pressure, the bifurcation phenomenon is connected with the formation of a large number n of waves (Fig. 4.2) that occur in a band near the edge where the circumferential stress are compressive. This is shown in Figure 4.3, case 3.

The influence of an additional edge load N to the case of internal pressure is examined in the next section.

b) Internal pressure η together with edge membrane force N.

We consider now the case of a cap with movably clamped edges ($l = 6$), the bifurcation pressures η of which are given in Table 4.1 for different values of the load N. The compilation of these bifurcation loads in terms of η was carried out as follows.

For each value of N, the pressure η was increased from zero up to a certain upper limit, and the loads at which bifurcation took place were tabulated together with the wave number n of the corresponding buckling mode. For example, for $N = -1$, the highest value of the buckling load considered was $\eta = 3.5$. The bifurcation points found in this interval were characterized by wave numbers varying between $n = 6$ and $n = 10$. For higher values of the load, other bifurcation points exist associated with buckling modes with wave numbers exceeding 10. These bifurcations were not further examined.

From the results that we obtained, it can be seen that the addition of tensile edge forces leads to an increase of the bifurcation loads for η. For example, in the "classical" case $N = 0$, bifurcation occurs at $\eta = 3.4$, with $n = 8$; while for $N = 2$, the bifurcation pressure is $\eta = 4.1$, also with $n = 8$. But the particular wave number n that corresponds to the critical (minimum) value of η is practically independent of N, as we can see from the data in Table 4.1, where it is fixed at $n = 8$. As can also be seen from this table, there are two cases where two modes (with different n) occur almost simultaneously.

Table 4.1

$\nu = 0.3$

	n										
N	5	6	7	8	9	10	11	12	13	14	15
−1	3.3	3.0	3.0	3.2	3.5						
0	5.4	3.8	3.4	3.4	3.5	3.8	4.2	4.7	5.3	6.0	6.8
1			3.8	3.7	3.9						
2		4.8	4.2	4.1	4.3	4.6	5.0	5.5	6.1	6.8	7.6

c) *External pressure η and edge membrane force N.*

We will now consider the behavior of a cap with clamped, but movable edge under the combined action of external pressure η and a uniformly distributed edge load N. In Figure 4.5, for some values of N, the limit pressures $\eta_c(\lambda)$ belonging to the first limit points of the axisymmetric equilibrium states are given by the solid curves, while the critical bifurcation pressures η_{cb} belonging to the first bifurcation points along these states are given by the dashed curves. For comparison, the critical pressures for the "classical" case of immovable clamped edges (see Section 3.4) are also shown. We will present a detailed discussion of the results for a particular geometry, $\lambda = 8$, that should furnish a better understanding of the intricacies of the solution of this problem.

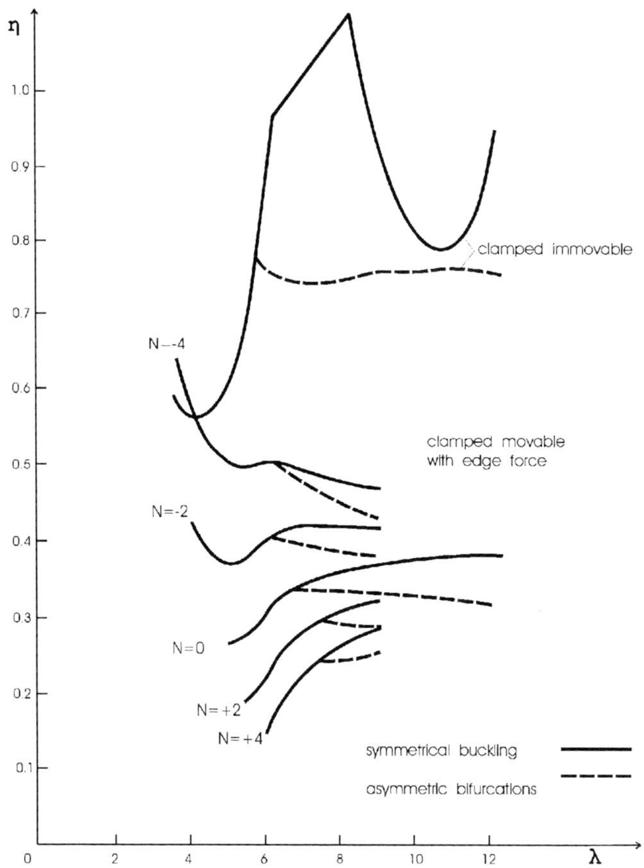

Figure 4.5 Limit and bifurcation loads for cap with movably clamped edge.

Figure 4.6 shows plots of $\eta(\overline{\rho})$ for some values of N, where $\overline{\rho}$ is the so-called "average" displacement given by formula (4.1).

$$\overline{\rho} = -\frac{\lambda^4}{2\overline{\xi}}\int_0^1 \overline{\rho}^2\theta d\overline{\rho}. \tag{4.1}$$

The presentation of the bifurcation points and the corresponding number n of the circumferential waves shown in Fig. 4.6 are based on results obtained in [Shilkrut, 1980; Shilkrut & Kochavi, 1983]. Some discrete values of bifurcation loads for larger n, which were found for $N = -20$ and -30, are left out in Figure 4.6. Figure 4.7 shows the circumferential stress σ_ϕ as a function of $x = \lambda\rho$ at some of these bifurcation states.

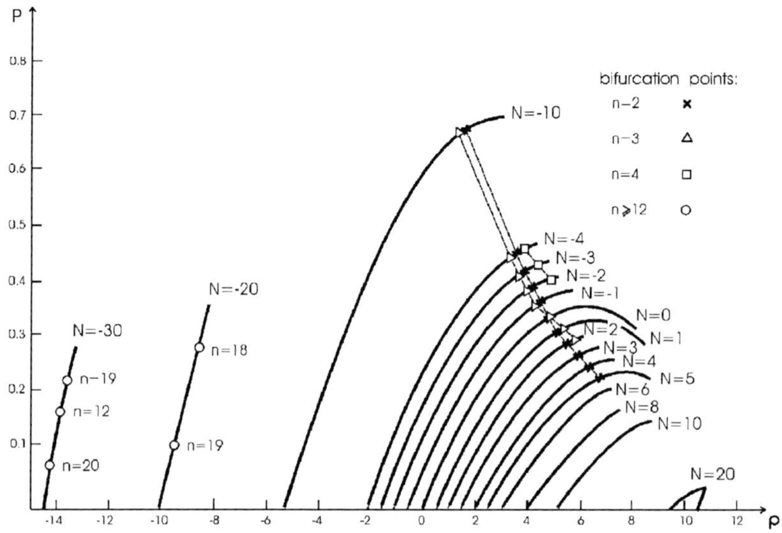

Figure 4.6 Some response curves, pressure η vs. average displacement $\overline{\rho}$ for various values of *N*.

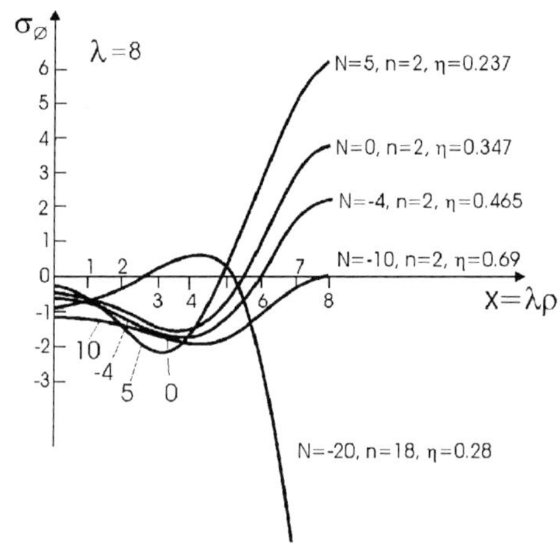

Figure 4.7 The hoop stress distribution at various bifurcation states.

The presented data allow us to make the following conclusions about the behavior of the cap when it is loaded by external pressure and an edge force:

1. A tensile (positive) edge force *N* lowers the pressure at which bifurcation buckling and snap-through occurs. This effect is the opposite of what is observed in the case of internal pressure (the previous case section 4.2).

(The conclusions that now follow are solely based on the results obtained for $\lambda = 8$, but similar qualitative behavior is also expected for other values of λ).

2. The bifurcation phenomenon disappears when N exceeds a certain positive value; for example, when $N \geq 6$ at $\lambda = 8$.

3. The wave number of the buckling mode associated with lowest bifurcation pressure η depends on N, as shown in Figure 4.6.

4. For positive as well as small negative values of N ($N \geq -10$ in the case $\lambda = 8$), bifurcations occur with a few waves ($n = 2$-3) at positive values of the average deflection $\overline{\rho}$. For large negative values of N ($N = -20, -30$), bifurcations appear, with a large number of waves at a negative value of the average deflection $\overline{\rho}$ (see Fig. 4.6).

5. In Figure 4.7, one can observe that for $N \geq -10$ the maximal compressive (negative) circumferential stresses occur somewhere between the pole and the boundary. The location of the highest value of the compressive hoop stress σ_ϕ moves toward the edge when the magnitude of N increases. For $N = -20$, σ_ϕ reaches a maximum near the edge. We can thus conclude that, when $N \geq -10$, the lobes of the buckled mode are located in a circular banded area that lies between the pole and the edge. When the compressive edge load N increases in intensity, the band of circumferential waves approaches the edge. For $N \leq -20$, the band of waves is already very close to the edge.

Thus, two basic forms of nonsymmetric buckling are encountered. The first corresponds to only a few waves (in circumferential direction) that are localized in belt-like region somewhere between the pole and the boundary. Such bifurcations occur for positive or small negative edge loads N added to the external pressure. This phenomenon is similar to the case of "bifurcation in compression" (see Section 2.1).

The second form of buckling corresponds to the appearance of a buckling mode with a much larger number of waves. This buckling pattern is found close to the edge and occurs when large negative values of N are combined with internal pressure (as in the case of "bifurcation in tension"; Section 2.1).

4.3 The deformation map—the analysis tool for systems under multi-parameter external loads.

The deformation map is one of the most convenient and efficient tools of investigation of the behavior of shells under multi-parameter loading (or multi-parameter excitations, in the general case). Its concept and use were first presented in Shilkrut [1974], while other applications of it can be found in several of his later works [1982, 1983b]. The map contains global information about the behavior of nonlinear structures under the given multi-parameter system of loadings. Of great importance is the fact that the topological (geometric) structure of the map is able to reveal the whole conglomerate of qualitative features that are essential for the understanding of the response of the nonlinear deformable system under consideration.

The algorithm for the construction of the map and the discussion of some of its applications is given below. In this discussion we will use a model that comprises all the parameters that are minimally required to expose the basic features of the map in terms of construction, applications, and quality of information. The model that we have chosen is the problem of the axisymmetric, thermo-elastic deformation of an orthotropic shell of revolution of uniform thickness h.

The governing equations for this case are derived from the basic equations (1.58)-(1.60) (see Section 1.2) and they are of the form:

$$\mathrm{L}_k(\omega) = k^2\left[-\frac{\theta^2 + 2\theta\bar{\theta}}{2\rho} + \left(\frac{1-\lambda^2}{\rho}N_T - \lambda^2 N_T'\right)\right]; \tag{4.2}$$

$$\mathrm{L}_k(\theta) = -m_a\left[\frac{1}{\rho}\int_0^\rho sq(s)\mathrm{d}s + \frac{P}{\rho} - \frac{\omega(\theta+\bar{\theta})}{\rho}\right]$$

$$+\left[\frac{\left(1-k^2\nu_r\right) - k^2\lambda^2(1-\nu_r)}{\rho}M_T + \left(1+\nu_r k^2\lambda^2\right)M_T'\right]; \quad (0 \le \rho \le 1). \tag{4.3}$$

where the operator L_k is given by

$$\mathrm{L}_k(\ ) = \frac{\mathrm{d}^2(\ )}{\mathrm{d}\rho^2} + \frac{1}{\rho}\frac{\mathrm{d}(\ )}{\mathrm{d}\rho} - k^2\frac{(\ )}{\rho^2}. \tag{4.4}$$

These equations are valid for caps that are closed at the apex. The dimensionless parameters appearing in (4.4), (see (1.61)-(1.65)), are defined by:

$$k^2 = E_\varphi / E_r;\ \lambda^2 = \alpha_\varphi / \alpha_r;\ m_a = 12\left(1-\nu_r\nu_\varphi\right) = 12\left(1-\nu_r^2 k^2\right). \tag{4.5}$$

$$N_T = \alpha_r \frac{a^2}{h^2}\int_{-1/2}^{1/2} T(\rho,\zeta)d\zeta;\ M_T = \alpha_r \frac{12a^2}{h^2}\int_{-1/2}^{1/2} \zeta T(\rho,\zeta)d\zeta;\ \zeta = z/h. \tag{4.6}$$

$$\nu_\varphi = k^2\nu_r; \qquad \nu_r + \nu_\varphi \le 1. \tag{4.7}$$

The unknown variables in this formulation are the membrane stress function $\omega(\rho)$ and the rotation $\theta(\rho)$ of the normal to the middle surface. All the other field quantities can be expressed in terms of these functions, as is shown below. Please note that the physical quantities are denoted by asterisks.

The listing that appears below concerns the membrane stress resultants, membrane stresses, the stretches of the middle surface, the changes of curvature of this surface and the bending moments.

$$N_r(\rho) = N_r^*(\rho)a^2 / E_r h^3 = \omega(\rho)/\rho; \; N_\varphi(\rho) = N_\varphi^*(\rho)a^2 / E_r h^3 = \omega'(\rho); \tag{4.8}$$

$$\sigma_r(\rho) = \sigma_r^*(\rho)a^2 / E_r h^2 = \frac{\omega(\rho)}{\rho} + \frac{1+\nu_r k^2\lambda^2}{1-k^2\nu_r^2}\left[N_T(\rho) - \alpha_r T(\rho,0)\frac{a^2}{h^2}\right];$$

$$\sigma_\varphi(\rho) = \sigma_\varphi^*(\rho)a^2 / E_r h^2 = \omega'(\rho) + \frac{k^2(\lambda^2+\nu_r)}{1-k^2\nu_r^2}\left[N_T(\rho) - \alpha_r T(\rho,0)\frac{a^2}{h^2}\right]; \tag{4.9}$$

$$\varepsilon_r(\rho) = \frac{h^2}{a^2}\left[\frac{\omega(\rho)}{\rho} - \nu_r\omega'(\rho) + N_T(\rho)\right];$$

$$\varepsilon_\varphi = \frac{h^2}{a^2k^2}\left[\omega'(\rho) - \nu_r k^2\frac{\omega(\rho)}{\rho}\right] + \frac{h^2}{a^2}\lambda^2 N_T(\rho). \tag{4.10}$$

$$\chi_r(\rho) = \chi_r^*(\rho)a^2 / h = \theta'(\rho); \quad \chi_\varphi(\rho) = \chi_\varphi^*(\rho)a^2 / h = \theta(\rho)/\rho; \tag{4.11}$$

$$M_r(\rho) = M_r^*(\rho)a^2 / D_r h = \theta'(\rho) + \nu_r k^2\theta(\rho)/\rho - \left(1 + k^2\nu_r\lambda^2\right)M_T(\rho);$$

$$M_\varphi(\rho) = M_\varphi^*(\rho)a^2 / D_r h = k^2\left[\theta(\rho)/\rho + \nu_r\theta'(\rho)\right] - k^2\left(\lambda^2 + \nu_r\right)M_T(\rho). \tag{4.12}$$

$$D_r = E_r h^3 / \left[12\left(1 - \nu_r^2 k^2\right)\right]. \tag{4.13}$$

The bending moments are considered to be positive if they tend to increase the curvature of middle surface. The curvature of the meridian is positive when the convex side of the shell is directed downwards in the positive direction of the z-axis.

The horizontal and vertical components of the displacement of a point of the middle surface (measured with respect to the undeformed surface) are given by:

$$u(\rho) = u^*(\rho)/h = \rho\varepsilon_\varphi(\rho)a/h = \frac{\rho h}{ak^2}\left[\omega'(\rho) - \nu_r k^2 \frac{\omega(\rho)}{\rho}\right] + h\lambda^2 \rho N_T(\rho)/a. \tag{4.14}$$

$$w(\rho) = w^*(\rho)/h = -\int_0^\rho \theta d\rho + \xi; \quad \xi = w(0). \tag{4.15}$$

where u is positive when it is directed outwards, and w is positive (as is ξ) when it points downwards (in the z-direction). We note that, when the edge is fixed,

$$w(\rho) = \int_\rho^1 \theta d\rho \text{ and } \xi = \int_0^1 \theta d\rho, \tag{4.16}$$

(see (4.15)). The shape of the undeformed meridian is given by the function $\overline{w}(\rho)$ and parameter $\overline{\xi}$, which are defined in the same way as $w(\rho)$ and ξ. These components are measured from the base plane of the shell. The angle $\overline{\theta}(\rho)$ is positive when the convex side of the undeformed shell is directed downwards.

The distributed and concentrated loads, which are positive when they are pointing downwards, are given in the dimensionless form:

$$q(\rho) = q^*(\rho)a^4/E_r h^4; \quad P = P^* a^2/2\pi E_r h^4. \tag{4.17}$$

The boundary conditions most frequently encountered are:

$$\omega(0) = \theta(0) = 0; \quad \alpha_1\omega'(1) + \beta_1\omega(1) = \gamma_1; \quad \alpha_2\theta'(1) + \beta_2\theta(1) = \gamma_2. \tag{4.18}$$

The conditions at $\rho = 0$ ensure the continuity and boundedness of the functions ω, θ in the vicinity of the apex. The condition at $\rho = 1$ for ω characterizes the degree of mobility of the boundary supports in the plane of the base of the shell. The last condition of (4.18) characterizes the degree of fixation against bending along the edge. The two conditions at $\rho = 1$ are usually independent of one another. The nine basic combinations of the boundary conditions are given in Table 4.2 (see also Table 1.1)

At first, we shall construct the deformation map for a relatively simple case. An orthotropic circular plate $\left(\overline{\theta}(\rho) \equiv 0\right)$ with movably hinged supports is loaded by an axisymmetric, two-parameter load system which consists of a uniformly distributed edge force $N = N_r(1)$ and a uniformly distributed edge bending moment $M = M_r(1)$.

Table 4.2

Nine basic combinations of the boundary conditions.

	Mobile support without normal boundary forces	Mobile support with active normal boundary force N	Immovable support
Conditions for ω	$\alpha_1=0;\ \beta_1=1;\ \gamma_1=0$	$\alpha_1=0;\ \beta_1=1;\ \gamma_1=N$	$\alpha_1=1;\ \beta_1=-\nu_r k^2;$ $\gamma_1=-k^2\lambda^2 N_T(1)$
Conditions for θ	$\alpha_2=0;\ \beta_2=1;\ \gamma_2=0$	$\alpha_2=1;\ \beta_2=\nu_r k^2;$ $\gamma_2=(1+k^2\lambda^2\nu_r)M_T(1)$	$\alpha_2=1;\ \beta_2=\nu_r k^2;$ $\gamma_2=M+(1+k^2\lambda^2\nu_r)M_T(1)$
	Clamping	Hinged support without boundary bending moment	Hinged support with bending moment M

The deformation map is a projection of the solutions of our problem in the plane spanned by the two load parameters N, M/m_a. The solutions can conveniently be obtained by considering the problem as a Cauchy (initial value) problem. Since the basic equations (4.2) and (4.3) are of the fourth order, we need four initial conditions to be able to solve Cauchy's problem. To formulate these conditions, we will make use of the property that in the vicinity of the initial point $\rho = 0$, the unknown functions $\{\omega(\rho), \theta(\rho)\}$ are of the form $\Delta\rho^k$ [Shilkrut, 1983b]. Consequently, we can write for any $\Delta\rho << 1$:

$$\omega(\Delta\rho) = \omega'(\Delta\rho)\Delta\rho / k; \quad \theta(\Delta\rho) = \theta'(\Delta\rho)\Delta\rho / k; \quad (0 < \Delta\rho << 1). \tag{4.19}$$

We shall now assume that the values of $\omega'(\Delta\rho)$, $\theta'(\Delta\rho)$ are known beforehand, so that the corresponding values of $\omega(\Delta\rho)$, $\theta(\Delta\rho)$ can be computed with (4.19). The corresponding Cauchy problem can then easily be obtained using numerical integration, with the method of Runge-Kutta, for example, or any other appropriate integration method. We start the solution of Cauchy's problem at the location $\rho = \Delta\rho <<1$, where $\Delta\rho \neq 0$. This is done because the operator L_K (see (4.7) is singular at $\rho = 0$, and we are thus forced to take a small sidestep away from this point. Using the integration process in this way, we can calculate the unknown functions and their derivatives in the interval $\Delta\rho < \rho \leq 1$ and determine the edge forces N and M, which correspond to these predefined initial conditions.

A combination of the load intensities N and M determine a certain point in the (N, M) plane. Having computed one particular combination of these parameters, we can repeat the computational procedure with a new set of initial values, for example with the same initial value $\omega'(\Delta\rho)$ but a slightly changed initial value $\theta'(\Delta\rho)$. Then, just as in the previous case, a next point will be generated in the plane (*N*, *M*). Continuing in this way, we obtain a curve in the (*N*, *M*) plane for $\omega'(\Delta\rho)$ = const., and for increasing or decreasing values of $\theta'(\Delta\rho)$. This whole process can, of course, be repeated with a new fixed value of $\omega'(\Delta\rho)$, and this will produce a new curve in the plane (*N*, *M*), and so on. (The maps in Figs. 4.8 and 4.9 are constructed in this way).

Thus, the deformation map can be viewed as a projection, of the two dimensional equilibrium surface parametrized by $\omega'(\Delta\rho)$ and $\theta'(\Delta\rho)$ onto the plane spanned by *N* and *M*. This equilibrium surface, and thus also its projection, represents the solutions defined by (4.2 to 4.18) for the range of values of $\omega'(\Delta\rho)$ and $\theta'(\Delta\rho)$ taken into consideration. The interesting feature of this (contour) map is that it provides a complete picture of the behavior of the plate under the action of the two parameter system of loads. We shall explain this in more detail below.

In the deformation map obtained for the plate (see Figs. 4.8 and 4.9), it can be seen that, along the *N*-axis there is a series of irregular points which are denoted by A, B, C,... . These points determine the values of the classical bifurcation loads *N* in "compression" at which the planar states of the plate are transferred into axisymmetric buckled states. As is well-known, these critical values of $|N|$ form an infinite discrete spectrum for each fixed value of $k = \sqrt{E_\phi / E_r}$ [see, e.g., Timoshenko & Gere, 1961].

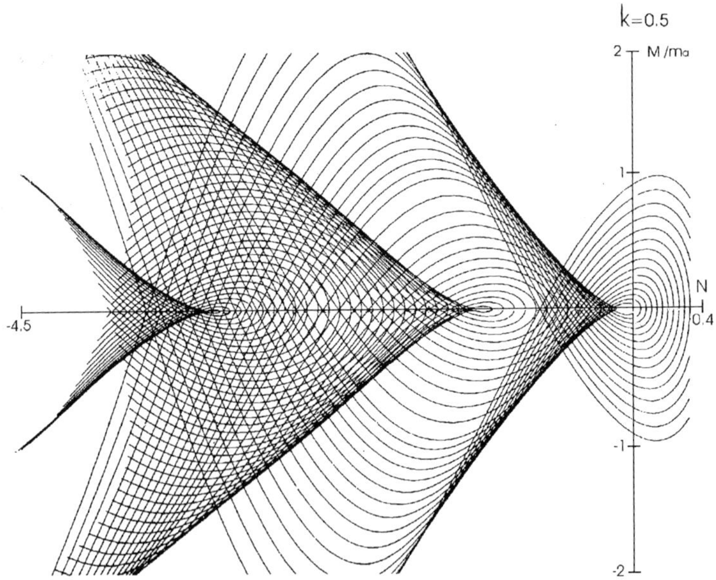

Figure 4.8 The deformation map for an orthotropic circular Plate 1.

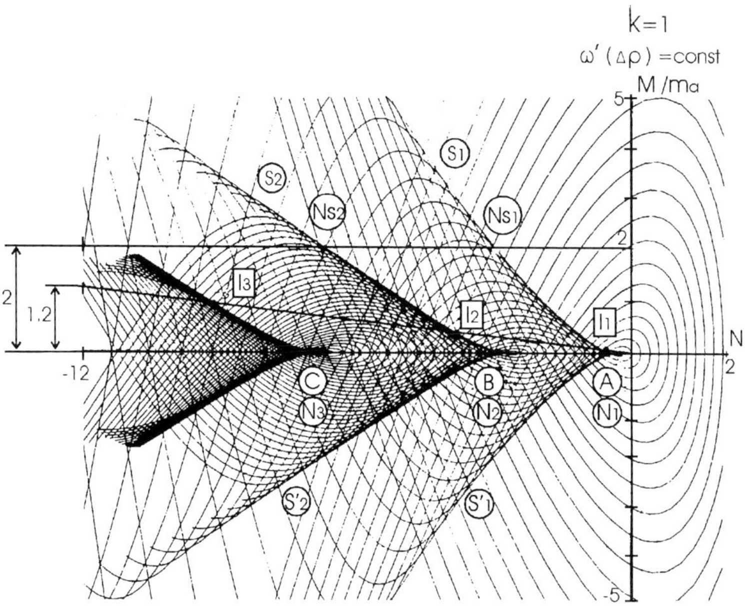

Figure 4.9 The deformation map for an orthotropic circular Plate 2.

From the standpoint of the theory of catastrophes and bifurcations [see, e.g., Woodcock and T. Poston, 1970; Thompson & Hunt, 1984], these singular points are typical cusps. The first three critical loads are presented in Table 4.3 in dimensionless form (see (4.8)) for three values of k.

Table 4.3

Values of critical loads.

k	$-N_1$	$-N_2$	$-N_3$
0.5	0.1063	1.1383	3.1127
1	0.3844	2.6598	6.7286
1.2	0.5671	3.5484	8.7996

All the numerical solutions described in this work are obtained for $\nu_r = 0.3$, so that $\nu_\phi = 0.3k^2$. The presented values of the critical load N for the classical (isotropic) case ($k = 1$) agree with the known solutions [see e.g., Timoshenko & Gere, 1961]. The figures in Table 4.2 show that the influence of k on N_1 is most noticeable when $k < 1$.

A family of curves symmetric with respect to the N-axis start to emerge at each of the above-mentioned irregular points A, B, C... . These curves, which are labeled by

with $\xi = \pm 8.7$ (point A in Fig. 4.10 and its corresponding symmetric image). When the moment is increased from zero in the state $\xi = 8.7$, loss of stability does not occur, because the deformation curve $M(\xi)$ beyond the starting point $(\xi = 8.7,\ M = 0)$ is monotonic. If we start from point A ($\xi = -8.7$) in the positive direction of M, then loss of stability by snapping will occur at the limit point C. In the latter case, the positive bending moment M counteracts the state of bending of the plate induced by the force $N = -2$, and therefore loss of stability may occur. In the other case, starting from $\xi = 8.7$, the positive moment M tries to increase the amplitude of the initially deformed shape and the phenomenon of snapping can then not occur.

Now let us investigate horizontal sections of the deformation map. The line $M = 0$ describes the behavior of the plate under the action of N alone. When $N > N_1$, the plate is in a state of plane tension or compression, and then $\xi = 0$. When $N \leq N_1$, the plate is in the critical or post-critical state of equilibrium. The corresponding deformation curves have the shape shown in Figure 4.11. Transition through any bifurcation point N_i changes the number of corresponding solutions of the basic equations by two. In Figure 4.11, the states marked by crosses are unstable in the small (see Section 6.2), as was proved by Shilkrut [1984].

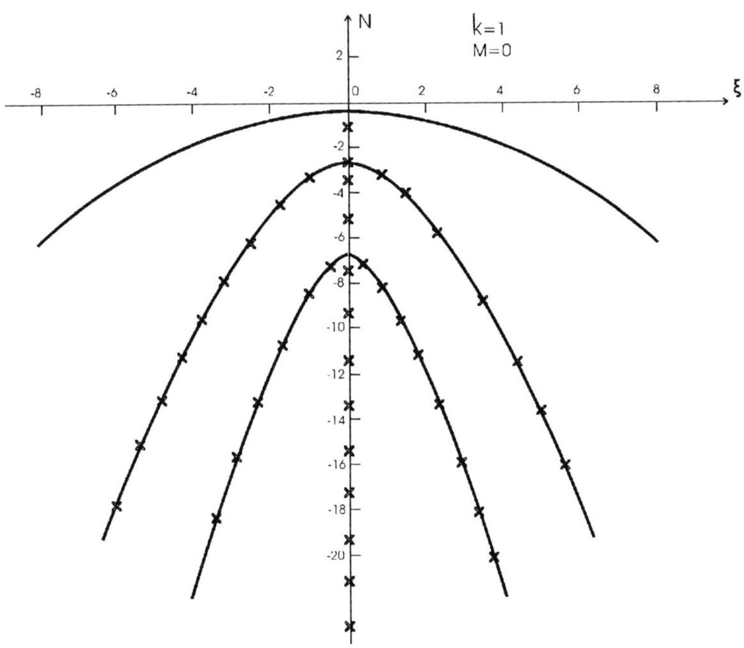

Figure 4.11 The section $M = 0$, the bifurcation diagram for the plate in pure compression.

The line M = const. $\neq 0$ provides the following information. For $N > N_{s1}$ (where N_{s1} is the intersection point between the line M = const. and the curve S_1), there is only one equilibrium state of the plate (see, e.g., Fig. 4.9). For $N_{s2} < N < N_{s1}$, there are already three solutions of the basic equations, and so on. The corresponding deformation curves (characteristics) are presented in Figure 4.12 for $M = 2m_a$ = const. > 0, and they prompt us to the following observation. When M = const. $\neq 0$, the bifurcation phenomena that are characteristic for the case $M = 0$ no longer exist (compare Fig. 4.11 with 4.12). The first bifurcation point N_1 in Figure 4.11 disappears altogether, while the other bifurcation points (N_2, N_3,...) are transformed to limit points. There is one stable monotonic branch and another stable branch that emerges from the first limit point. (Both of these branches are drawn by solid curves in Fig. 4.12). A continuous transition from the stable monotonic branch of the deformation curve to other branches does not exist. A transition (transient) can only be effectuated if an additional special impulse is applied. Thus the addition of the moment M in fact eliminates the danger of loss of stability (in the small) of the plate. The crosses in Figure 4.12 mark the axisymmetric equilibrium states that are unstable (in the small). The same unfolding takes place when $M = -2m_a$ = const. < 0, but the deformation curves are then the mirror image of the former (mirrored with respect to the N-axis in Fig. 4.12).

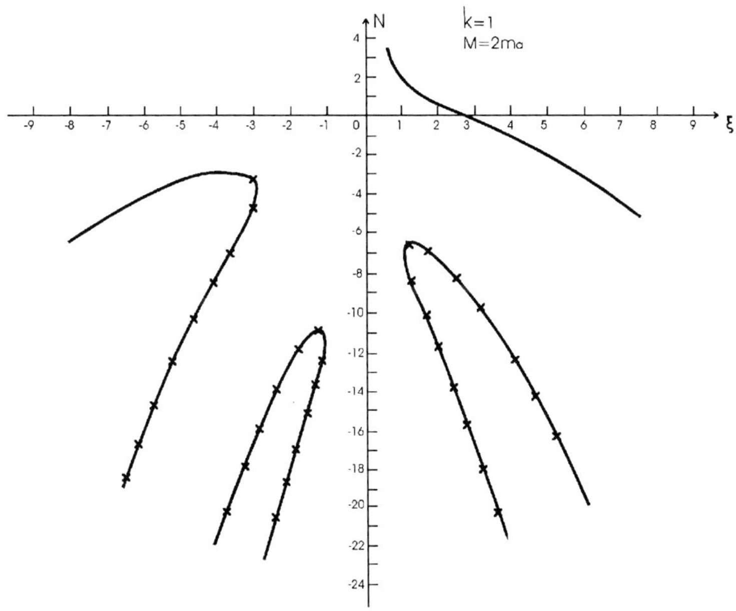

Figure 4.12 The section $M = 2m_a$, the branching diagram for the plate when the additional load M is present.

Let us now look at the influence of the eccentricity of the compressive force N on the behavior of the plate; that is, when the load combination (N, M) is replaced by (N, eN) so that M

takes the special value

$$M = eN \tag{4.20}$$

The eccentricity e is here defined in such a way that e is positive when the force N acts in a plane below and parallel to the middle surface.

To investigate this important problem, we must focus on the intersections of the lines represented by the relation (4.20) with the curves of the deformation map. Let us take, for example, a case where $e < 0$. $I_1, I_2, \ldots$, denote the points of intersections of the line $M = eN$ with the curves $S_1, S_2, \ldots$, of the deformation map (see Fig. 4.9). The stretch of this line between the origin and I_1 lies in the region of the map where there is only one equilibrium state of the plate belonging to one combination of M and N. If $M = eN$ exceeds the value that belongs to I_1, the line (4.20) enters the area of the map where there are already three equilibrium states. After the intersection with curve S_2, the load combination (N, eN) possesses five solutions, and so on. The deformation curves for a specific case $k = 1$, $e = -0.1m_a$, and movable hinged support are given in Figure 4.13. The behavior it portrays is similar in its qualitative properties to the case M = const. (see Fig. 4.12). It is also similar to the behavior of a compressed bar with imperfections [e.g., Thompson & Hunt, 1984], and it agrees (qualitatively) with the predictions of the theory of W. T. Koiter [1945, 1963b].

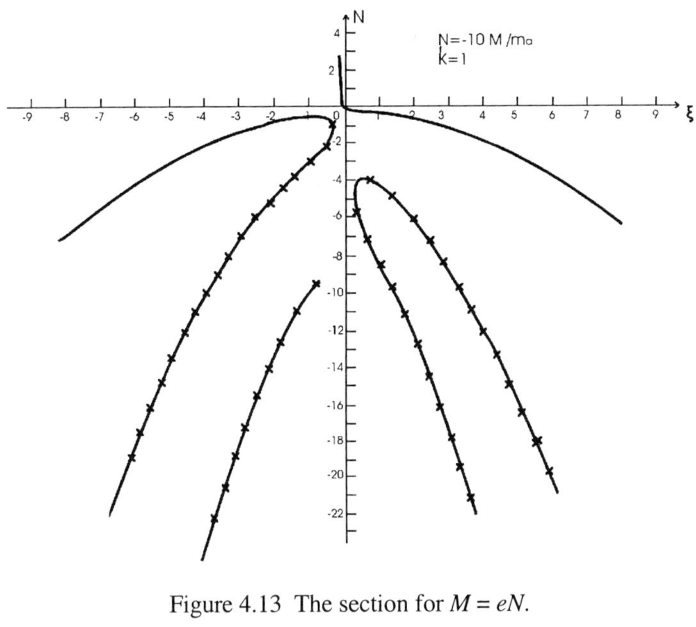

Figure 4.13 The section for $M = eN$.

Please note that the behavior of the plate under any other type of loading $M = f(N)$ can be examined in the same way, without any difficulties, once the solutions for a range of M, N

values have been obtained.

The numerical data used to construct the deformation map contains not only all the information about the behavior of the plate under forces M and N when the edge is movable hinged, but, in fact, also the information for any other edge condition. Indeed, each point on the map corresponds to some value of M and N, and, these include the case $\theta(1) = 0$ (the edge is clamped). Thus the locus of these points represents the case of movable clamped supports where N is a one parameter load, and the corresponding moment M is then the reaction moment.

But, the map of the clamped plate just mentioned can also be obtained in another more formal way. During the construction of the basic map in the plane (N, M) for the movable hinged edge, one can also plot the corresponding solutions in the plane $(N, \theta(1))$. After all, we know the value of the unknown functions and their derivatives in the interval $\Delta\rho \leq \rho \leq 1$. The points of this plot $(N, \theta(1))$ that lie along the horizontal axis (axis of N) then make up the map for movable clamped plates (see Fig. 4.14a). It is understood that the map is now a one-dimensional representation, because there is now only a one load parameter.

The same approach can be used to construct the map for the case of immovable hinges (the external force is M is now acting by itself). This is the locus of points of the basic map for which $u(1) = 0$ ($u(1)$ is the horizontal displacement of the edge, which is proportional to the expression $\omega'(1)$-$\nu\omega(1)$) when temperature effects are neglected (Table 4.1). In other words, the case of immovable hinges is given by the points along the M-axis in the plot $(u(1), M)$ (see the qualitative picture in Fig. 4.14b). The case of the additional influence of a temperature field is considered in the next section. See also the map given in Figure 4.26.

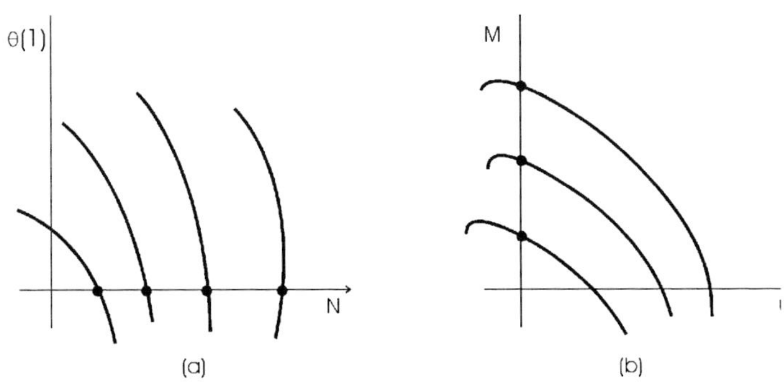

Figures 4. 14 a,b Alternative representations of the deformation map.

In the same manner, the maps for all types of supports can be investigated, as, for example, for nonlinear springs: $M = f(\theta(1))$ (or $N = F(u(1))$, where f and F are given functions.

The fact that all the additional maps mentioned above are parts of the "basic" map in the (M, N) plane leads to the following natural, but important, conclusion. The areas of global or absolute stability (where only one solution exists) of all the specific maps mentioned above cannot be smaller than the area of stable states of the basic map for the movable hinged plate.

The deformation map that we presented here is constructed by plotting the solution curves parametrized by $\theta'(\Delta\rho)$, while $\omega'(\Delta\rho)$ is kept fixed. This is of course not the only possible way. We can also plot the solutions for varying values of $\omega'(\Delta\rho)$, while we keep the value of $\theta'(\Delta\rho)$ fixed. Figure 4.15 shows such an alternative version of the map for a mobile, simply supported plate with moveable edges subjected to compound loading M, N. It is obvious, however, that the topological structure of both these maps, as well as other versions, must be identical, because they all reflect the same physical phenomena related to the same object under the same conditions.

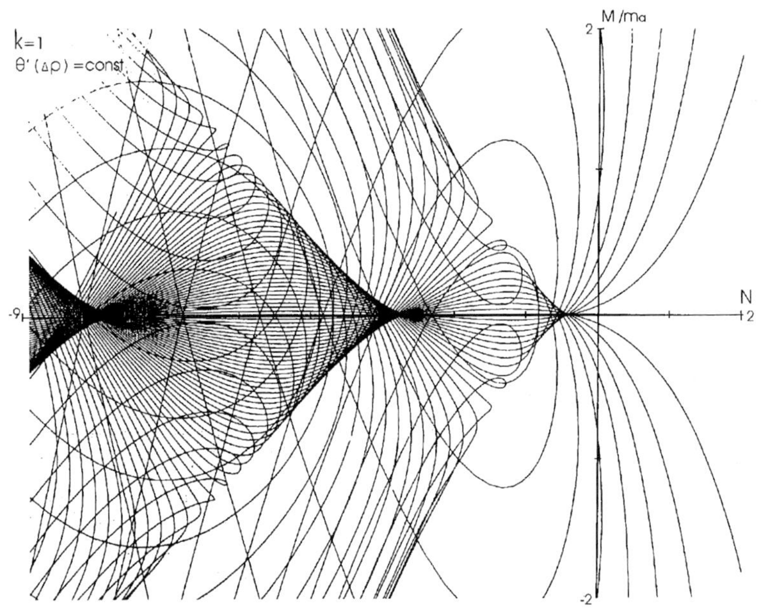

Figure 4.15 An alternative version of the map for a mobile, simply supported, circular plate.

We shall now present some results in terms of sections of the corresponding maps for orthotropic plates for different values of the orthotropy coefficient k. The deformation curves $M(\xi)$, in the case $N = 0$, for three values of the parameter k are shown in Figure 4.16. When k increases, the stiffness against the bulging of the plate increases. Figures 4.17-4.19 and Figure 3.5 also present relations that permit a comparison of behavior for different values of k when $N = 0$. Figure 4.20, for example, shows graphs of $\omega'(\rho)$ for different values of k for an immovable hinged support. The same qualitative picture is shown in Figure 4.18 for $\omega(\rho)/\rho$.

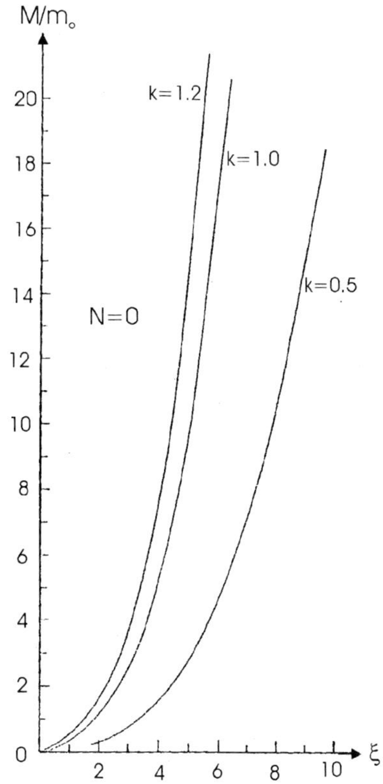

Figure 4.16 Response curves $M = M(\xi)$ for orthotropic circular plates, $N = 0$.

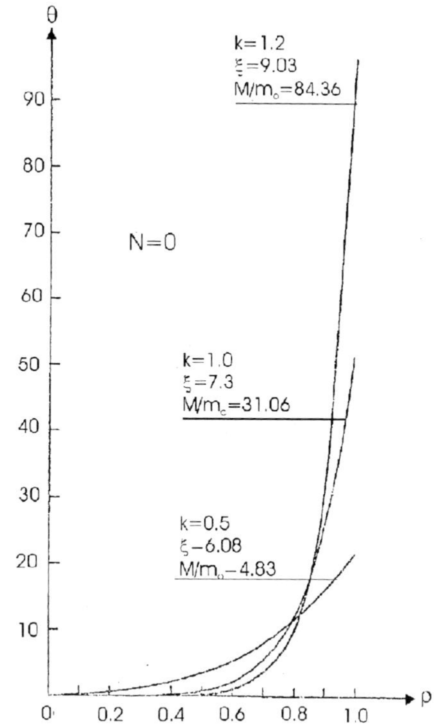

Figure 4.17 Some particular solutions $\theta = \theta(\rho)$.

In order to understand these results, it is necessary to take into consideration one of the basic features of the governing functions $\omega(\rho)$ and $\theta(\rho)$ which we already alluded to earlier. This is that, in the neighborhood of the apex, that is near the singular point $\rho = 0$, the solutions of the governing equations for $\omega(\rho)$ and $\theta(\rho)$ are generally of the form

$$\begin{matrix}\omega(\rho)\\ \theta(\rho)\end{matrix} = A\rho^k + \rho^k \sum_{i=1}^{\infty} a_i \rho^{i(k+1)}, \tag{4.21}$$

which is a consequence of the properties of the differential operator (4.2), (4.3) and (4.7), see [Shilkrut, 1983b].

Figures 4.21-4.23 show some characteristic deformation states of the movably hinged plate for different values of N that belong to the case $k = 0.5$. The corresponding deformation curves $M(\xi)$ for different N are displayed in Figure 4.10. These results clearly illustrate the qualitative features of the behavior of plate as discussed earlier for the isotropic case ($k = 1$), and as it follows from (4.21) when $k \neq 1$.

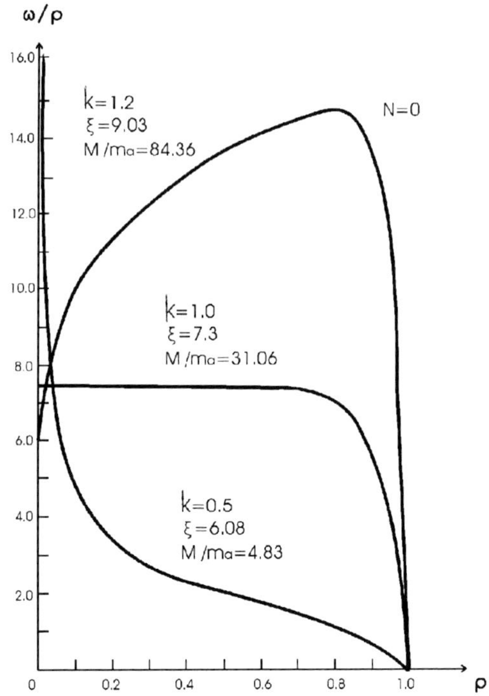

Figure 4.18 Some particular solutions $\omega/\rho = \Phi(\rho)$.

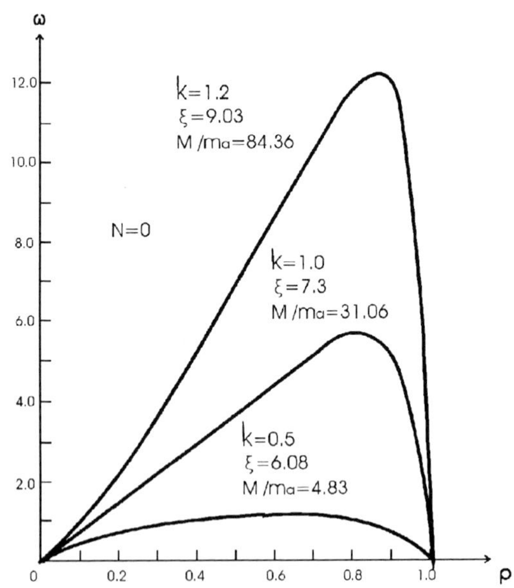

Figure 4.19 Some particular solutions $\omega = \omega(\rho)$.

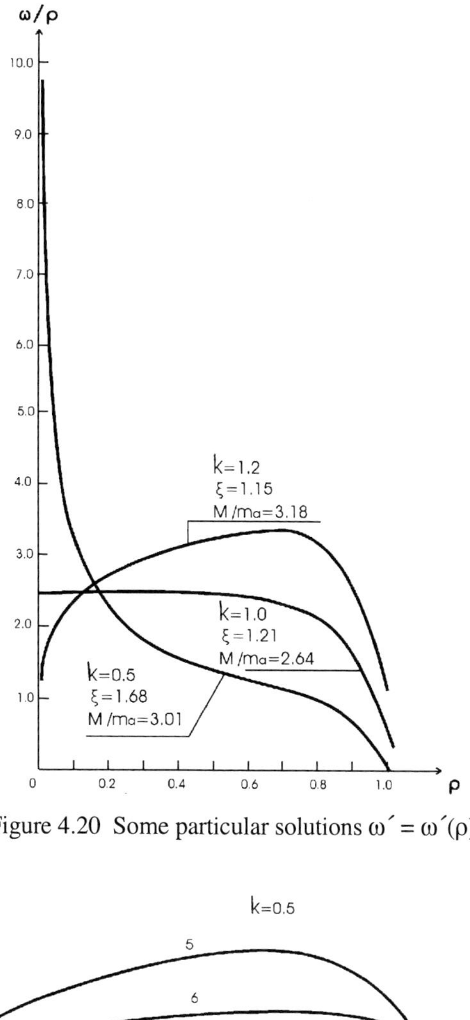

Figure 4.20 Some particular solutions $\omega' = \omega'(\rho)$.

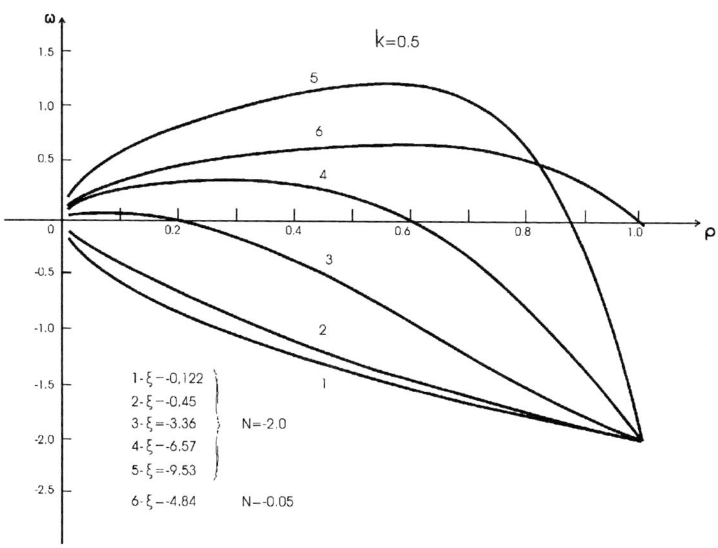

Figure 4.21 Some deformation states for the mobile, hinged circular plate, $\omega(\rho)$.

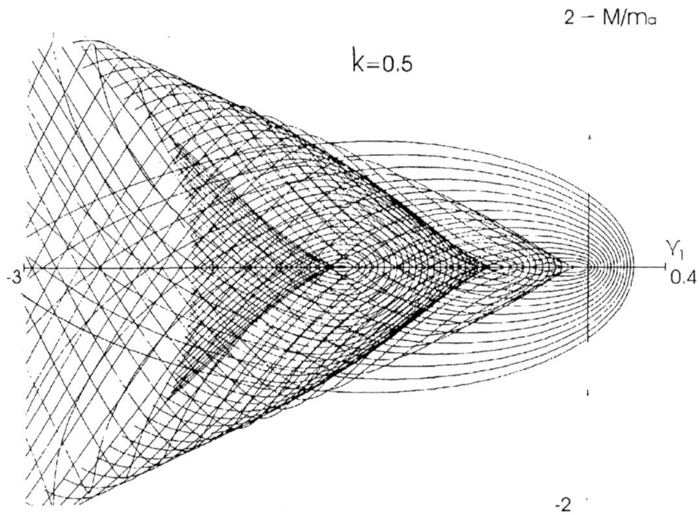

Figure 4.25 Alternative map for the case $k = 0.5$.

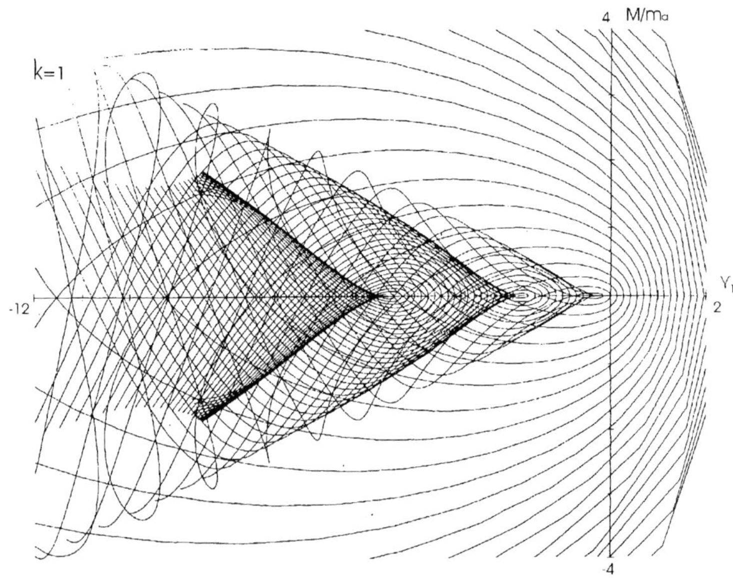

Figure 4.26 Alternative map for the case $k = 1$.

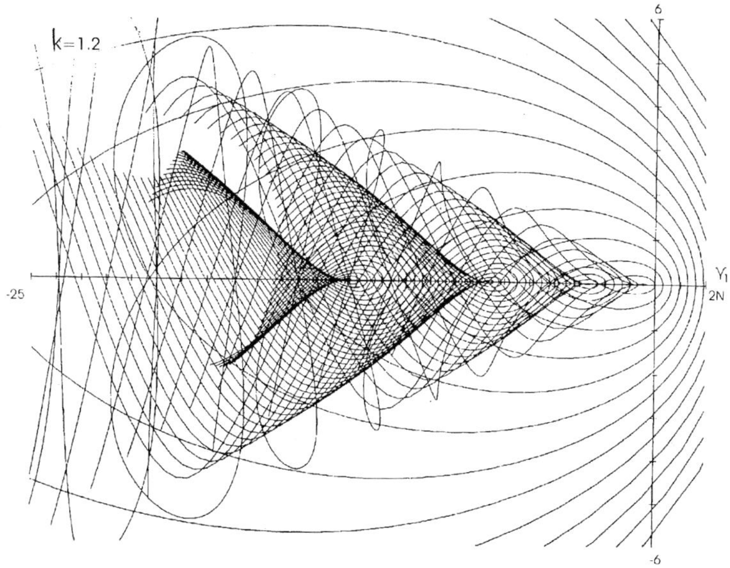

Figure 4.27 Alternative map for the case $k = 1.2$.

To conclude this section, we add some further comments about the deformation map that may clarify its significance as a tool for the investigation of the static solutions of any nonlinear structure or system. First of all, as was already mentioned, the deformation map gives a global picture of the behavior of the investigated systems. Its topological structure gives a direct and full description of the system controlled by two or three parameters. (If there are three, one should construct a series of plane maps, as will be demonstrated in the next section). The behavior of the system for a given loading path is defined by the corresponding cross-section of the map. The specific numerical solution can then be obtained directly without having to rely on incremental procedures that are often used in FEM and FDM codes. It is very important to note that the initial value problem (of Cauchy's problem) that is used to construct the maps has a unique solution, whereas the boundary problem of a nonlinear system, as a rule, has many solutions. This circumstance substantially simplifies the former procedure.

Besides the qualitative information, the map is also a "data base" of quantitative information on the behavior of plates and may be used in particular, as a nomograph. For each point of the loading path given by the relation $F(M, N) = 0$, the map reveals the initial parameters $\omega'(\Delta\rho)$ and $\theta'(\Delta\rho)$ of all equilibrium states that exist at this point (the corresponding values of $\omega(\Delta\rho)$ and $\theta(\Delta\rho)$ follow from (4.19)). For this, it is often convenient to use, both versions of the map — the one that is based on contours for $\omega'(\Delta\rho)$ = const., and the other, which is based on contours for $\theta'(\Delta\rho)$ = const. (See Shilkrut, 1983b, and Fig. 4.15, above).

Given these initial parameters, all the data of the considered equilibrium states can easily be found by the numerical solution of the corresponding Cauchy problem. Thus, we have here a direct link between boundary and initial value problems. Furthermore, it is interesting to compare the technique described here with the well-known "shooting" method, which can also be used for the numerical solution of equations of the type (4.2) and (4.3) [e.g., Shilkrut, 1980; Shilkrut & Vyrlan, 1977; Shilkrut, Shevandronov, Morar & Maksimov, 1969; Keller, 1968; Keller & Wolfe, 1967; Valishvili, 1976]. Both methods use the Cauchy problem as a basis from which to start, but there is an essential difference between the two approaches.

In order to solve the given boundary problem by the shooting method, it is necessary to test a series of solutions of initial value problems. Those that do not correspond to the given boundary conditions are removed and not used, even though they represent a large amount of information about the behavior of the investigated structure. In contrast to the shooting method, in our approach, all of this valuable information is preserved and stored in concentrated form in the deformation map, where it can be used for the formation of the global picture of the behavior of the system. Such a global, general, picture is not only very important from a theoretical point of view, it is also very useful for engineering applications. Of course, the construction of a deformation map demands more computer time than the numerical solution of a specific boundary problem, but we obtain, as a reward, a general picture of the structure's behavior, in both qualitative and quantitative sense, as was shown above.

Comparing the deformation map approach with that of the shooting method shows another substantial advantage of the former over the latter. The shooting method directly solves the specific boundary problem and is, as a consequence, connected with the solution of transcendental or nonlinear algebraic equations. Our map technique is much simpler and does not use such complex means. In fact, it solves only the initial value problem, which has always a unique solution. This is a factor that contributes to the robustness of the procedure. (The proof of the uniqueness of the solution of the Cauchy problem is given in Keller and Wolfe [1967]).

It is also important to note that the initial value problem is the simplest way in which a problem in terms of differential equations can be posed, and, because of this, the corresponding numerical algorithms for its solution are the simplest and most accurate. The deformation map can be used for any physical system described by ordinary differential equations — in particular, physically nonlinear problems in the elastic range. It also allows a generalization to problems described by partial differential equations.

4.4 Three-parametric loading systems and the manifestation of chaos-like phenomena in statics.

In the previous section, we considered the basic features of the deformation map, using the simplest case with only two control parameters active, as an example. Shilkrut [1982, 1992a] discovered that the extension to three control parameters leads to the appearance of new phenomena. Let us now consider such a three-parameter loading problem. As a basic model, we take the geometrically nonlinear, circular, orthotropic plate, supported at the edges by immovable hinges, which undergoes axisymmetric thermo-elastic deformations. The bending stiffness of the plate is characterized by a coefficient of orthotropy of $k = 0.5$. The plate is subjected to a temperature distribution $T(\zeta)$, which is an even function of ζ but does not depend upon ρ. In addition to this type of loading, the plate is subjected to normal pressure — the intensity of which is defined by the dimensionless load parameter q — and to a uniformly distributed edge moment M. The temperature field $T(\zeta)$ is isotropic; that is, $\lambda = 1$ (see basic relations (4.2)-(4.18)). In this case, we have $N_T(\rho) \equiv \text{const.} \neq 0$ and $M_T(\rho) \equiv 0$.

Thus defined, the problem is one of thermo-elasticity, which is governed by three control parameters, the temperature parameter γ_1 (see Table 4.1), the edge moment M, and the pressure q. The solution of it can be displayed using a series of deformation maps, each of which is constructed for a fixed value of one of the control parameters. The choice of the latter is a matter of convenience, and we shall here select the pressure q for this purpose. The map in Figure 4.28 is constructed for $q = 30$ in the plane γ_1 (determined by (4.23)), where $u(1) = 0$) and M/m_a.

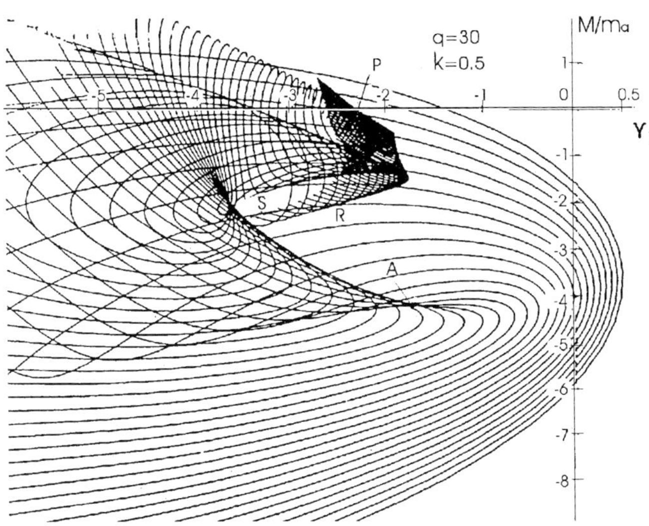

Figure 4.28 The deformation map for the thermo-elastic plate problem.

It is useful to compare the map of Figures 4.9 ($q = 0$, $k = 1$) with that of Figure 4.28 ($q = 30$, $k = .5$). In this comparison, it is immaterial that the map for $q = 0$ in Figure 4.9 refers to the case $k = 1$, since our investigations have shown that this difference does not cause any qualitative changes [Shilkrut, 1983b] (see also Fig. 4.25). First of all, the distribution of the cusps at A, B, C,... for $q = 0$ along the horizontal axis $M = 0$, shown in Figure 4.9, is transformed to a distribution in the plane $\gamma_1, M/m_a$ whereby the original regularity of the topology of the surroundings of these cusps is destroyed (Fig. 4.28). When q is increased from zero, the first cusp (point A in Fig. 4.9) and its neighborhood move downwards (compare the two maps), but though there is some distortion of this neighborhood, the cusp remains topologically of the same class (see region around A in Fig. 4.28). Another type of change occurs with the "higher" cusps (B, C,... in Fig. 4.9) and their neighborhoods. Figure 4.28 is an indication that accumulation of the cusps (B, C,...) occurs, accompanied by a geometrical distortion and superposition of their neighborhoods in a bounded region around P of the map above point A. Due to this accumulation phenomenon, new singular points appear (e.g., labeled in Fig. 4.28 by S). We believe that these singular points represent what is called a hyperbolic umbilic type of catastrophes [Woodcock & Poston, 1970].

Some of the details of the region P are left out of Figure 4.28, since the contour plot would become too crowded otherwise. This region around P has an extremely complex structure because the cusps that come together there undergo a change in character as a result of the nonlinear interaction that occurs between these solutions when they approach each other. It is fairly clear that not all cusps of the (infinite) spectrum A, B, C,..., are clustered in the region P (Fig. 4.28). In other words, it seems that the further the cusps are removed from A when $q = 0$, the less they are affected by the pressure q, but that, when q is increasing, its influence also increases, with the outcome that more and more cusps are drawn to the region P.

In investigating the main features of the map of Figure 4.28 — in particular the significance of the solutions in the region of the accumulation of cusps — it should first be noted that the boundaries of the areas of the cusp A and P that we see in Figure 4.28 determine the plate's stability boundaries (of course, only in the subspace of axisymmetric deformations). Outside these regions (see, e.g., the area R in Fig. 4.28), there is only one equilibrium state for the plate for a given combination M and N. The topography of the map shows that each cross-section of the map, defined by γ_1 = const., in R will show a monotonic deformation curve (load-displacement curve) in the plane ξ, M (where, as usual, ξ is the dimensionless displacement of the plate's center). As soon as the plane γ_1 = const. cuts through the boundaries of the cusp A and the area P (that is, when the value of γ_1 lies approximately in the interval $(-3.4 < \gamma_1 < -1.6)$, the existence of a separate deformation curve that forms a closed loop is expected. This follows from the presence of a separate bundle of solutions between the regions A and P). If $\gamma_1 < -3.4$, then the corresponding deformation curves in the sections γ_1 = const. will have a complicated, but, again, continuous, character. Some particulars of these features are illustrated in Figure 4.29, where the equilibrium curves are given in a number of planes γ_1 = const.

When $\gamma_1 = -1.0$ and -1.5, the equilibrium curves are monotonic. The plane $\gamma_1 = -1.5$ almost passes through point A, and the corresponding deformation curve in Figure 4.29 has a point of inflection with a horizontal tangent. When the plane γ_1 = const. moves further to the left in Figure 4.28, the deformation curve shows a part where the degree of non uniqueness of the solution is three (curve 0 in Fig. 4.29, for $\gamma_1 = -3$); i.e., the former inflection point splits into two extremal points. Moreover, a separate closed branch appears (see Fig. 4.29, curve L). This isolated branch with a complicated shape is the intersection of the plane γ_1 = const., with the equilibrium surface in the area P of the map in Figure 4.28), where the clustering of cusps has taken place.

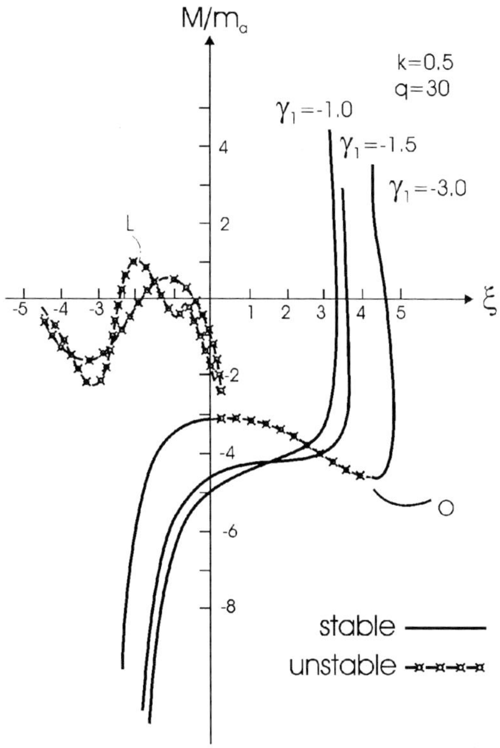

Figure 4.29 Some cross-sections of the map of Figure 4.28.

Two other subsections of the map of Figure 4.28 ($\gamma_1 = -2; -3.5$) are presented in Figures 4.30 and 4.31. It can be seen that, in the section $\gamma_1 = -3.5$, the two branches, which were previously isolated, are now merged into one, resulting in very complicated behavior. This happened because the section $\gamma_1 = -3.5$ cuts the equilibrium surface left of the singular point S in Figure 4.28. The singular point S is the boundary where the separate isolated loops "connect" with the other branches.

Thus, in the right side of point A in Figure 4.28, there is only one unique solution for each value of M/m_a, and the corresponding deformation curves are monotonic. Between points A and P, the plane sections γ_1 = const. contain deformation curves with a maximum and a minimum. Between points P and S, the plane sections γ_1 = const. contain two separate branches. One of these branches has a maximum and a minimum with respect to M/m_a and the other is closed; i.e., it forms a loop. In Figure 4.28 to the left beyond S, the separate loops disappear until a new singular point is encountered. There are thus two values of γ_1 (about $\gamma_1 = -1.6$ and -3.4) at which a single branch of solutions in the plane γ_1 = const. bifurcates into two separate branches, one of which is a closed loop.

By cutting cross sections of the maps, the deformation curves (such as those in Figs. 4.29-4.31) can be generated automatically, whereby each discrete point of the curves is found

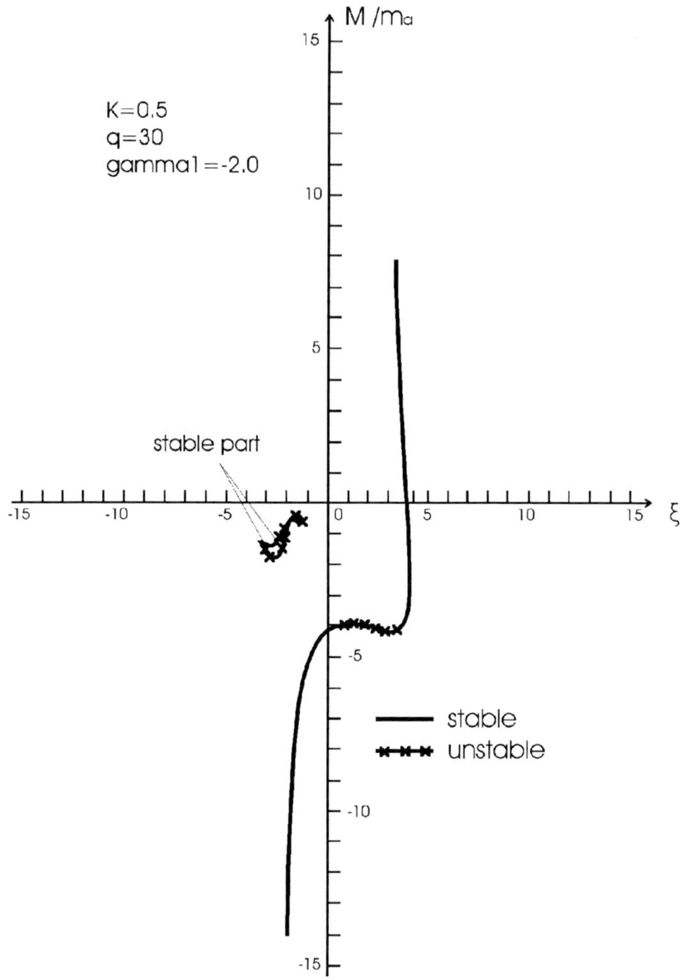

Figure 4.30 The subsection $\gamma_1 = -2$.

independently from any other point. Thus the generation of points of the curves in question does not demand any "step-by-step" procedure along these curves. This also counts for the separate loops, which are obtained in the same direct way. In Figures 4.29, 4.30, and 4.31 the unstable parts of the equilibrium branches are marked by crosses. The method and the algorithm for the determination of the stability of the equilibrium states are described in Section 6.2 [see also Shilkrut, 1984].

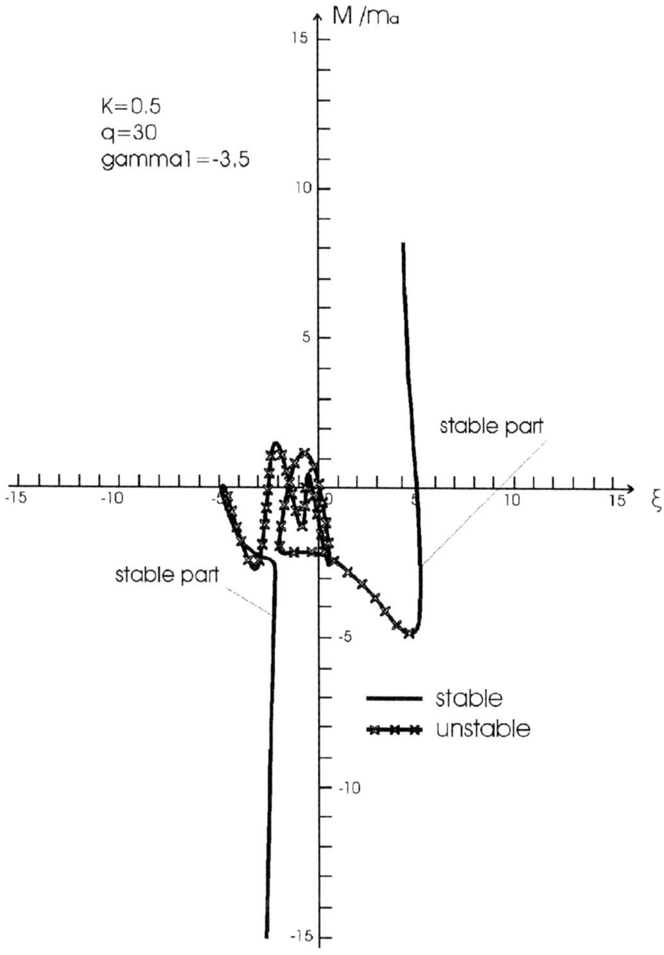

Figure 4.31 The subsection $\gamma_1 = -3.5$.

The existence of the separate loop *L* for the case $\gamma_1 = -3$ in Figure 4.29 implies that an equilibrium state that corresponds to the branch 0 cannot be changed into one that belongs to the loop by a quasi-static process. The possibility of loss of stability of the initially stable equilibrium state of the plate (as represented by a point of the lower part of the branch 0), can only occur at the first limit point of this branch by snapping. Thus, at this particular point, a jump, which is a transient process, will occur and this will most likely end at the second stable part of 0. A transition from a state of this second part of the branch 0 (for higher values of

M/m_a) to a state of the branch L cannot occur unless a sufficiently large, appropriately directed, additional disturbance in the form of an impulse is applied to the system.

The possibility described above can sometimes lead to dangerous situations, because it is not always realized that separate loops with stable solutions exist. The incremental solution methods of the Finite Element Codes that are used to do the structural analysis are generally not capable of detecting such separate deformation states. The danger can be avoided when one can construct a global map that contains the relevant qualitative, as well as the quantitative, information of the behavior of the system.

It can also be concluded from the results that lead to Figure 4.29 that, in the region P of the map, at least four cusps are concentrated. The number of equilibrium states that correspond to a combination of M and γ_l in this zone add up to nine: eight corresponding to points belonging to the loop, and one to the curve 0.

The comparison of the cases $q = 0$ with that of $q = 30$ is a clear indication that q is the main control parameter that determines the behavior of the plate. For this reason, it is important to determine the way the deformation map changes when the value of q increases. The maps for $q = 5$, 10, and 15 are presented in Figures 4.32, 4.33, and 4.34, and they show that, when q increases, the first cusp descends, while the third and subsequent cusps approach the second one — a process that indicates the beginning of accumulation of higher cusps. This accumulation effect can also be seen in Figure 4.35 for $q = 30$ (this figure differs from Fig. 4.28 by its scale, which makes it possible to highlight some further details of the map connected with the accumulation process).

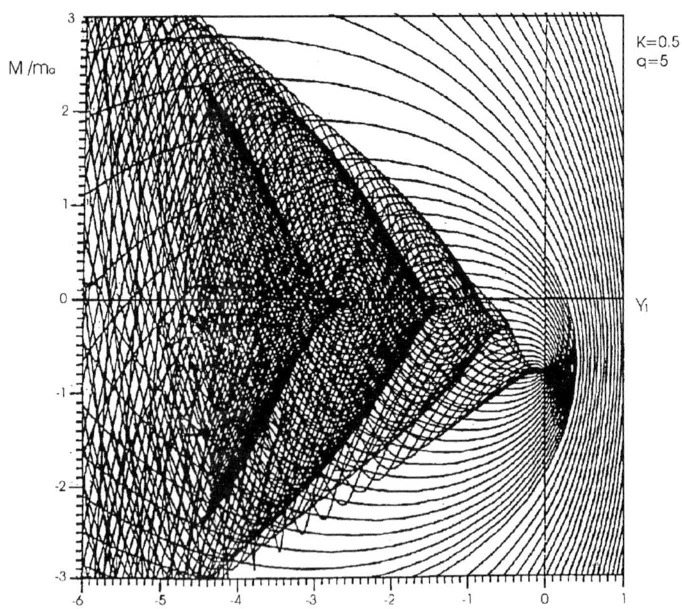

Figure 4.32 The influence of q as a control parameter; the case $q = 5$.

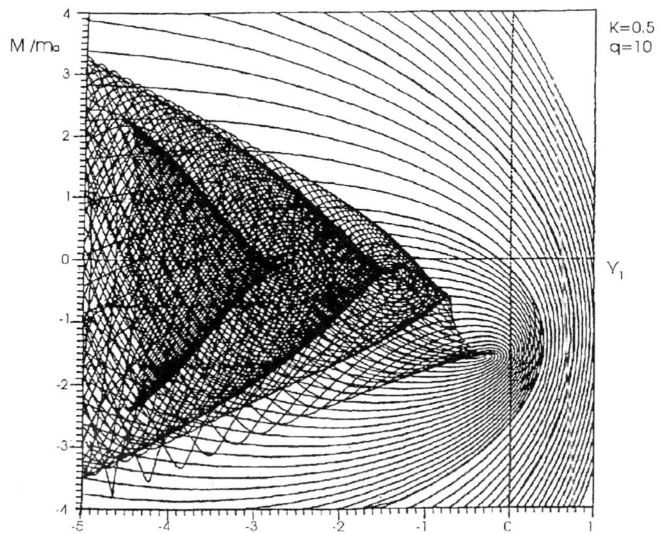

Figure 4.33 The influence of q as a control parameter; the case $q = 10$.

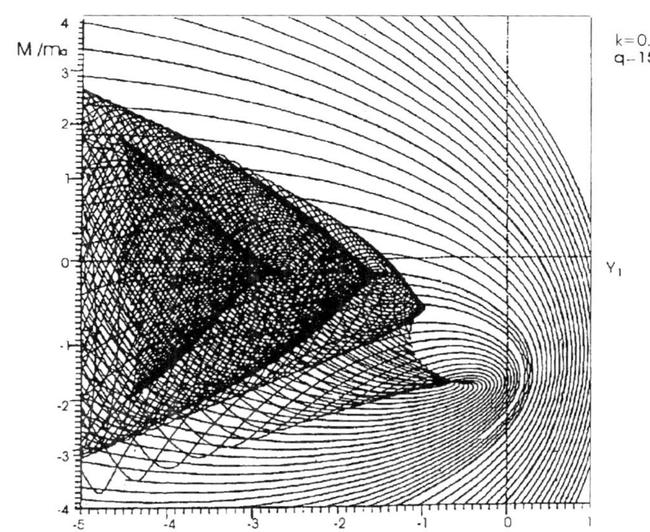

Figure 4.34 The influence of q as a control parameter; the case $q = 15$.

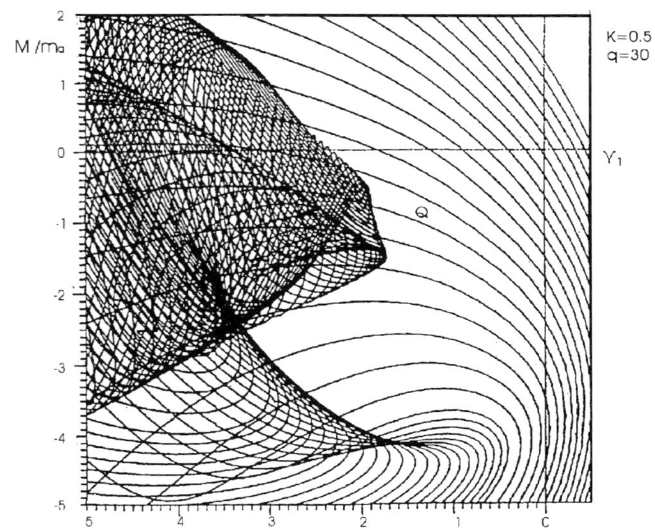

Figure 4.35 The influence of q as a control parameter; the case $q = 30$.

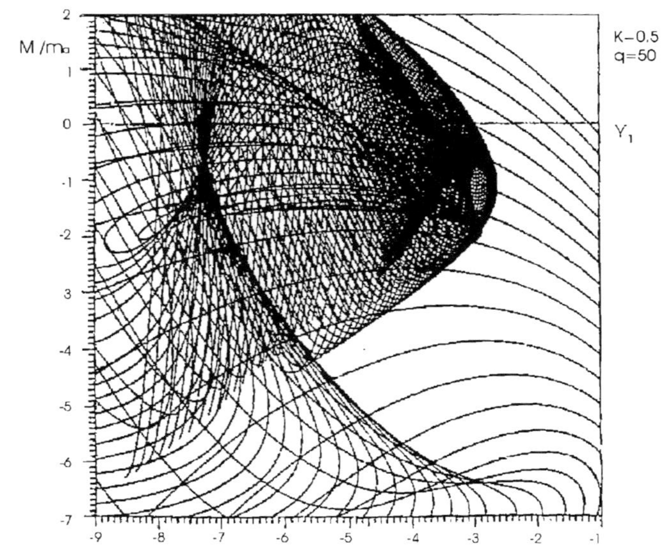

Figure 4.36 The influence of q as a control parameter; the case $q = 50$.

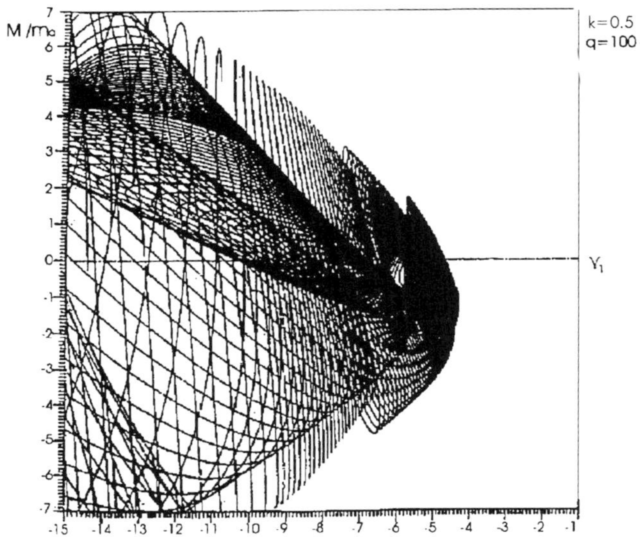

Figure 4.37 The influence of q as a control parameter; the case $q = 100$.

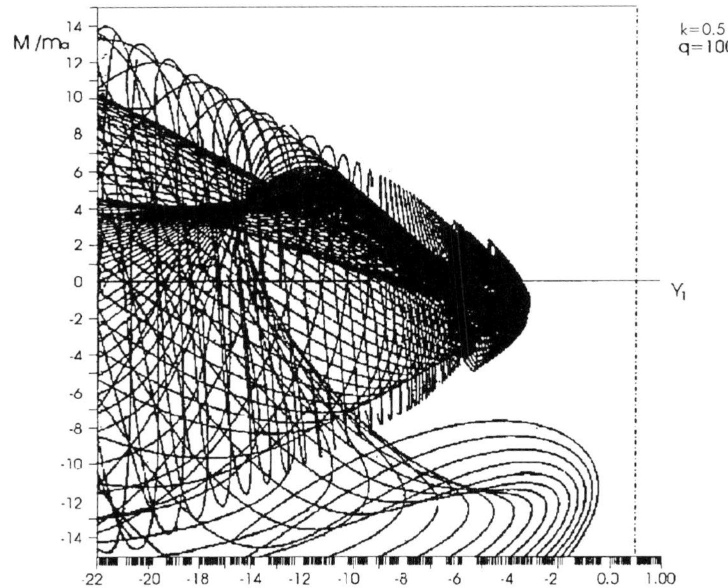

Figure 4.38 The influence of q as a control parameter; the case $q = 100$, to a different scale.

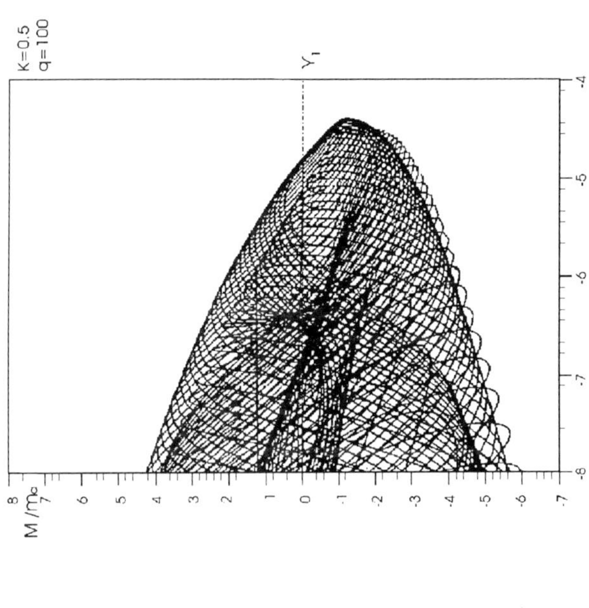

Figure 4.39 The case $q = 100$, first enlarged detail.

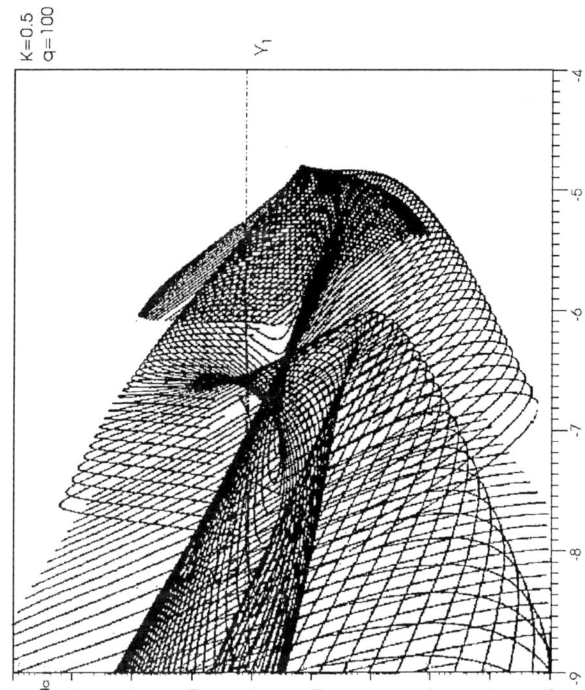

Figure 4.40 The case $q = 100$, second enlarged detail.

A map for $q = 50$ is presented in Figure 4.36, while the case $q = 100$, is drawn in Figure 4.37 and, to a different (smaller) scale, in Figure 4.38. It is noted, that these presentations are far from complete to avoid unnecessary loss of the most important parts of the details. (For example, the series of parabola-like curves of the type labeled by letter Q in Fig. 4.35 is not shown in the zone of the cusps' accumulation in the maps for $q = 100$.)

The pictures show that the increase of q generates a significant intensification of the process of accumulation and nonlinear interaction of the higher cusps. These maps exhibit an increasing degree of complexity in structure. For $q = 100$, it appears that the corresponding map is already attaining an unpredictable, chaotic character. To illustrate this point further we also added Figures 4.39 and 4.40, which give different details of the region of accumulation for $q = 100$.

The enormous complexity of the plate's static solutions is also demonstrated by the different parts (also incomplete) of the deformation map for $q = 100$ presented in Figures 4.41, 4.42, and 4.43.

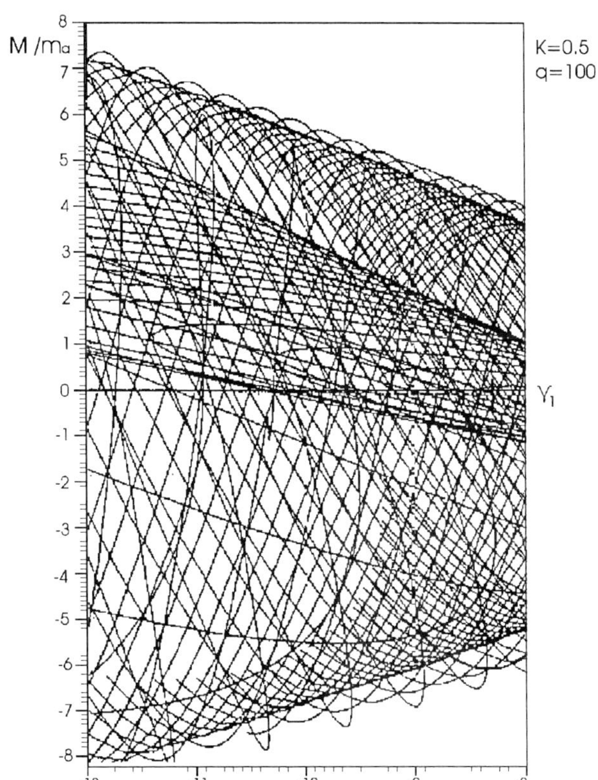

Figure 4.41 The case $q = 100$, third enlarged detail.

Figure 4.41 is associated with the interval ($-12 < \gamma_l < -8$). The part of the map for ($-16 < \gamma_l < -12$) is shown in Figure 4.42, and Figure 4.43 represents the behavior of the system when ($-20 < \gamma_l < -16$). In our opinion, the chaotic behavior of the static solutions is evident.

The solutions considered here, although very complex, belong in fact to one of the simplest nonlinear problems; i.e., the axisymmetric deformation of a geometrically nonlinear circular plate. It can therefore be expected that, in other cases, the behavior will be much more complicated. Take for example, the case of nonsymmetric deformations of plates, or that of nonsymmetric deformations of shells of revolution, where an additional geometrical control parameter $\bar{\xi}$, which characterizes the initial geometry of the shell, is introduced.

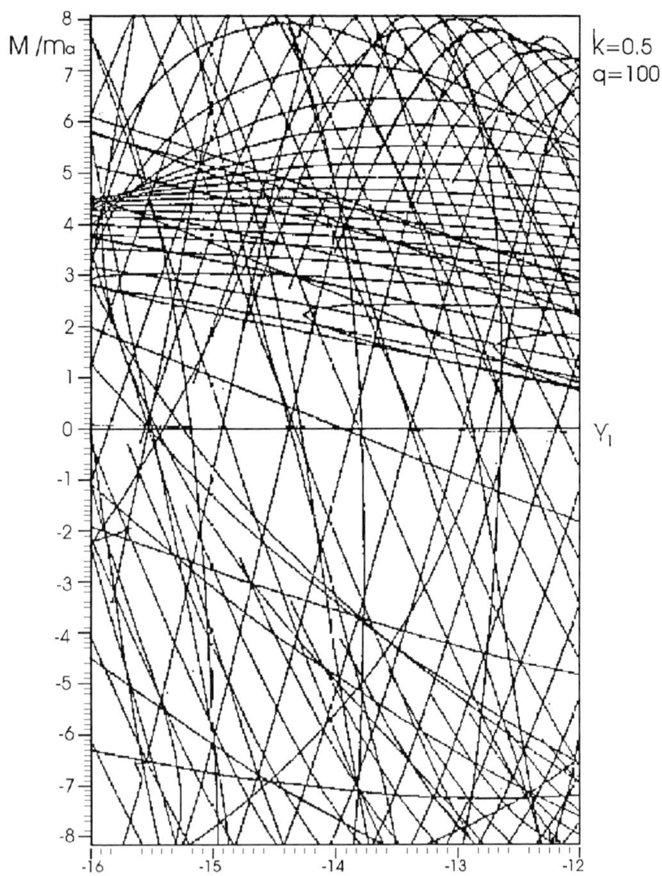

Figure 4.42 The case $q = 100$, fourth enlarged detail.

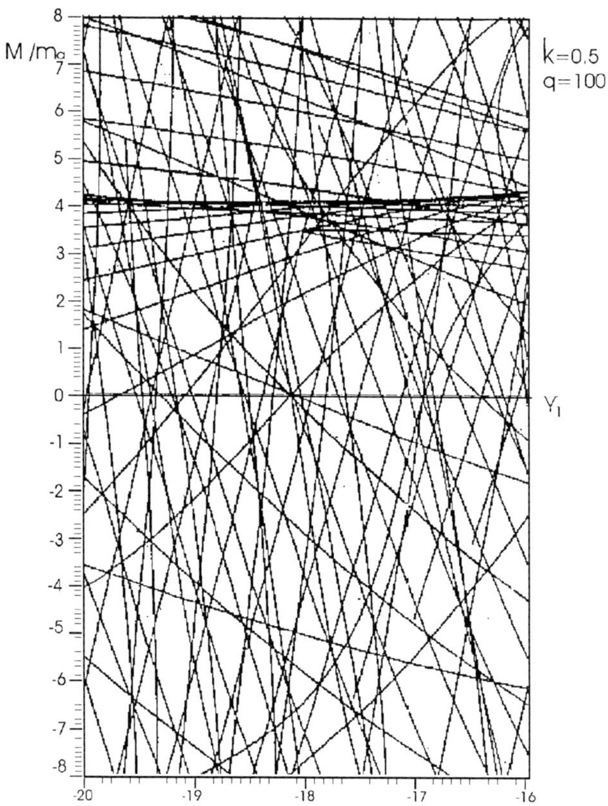

Figure 4.43 The case q = 100, fifth enlarged detail.

The theory of chaos for deterministic dynamic systems is already quite well established [see e.g., Moon, 1987; Seydel, 1988]. Recently some attempts were made to find static analogies [El Naschie, 1990; Hunt, Bolt & Thompson, 1989; Thompson & Virgin, 1988], but these considerations were confined to structures subject to special artificial conditions (e.g., infinitely long bars with sinusoidal geometric imperfections), whereby the boundary value problem (which describes the static deformation of a structure) is in fact transferred into an initial value problem (which describes motion).

The chaotic (unpredictable) behavior that we encountered, however, is generated by an ordinary nonlinear structure of finite dimensions under quasi-static loading conditions, a structure that is governed by a boundary value problem. In our case, where several external control parameters come into play, the plate's chaotic-like behavior is revealed by the complexity of the map's structure. This condition is caused by the circumstance that, for a given combination of the control parameters, many equilibrium states exist, and that the multiplicity of these solutions can grow and become practically unpredictable when the control parameters come within some specific range of values.

A better insight into the static behavior of plates can be obtained by an investigation of the stability of the whole static solution set that belongs to the range of parameters under consideration. It is then only necessary to look at the stable states. In this way, the multitude of equilibrium states can drastically be reduced. It was recently shown [e.g., Shilkrut, 1980; Shilkrut and Vyrlan, 1977, 1975] that, for a single-parameter loading, or sometimes for a two-parameter loading, plates and shells possess no more than two stable, axisymmetric equilibrium states for any selection of the control parameters, independently of the number of solutions that the corresponding boundary value problem admits (see Chapter 6).

How many stable equilibrium states exist in relation to the number of control parameters (when this number exceeds two) is a question that cannot be answered yet. It is to be expected, however, that the more control parameters come into play in a given problem, the higher the chances are for a chaos-like behavior of its static solutions.

Another factor that can directly contribute to the generation of chaos is the choice of the loading sequence in the multi-parameter case. Consider, for example, a case of two-parameter loading (N and M) acting on the plate. Suppose that the task is to find the equilibrium state of the plate at $N = 10$ and $M = 5$. This problem cannot be solved by the superposition of solutions that belong to two separate and independent load cases $N = 10$ and $M = 5$, because the problem is nonlinear.

A solution of the corresponding boundary value problem (4.18), (4.2), and (4.3) for the given N and M, could be considered, but this approach cannot, in principle, reveal the influence of the loading paths on the final equilibrium state of the structure. Indeed, there are infinitely many ways to load the plate by a multi-parameter loading system. One may, for example, simultaneously apply the whole load system, starting from zero and increasing it very slowly (i.e., quasi-statically) until its final value. Or, one can load the plate by first using only a part of the system, again by slowly increasing its intensity to its final value, and, subsequently activate the second part, and so on, until the whole load system is applied. There is absolutely no guarantee that, for these two different loading procedures, the same equilibrium state will be reached. On the contrary, the solutions in these two cases will generally be different, despite the fact that the equations of equilibrium and the corresponding boundary conditions are devoid of any terms that signal the way the loading procedure is carried out.

The concept of a quasi-static loading process, which stands at the basis of our considerations, implies that the loads are varied infinitely slowly with time. This does not mean, however, that the actual response of the structure to this slowly varying load will be determined by the corresponding slowly varying solutions of the static equations of equilibrium. The reason for this is that a great many of these solutions are unstable, so that, as soon as such a state is reached, the equilibrium in this state can no longer be maintained. At that instant, a transient change to another stable state will take place, which is no longer described by the static equations of equilibrium.

Thus, to study the influence of the loading procedure, it is necessary to take the transient nature of the structure's response into account, even when the loads are changed in a quasi static fashion. It is by means of such an approach that the influence of the loading path will be investigated in the next section [see also Shilkrut, 1988, 1992b].

The main points surrounding the question of the influence of the loading path can now be summarized as follows:

1. The set of the solutions of the governing equations of equilibrium does not depend on loading procedure. For an actual structure, however, equilibrium can only be maintained at a stable state no matter how slowly the load is applied.

2. Different loading procedures leading to the same final value can generate either the same stable state or a different one that belongs to the set of solutions mentioned above. The question of how the selection of the final stable state is influenced by the loading procedure cannot be answered, in general, at this time. Sometimes partial answers can be found for specific cases, but at present, it is not clear that an answer for the question of this selection exists, *a priori*.

It has been shown [Shilkrut, 1992b] that a small deviation in the loading procedure (e.g., deviation due to numerical errors) can lead to a very large deviation in the final result. This phenomenon points at a degree of unpredictability in the response of the system, which is otherwise associated with chaotic behavior. When many control parameters are involved, the system may become more and more unpredictable in the sense just mentioned. In this preliminary stage of our investigations, the physical nature of chaos-like phenomena in statics cannot yet be clarified to a satisfactory degree.

In our case, as was demonstrated above, the main source of the advent of unpredictable behavior is the accumulation of higher cusps and their nonlinear interaction. It is apparent that this interaction generates singular points of increasing complexity, which invoke this unpredictable behavior. In order to obtain some additional insight into this mechanism, it is useful to look at what is happening with the topology of the map when interaction intensifies. During the process of accumulation, strong geometric transformations of the curves that represent this topology take place. When $q = 0$, they have still a very regular structure, as shown, for example in Figure 4.44. Each curve is formed by a sequence of symmetric loops, which go around the

internal (elastic) forces in the body. This means that, during the loading process, the inertial (d'Alembert) forces that are acting on the body's particles are equal to zero (or are negligible) during the loading process. This is the commonly accepted notion of the process of static loading, or, better, the notion of quasi-static loading [see, e.g., a relatively recent article by Itzhak and Baruch, 1986]. It is a concept based on intuition, which is often, but wrongly, viewed as obvious.

The first effort to query this concept immediately shows a serious defect. First of all, quasi-static loading is a process that is evolving in time, but each equilibrium state along the equilibrium path that is reached in this way is defined as a static state that is attained when the time $t \to \infty$. Thus, the "classical" notion of quasi-static loading must be corrected by adding the stipulation that equilibrium states along the equilibrium path are *static* solutions obtained by a *quasi-static loading process*.

But this modification does not help us any further. In order to see this, it is enough to consider the example of a snap-through process. This is a transient process — a jump that takes place as a result of a quasi-static loading procedure — while pre-snapping and post-snapping states are *de facto* static states of equilibrium. (For a clear understanding of this example, we must introduce a small amount of damping to the system, as we mentioned in Chapter 3.) The process can be demonstrated by a series of examples, where large velocities and accelerations occur during the snapping motion, but where, when $t \to \infty$, a final state of static equilibrium is reached in the sense of the definition given above [see, e.g., Shilkrut, 1988].

By using the equations of motion to explore the influence of the loading path on the response of a given system, we will now introduce a notion that will be of help in the identification and determination of (classical) static equilibrium states. This is the notion of a "steady state." A structure has reached a *steady state* (often called equilibrium state) if, for $t \to \infty$,

1) this state is not accompanied by accelerations, (the classical notion of a static equilibrium state).

2) the state is characterized by a periodic motion.

3) the state is characterized by an almost-periodic motion. In this case the motion is the sum of periodic motions where the ratios of the frequencies are rational but not whole.

The average of each of the two latter cases can be identified as the classical, static equilibrium state. If the structure exhibits damping so that energy dissipation takes place, the last two cases simply disappear.

Thus, in the approach that we will adopt here, the applied external excitation varies with time in an arbitrary way, but slowly. At $t \to \infty$, it must have arrived at its final constant value. The path the loading is forced to follow can be arbitrary. Once the motion of the structure reaches a steady state, this state determines the static equilibrium state. With these additional stipulations, the classical notion of a quasi-static deformation process loses its rather restrictive significance and becomes applicable in numerical computations.

Having clarified the notion of the equilibrium states, we are now ready to enter the discussion of how the equilibrium states of a given structure are influenced by the sequence by which the loads are applied. To elucidate the physical picture of the phenomenon in question, we will take as an example the axisymmetric deformation of a mobile-hinged, isotropic, geometrically nonlinear circular plate subjected to independent edge loads M and N (see Fig. 4.45). Thermal loads are not considered in this case.

Figure 4.45 The problem of a circular plate loaded by two edge loads, N and M.

We are now faced with the problem of determining the equilibrium states of the plate for specific values of forces $M = M_0$ and $N = N_0$, whereby two different loading paths will be used. The analyses will be carried out by the solution of the (nonlinear) equations of motion, where the loading is varied very slowly with time. In this a way, the actual physical process of the plate under a slowly varying load will be simulated numerically. This means, for example, that if the real structure suffers a snap-through at some instant during this process, snap-through will also occur during the simulation. The two loading paths that we consider here are described as follows.

1) At the beginning, only the boundary moment is applied, the value of which increases very slowly from zero up to its final magnitude M_0, while N is kept at $N = 0$. When M reaches its final value M_0, it remains fixed, and subsequently the edge force N is applied, in the same way as M (i.e., very slowly) until its final value N_0. As soon as $N = N_0$, the loading process is finished. The steady state that is found after this instant determines the final static equilibrium state.

2) The load sequence is now carried out in the reverse order; that is, from the beginning, N acts alone and after this, M is applied. The rate of change of loading parameters is the same as in the previous case. The comparison of steady (equilibrium) states that are reached by the plate via these two different ways of loading can reveal the influence of the paths on the plate's behavior, as mentioned above.

We shall now assume that the solutions of this problem possess the following feature.

The steady states (established when $t \to \infty$; $\lim_{t\to\infty} M = M_0$ and $\lim_{t\to\infty} N = N_0$) are equilibrium states represented by solutions of basic static equations that correspond to the following edge conditions:

$$\theta(0) = \omega(0) = 0; \quad \omega(1) = N_0; \quad \theta'(1) + \nu\theta(1) = M_0, \tag{4.24}$$

and this *outcome* is independent of the initial conditions and the specific loading path. This assumption is not unreasonable. It means that the steady states of the equations of motion (when external friction is available) are members of the set of solutions of the static equilibrium equations that are determined by the given values of the external forces.

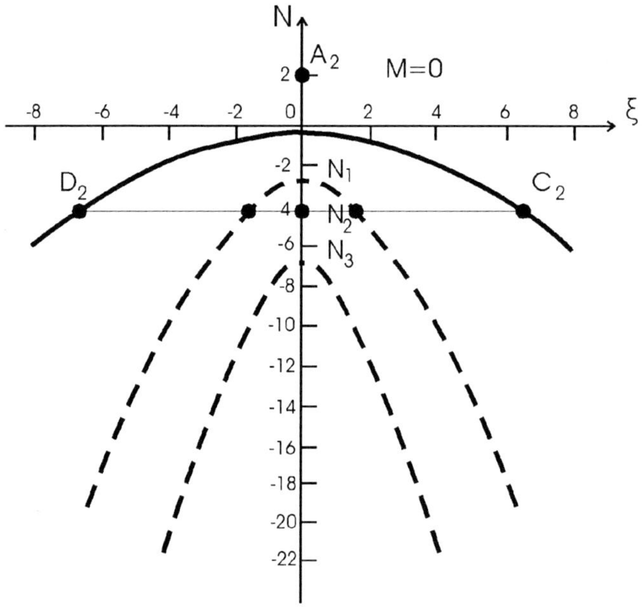

Figure 4.46 The solutions for M = 0.

In order to investigate the influence of the two different loading paths described above, we use the available solutions of the static equations when: a) only N is acting (Fig. 4.46); b) only M is acting (Fig. 4.47); and, c) both M and N are active (Fig. 4.48) (see Section 4.3). The dashed curves in these figures denote unstable branches.

We shall start with the first loading sequence described above. Let, for example, the two final values be given by $M_0/m = 2$ and $N_0 = 2$. In the first part of sequence 1, M acts alone, and when $M = M_0$, the plate will have reached the state marked A_1 in Figure 4.47. After this, N is applied until it attains the value $N_0 = 2$. The state of the plate will then reach the position marked by B_1 in Figure 4.48.

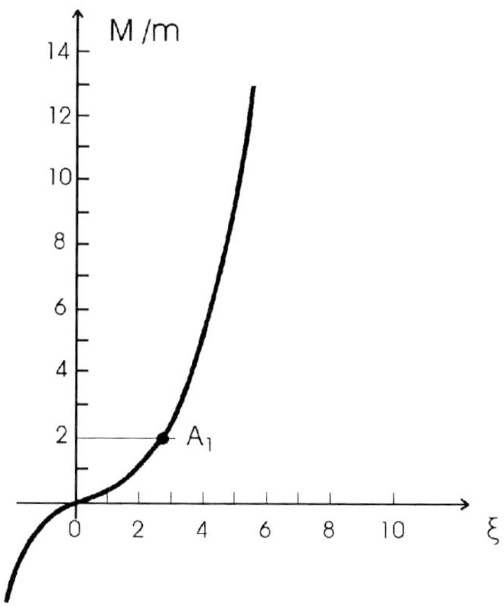

Figure 4.47 The solution for $N = 0$.

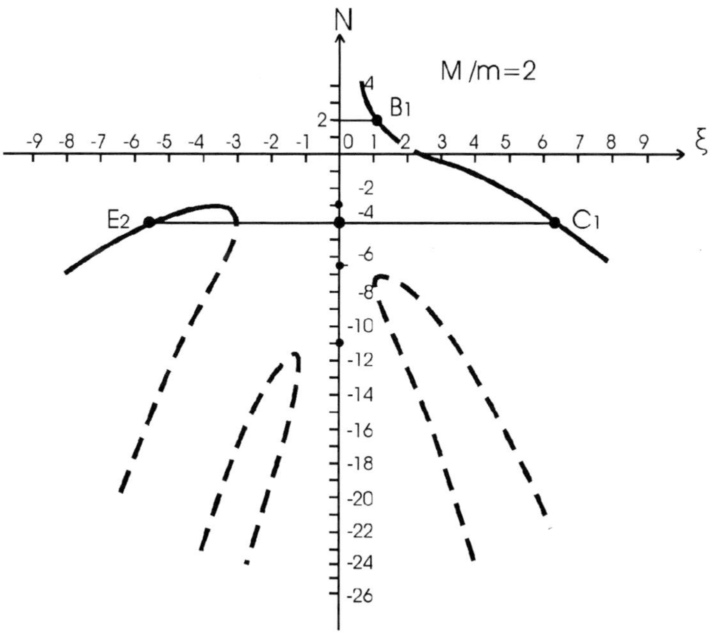

Figure 4.48 The case that both M and N are acting.

In the second loading sequence, at the end of the first stage, the system will arrive at the position marked by A_2 in Figure 4.46. At the end of the second stage, the plate again reaches position B_1 in Figure 4.48. Thus, for this particular problem, both loading paths lead to the same final equilibrium state. This result is not very surprising, since, for the given values $M_0/m = 2$ and $N_0 = 2$, there is only one solution (point B_1 on Fig. 4.48) of the static equilibrium equations.

However, we anticipate a different outcome, if, for the given combination of M_0 and N_0, more than one static equilibrium state exists. Let, for example, the load combination be given by $M_0/m = 2$ and $N_0 = -4$. In that case (see Fig. 4.48), there are three static solutions, two of which (E_2 and C_1) are stable. For this combination, after the end of the first stage of the first loading sequence, the plate will be in the same situation (point A_1 on Fig. 4.47) as in the previous case. The second phase of the first loading sequence will take the plate to the state that corresponds to point C_1 in Figure 4.48.

Indeed, at point A_1, the plate gains the shape qualitatively shown in Figure 4.49, with $\xi > 0$, where ξ can be seen as the amplitude of the deformation mode. If the compressive load N is applied and reaches $N_0 = -4$, the amplitude ξ will further increase. Thus, the plate will be deformed to the state that corresponds to the position C_1 in Figure 4.48, and not to one of the other two possible solutions of the equilbrium equations for the combination $N_0 = -4$ and $M_0/m = 2$.

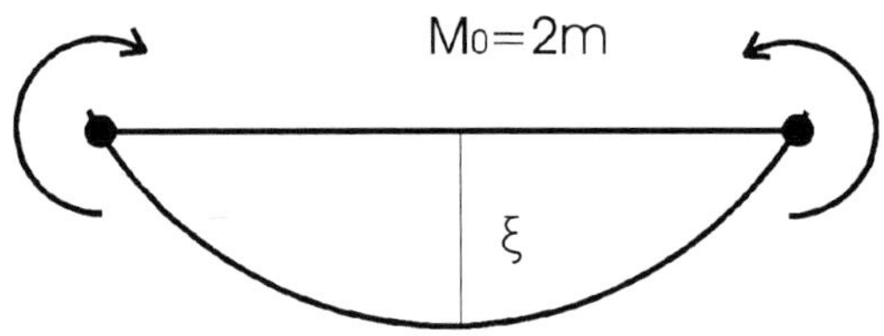

Figure 4.49 The deformed state A_1 in Figure 4.47.

In the second loading sequence. There are five possible equilibrium states that correspond to the level $N_0 = -4$ (see Fig. 4.46). Three of these are unstable, while the other two, marked by C_2 and D_2 in Figure 4.46, are stable. The unstable states are physically not attainable. This leaves us with the two remaining possibilities: either C_2 or D_2. We choose C_2 as an example, so that the corresponding deformed shape has a positive central displacement ξ. In this case, after the end of the second loading phase, the plate's final state will be the one that corresponds to point C_1 (see Fig. 4.48), and this is the state that was also reached in the previous loading sequence. It happens this way because the additional positive moment M_0 increases the ampli-tude ξ of the deformation mode. Thus, so far, both sequences lead again to the same result.

Now, if, at the end of the first step of the second loading sequence, the plate takes the state that corresponds to D_2 so that $\xi < 0$, instead of the state C_2 with $\xi > 0$, application of M will lead to the position marked by point E_2 in Figure 4.48. (The intermediate equilibrium state for $N_0 = -4$ in Fig. 4.48 is unstable.) In this case, a change in the loading sequence brings about a difference in the final equilibrium state.

From this qualitative analysis of the investigated phenomenon, one arrives at the following conclusions:

1. Different loading paths may lead to different responses.

2. Different responses can be expected when more than one equilibrium state exists for at least one of the components of the two force system (M_0 or N_0), as, for example, in the case shown in Figure 4.46. (The number of possible equilibrium states corresponding to a fixed value of the parameter of loading is always odd.)

3. The equilibrium state that the structure eventually will reach is always stable. Because the process of loading is considered to be time-dependent, the structure cannot remain in equilibrium at an unstable state.

4. The set of stable equilibrium states does not depend on the way the loading is varied. They belong (invariably) to the solution set of the static equilibrium equations.

5. The question of how the plate selects the deformation path to either C_2 or to D_2 (Fig. 4.46), the path that determines the final result, has no answer in the framework of an idealized mathematical model. It can only be answered if we consider an imperfect model, which removes the indeterminacy of the selection between C_2 and D_2. (Incidentally, an imperfection can be introduced unwillingly through errors produced by the numerical method that is used to solve the problem. In such a case, the numerical errors can steer the deformation process to the final result. An example of such a process will be shown below.)

6. The outcome of the problem investigated here has implications for the use of numerical solution methods. Numerical solutions of nonlinear problems are often obtained by incremental procedures (also called "path following" methods). The phenomena discussed in the foregoing show that great caution should be exercised when using these methods.

7. The dependency of the final equilibrium state on the process of loading introduces a degree of unpredictability in the expected response of the structure that could be a source of chaotic-like behavior.

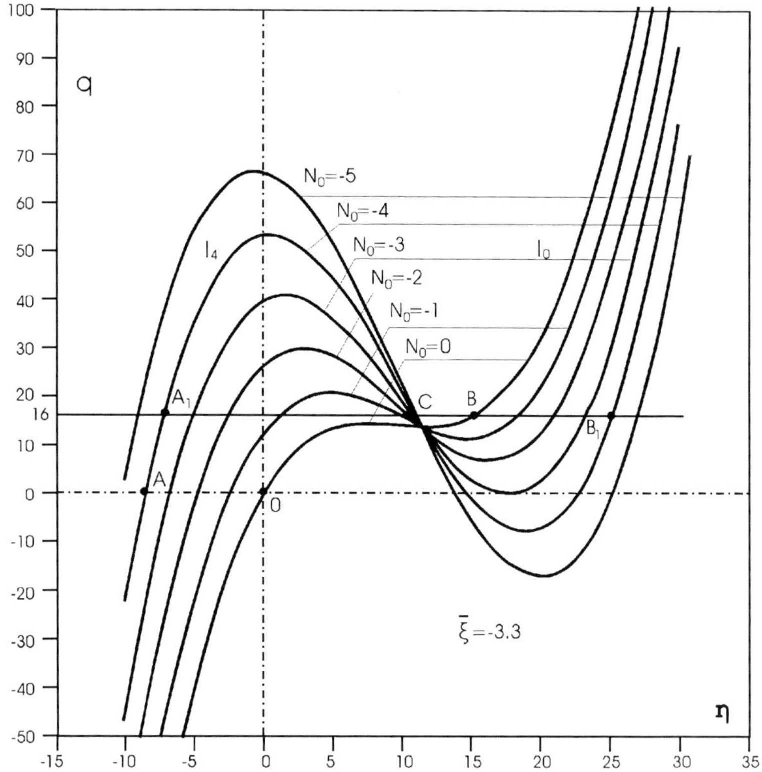

Figure 4.53 Results for the third loading path.

When the loading q is activated, the deformation of the cap follows the curve A-A_1 in Figure 4.54, corresponding to the trajectory A-A_1 along curve l_4 in Figure 4.53.

In the case of the second loading path II, at the time $\tau = \tau_1$, point B in (Fig. 4.54) is reached. From $\tau = \tau_1$ on, and until the establishment of the steady state, the path of deformation follows the branch B-B_1 (Fig. 4.54). In Figure 4.53, the first part of the process is along curve l_0 from point O until point B, the equilibrium state corresponding to the steady (final) state, is denoted by B_1. Thus, it is clear that, if one of the loading procedures leads to a snap-through event, then the final steady state is different in comparison to the procedure where no snap-through occurs.

Figure 4.55 illustrates a case for which the damping is removed. This corresponds to the choice $N_0 = -5$, with initial conditions $\eta(0) = 9$, $\dot{\eta}(0) = 0$. As can be seen, the response of the cap ends in strong oscillations around the steady state, shown in Figure 4.54 for $N_0 = -5$. The same results are obtained when the initial displacement is set to zero, but the initial velocities are taken as $\dot{\eta}(0) = 5$ or 15. Thus, the change in the initial conditions does not have any influence on the final result in this case; i.e., the steady states that are reached are the same after this change.

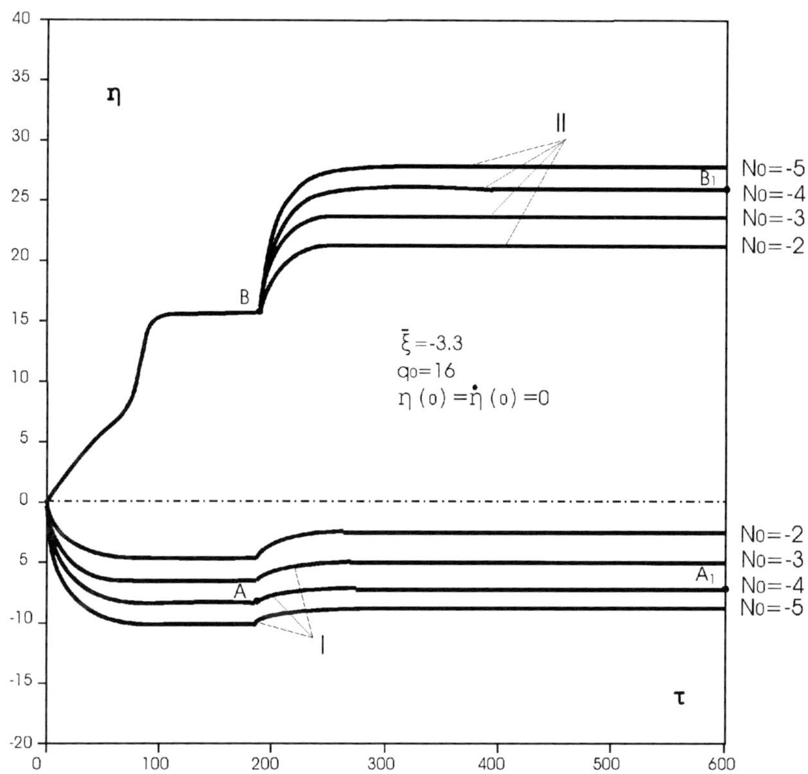

Figure 4.54 Transient solutions for the case: $q_0 = 16$.

A completely unexpected behavior was observed for the case that is defined by the initial conditions $\eta(0) = 9$ and $\dot{\eta}(0) = 0$ ($N_0 = -5$, $q_0 = 16$), where again no damping is taken into account. This occurred when the integration step of the integration process was taken as $\Delta\tau = 0.1$ instead of the step $\Delta\tau = 0.01$, which was used before. The result (shown in Fig. 4.56) is different from the one depicted in the previous figure in the sense that the steady states obtained here are the same for both loading paths I and II. This means that numerical errors can drastically change the final results, showing that the response of the shell is very sensitive to the loading process.

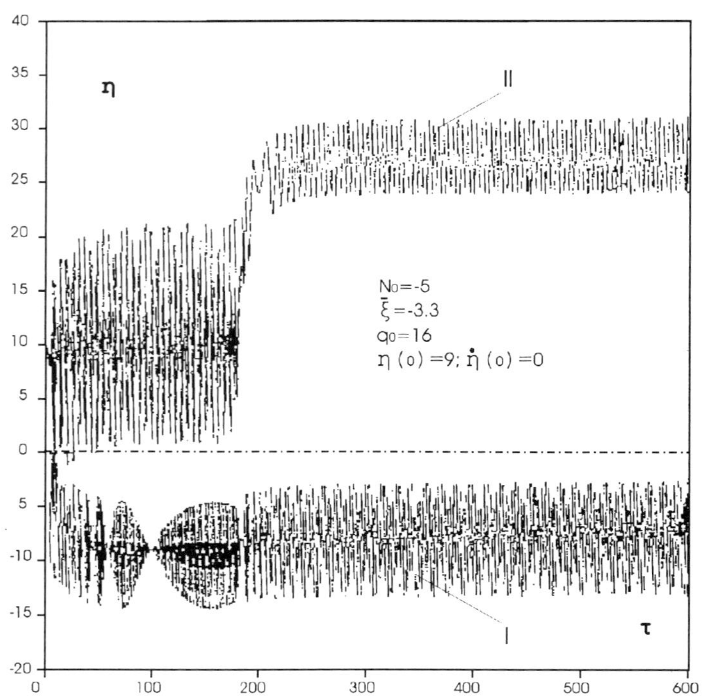

Figure 4.55 The case $q_0 = 16$, different solutions for different loading paths; no damping.

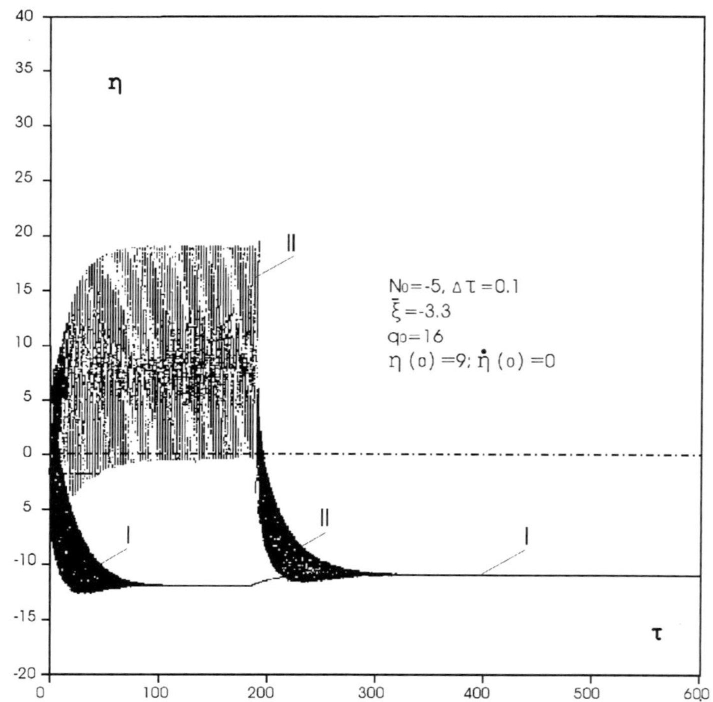

Figure 4.56 The effect of a relaxed error tolerance.

In Figure 4.57, the behavior of the cap is shown for the case that both loads grow simultaneously during the entire time interval ($0 \leq \tau \leq \infty$). The other specifications are: $N_0 = -5$, $q_0 = 16$, with the natural initial conditions $\alpha = \beta = 0$, and damping taken into account. In this case, the steady states were reached that were found for the loading sequence I given in Figure 4.54. It also seems that the edge load N is the dominant factor in the response of the cap. The removal of damping results in a response with only small additional oscillations. The case $N_0 = -5$, $q_0 = 24$ was also investigated, but, in comparison to the previous case ($N_0 = -5$, $q_0 = 16$), no new results were obtained.

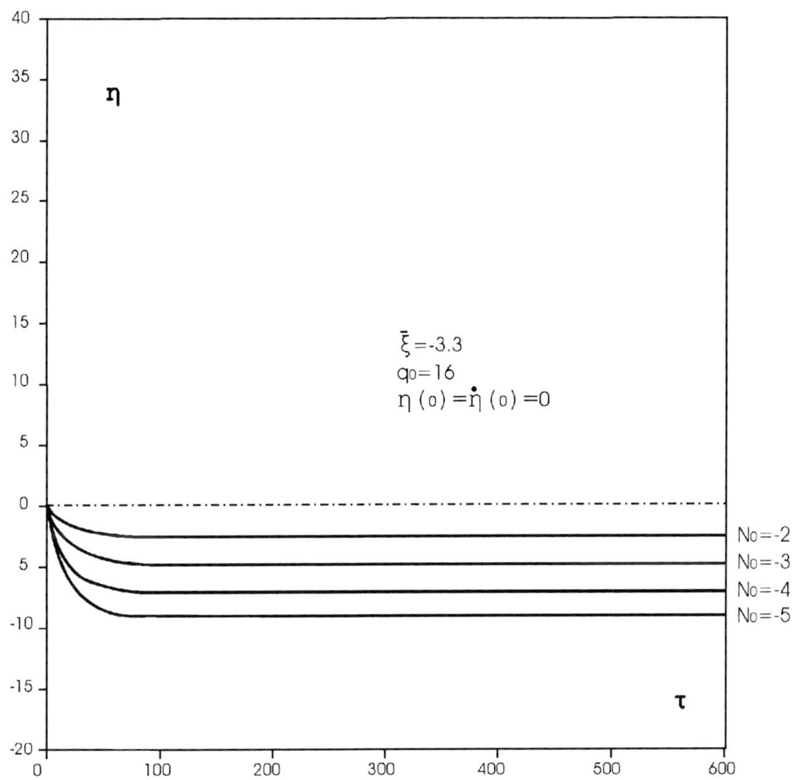

Figure 4.57 The effect of both loadings applied concurrently.

The results presented above correspond to caps with an initial geometry given by the rise parameter $\bar{\xi} = -3.3$. The same qualitative results, however, were obtained for a cap with $\bar{\xi} = -9.3$. It is also noted that shallow, geometrically nonlinear arches show a similar behavior. We will now formulate some conclusions that we can draw from the analysis of this model and add some further additional observations.

1. The numerical experiments described above confirm our expectations formulated earlier regarding the behavior of geometrically nonlinear shell structures.

2. The introduction of a little damping does not seem to influence the final result. This outcome can be explained in light of the basic assumption (introduced in the beginning of this section) about the coincidence of the steady state solutions (of the equations of motion) with the static solutions (of the equilibrium equations). There are many classes of differential equations that possess this feature. Examples of such equations were discussed by Shilkrut [1988].

3. There are a number of interesting cases that were not considered above; for example, a cap subjected to a two-parameter loading with an edge moment M, and pressure q. Let the final values M_0 and q_0 be of such magnitude that they generate only a single equilibrium state when they act alone. In spite of this restriction, there are still combinations of M and q possible for which at least three equilibrium states can occur. Such an example is presented in Figures 4.58 and 4.59, which are related to the axisymmetric deformation of spherical caps with an apex rise of $\overline{\xi} = -4$ with a mobile hinged edge.

   In Figure 4.58, a series of equilibrium curves $M = M(\xi, q = \text{const.})$ is presented (see also Section 3.7). Consider the case $M_0 = 40$ and $q_0 = -20$ (the negative sign means internal pressure). The unique equilibrium state that corresponds to $M_0 = 40$ and $q_0 = 0$ is marked by point A on Figure 4.58. The unique equilibrium state that belongs to the specification $M_0 = 0$ and $q_0 = -20$ is marked by B in Figure 4.59. Equilibrium states generated by the combination $M_0 = 40$ and $q_0 = -20$ are the three (points C, D, and E in Fig. 4.58), where the intermediate point D (marked by $\times$) is unstable.

   Thus, despite the fact that equilibrium states generated by each component of the loading (when they act separately) are unique, uniqueness of the equilibrium states cannot be guaranteed when the loads are not together. The final state then depends, of course, on the specific way the multi-parameter load system is applied. This is the distinguishing feature of this case, which is not widely known in the literature.

4. The class of problems dealt with here concerns a two-parameter load system. We believe that no new, qualitatively different results will emerge when three or more parameter load systems are active, despite the fact that these cases are considerably more complicated than the two-parameter case (see, e.g., the results for the specific case when a plate is subjected to pressure, an edge moment, and a thermal load [Shilkrut, 1982, 1984], which were discussed in the previous section).

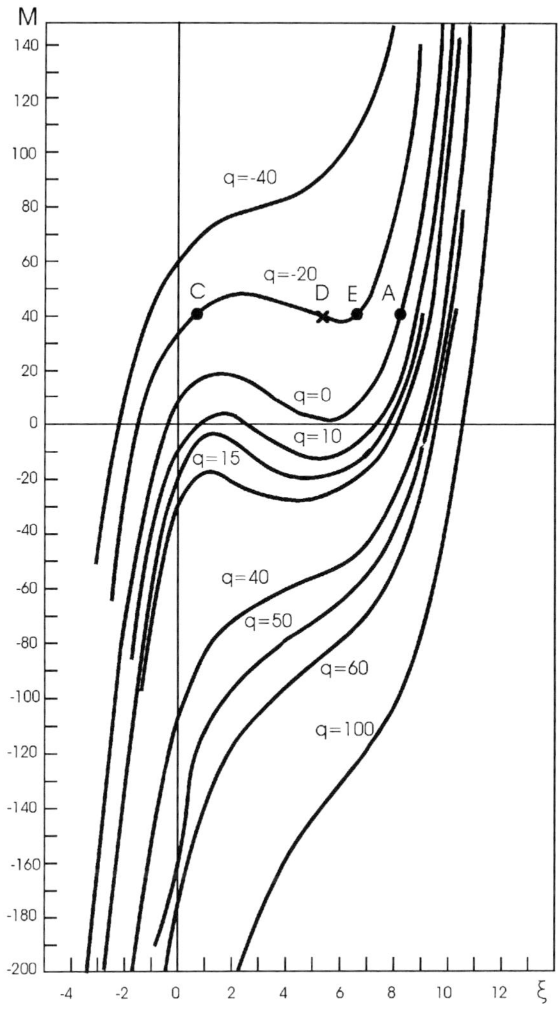

Figure 4.58 Static solutions for a cap loaded by uniform external pressure and an edge moment.

5. It is useful to understand the similarities and differences between the (transient) computational procedure used here and the step-by-step (incremental) approach that is available for the solution of the static equilibrium equations. The incremental approach in the case of multi-parameter loading system can, of course, be attempted using a series of different loading sequences.

The incremental procedure, which computes the static response of the structure for a certain loading sequence, is quite similar to our transient approach as far as the stable

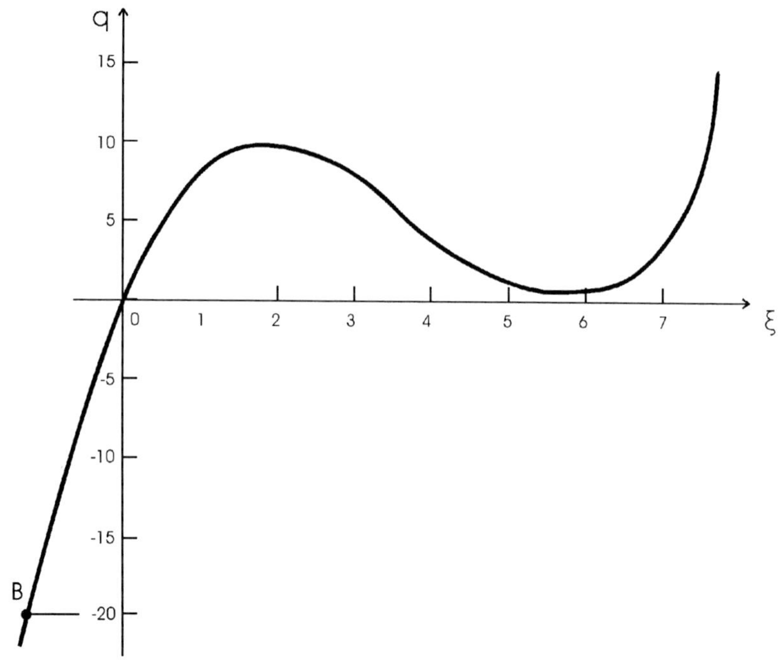

Figure 4.59 A particular solution.

parts of the solution branches are concerned. But in order to reach the final stable state of the problem, the incremental procedure must be guided along an unstable branch, which is connected with the stable final state. Such a connection, if it exists, is very difficult to find in the general case and is definitely not known *a priori*. The situation is particularly complicated because the number of unstable equilibrium states is very large as compared to the number of stable states (see Section 2.1). Moreover, even when the procedure is successful and a new stable state is found, there is no guarantee that this state corresponds to the one to which the structure would move in the transient case; the actual final state that is reached via a transient process might belong to a separate loop. This means, that there is no absolute certainty that the step-by-step (incremental) procedure employed can obtain the correct result for the given problem. (A better and more promising approach is the use of the incremental, path-following method in conjunction with a transient method, such as proposed by Riks [1979; Riks, 1998, Riks & Rankin, 1997].

6. As was shown in the previous section, the dependence of the final steady state of a structure on the loading path and impulsive additional excitations (perturbations) can result in an unpredictable response.

7. Without much additional trouble, the approach used here can be generalized to problems with nonlinear material behavior; internal friction (viscoelasticity), etc.

4.6 Multi-parametric thermo-elastic problems: Their analogy with pure elastic problems and some additional numerical results.

Because the problems cited in the title are important from a theoretical, as well as a practical, point of view, we shall continue with a discussion that we started in the previous sections of the present chapter. We shall begin with the analogy that exists between thermo-elastic and pure elastic problems, an analogy that we mentioned in Section 2.7, but which we will now discuss in greater detail. The governing equations and other relations (4.2)-(4.18) of the von Karman-Marguerre type for thermo-elastic problems will serve as the mathematical basis of our considerations.

We shall first consider the particular class of thermo-elastic problems that have no temperature-dependent terms in the equilibrium equations for the interior of the shell, equations (4.2) and (4.3). The cases that belong to this special class of problems are given by:

(i) $\lambda = 1;\ N_T \equiv \text{const.} \neq 0;\ M_T \equiv 0;$ (4.41)

(ii) $\lambda = 1;\ k = 1;\ N_T \equiv \text{const.};\ M_T \equiv \text{const.}$ (4.42)

If the support is movably clamped (see Table 4.1), the corresponding solutions for $\omega(\rho)$ and $\theta(\rho)$ do not depend on the temperature field T, while the strains, stresses, and displacement u are directly dependent on T (see the corresponding relations (4.9)-(4.15). For other types of supports, the basic unknowns $\omega(\rho)$ and $\theta(\rho)$ become dependent on the temperature field T through the dynamic boundary conditions, which are determined by $M_T(1)$ and $N_T(1)$.

When restriction (**i**), (4.41), applies, the thermo-elastic problem for the cases of mobile hinged supports is reduced to a pure elastic problem with the same kinematic, and dynamic boundary, conditions (see Table 4.1).

$$\omega(1) = N;\ \ \theta'(1) + \nu_r k^2 \theta(1) = M, \tag{4.43}$$

where the functions $\omega(\rho)$ and $\theta(\rho)$ are not dependent on T. This condition is a consequence of the restriction $u(1) = 0$ (see (4.14, 4.15)). For immovable (hinged or clamped) supports, case

(4.41) is equivalent to a pure elastic problem with the edge condition

$$\omega'(1) - \nu_r k^2 \omega(1) = -k^2 N_T(1) \tag{4.44}$$

for ω. For immovable hinged supports, the condition for θ remains identical to (4.44), while, for clamping, we have $\theta(1) = 0$.

When restriction (**ii**), (4.42) applies, and the material is isotropic (i.e., $k = 1$), the boundary conditions for ω at $\rho = 1$ can be either of type (4.43) or type (4.44). With regard to the condition for θ at $\rho = 1$, this can be either $\theta(1) = 0$ or the condition given at (4.43) (for $k = 1$), or, it can be of the following form:

$$\theta'(1) + \nu\theta(1) = M - (1+\nu)M_T(1). \tag{4.45}$$

Thus, both cases (4.44) and (4.46) are reduced to pure bending (see Section 3.7), or to a two-parameter loading elastic problem, where M and N act as a set of edge loads.

The case (4.42) can be realized if $T(\rho, \zeta)$ is an even function on ζ and does not depend on ρ. The case (4.43) can occur if T is an arbitrary function of ζ only.

We will now consider some additional cases where a thermo-elastic problem can be transformed to a pure elastic problem. Let the thermal load be characterized by

(**iii**) $\lambda = 1$; $N_T \equiv$ const.; $M_T \equiv$ const., but $k \neq 1$. (4.46)

The action of the temperature field is then reduced to an additional equivalent concentrated force $P = -(1 - k^2)\, M_T/m_a$ (see the corresponding terms in (4.3)).

(**iv**) $\lambda = k = 1$. (4.47)

In this case, the temperature distribution is given by the function $T(\rho, \zeta) = R(\rho)Z(\zeta)$, where Z is an odd function of ζ. Then $N_T \equiv 0$ and $M_T = f(\rho)$ is a function of ρ. The thermo-elastic problem is reduced to an equivalent elastic problem with a distributed load of type

$$\rho q(\rho) = -(1+\nu)\left[\rho M_T'\right]' / m_a \tag{4.48}$$

(see (4.3)). It is clear that $R(\rho)$ must be of such a type that $\rho q(\rho)$ in (4.48) will be an integrable function in the interval [0, 1].

(**v**) $\lambda = k \neq 1$. (4.49)

This is the same case as the previous one (**iv**), but $k \neq 1$ produces an additional equivalent concentrated force P, of the type mentioned above in case (**iii**).

(**vi**) $T(\rho, \zeta) = R(\rho)Z(\zeta) + A,$ (4.50)

where Z is an odd function of ζ and A is a constant. Consider also $\lambda = 1$; $k \neq 1$. Then we have the previous case (**v**) with the additional stipulation $N_T \equiv$ const. $\neq 0$. In the last four cases, it was necessary to write down the corresponding edge conditions at $\rho = 1$, as was done in the cases (**i**) and (**ii**), using the information provided in Table 4.1. Thus, there is a large class of thermo-elastic problems that can be reduced to an equivalent class of standard, elastic problems. By solving these, we solve the equivalent thermo-elastic problem at the same time.

Before we proceed with a survey of numerical solutions to some of these problems, it is worth considering the properties of $\omega(\rho)$, which are related to loss of stability using the qualitative analysis given in Chapter 2. As was stated in Section 2.8, Property 12, stability loss by snap-through in the space of axisymmetric deformations is impossible if $\omega(\rho) \geq 0$; $(0 \leq \rho \leq 1)$. We consider here the isotropic case ($k = 1$) for which these qualitative results were obtained. Applying Property 6 (Section 2.8) to equation (4.2) for one of two boundary conditions,

$$\omega(1) = 0; \;\; \omega'(1) - \nu\omega(1) = -\lambda^2 N_T(1) \tag{4.51}$$

(see Table 4.1), we observe that the sufficient conditions that ensure that $\omega(\rho)$ will be non-negative are:

$$\frac{\theta(\rho)\bar{\theta}(\rho)}{\rho} + \lambda^2 N_T'(\rho) - \frac{1-\lambda^2}{\rho} N_T(\rho) \geq 0; \;\; (0 \leq \rho \leq 1), \tag{4.52}$$

$$N_T(1) \leq 0 \tag{4.53}$$

for both conditions (4.51). The physical meaning of the last condition is that the edge $\rho = 1$ is cooled down. If at least one of the inequalities (4.52) or (4.53) is not satisfied, Property 6 loses its validity. In this case, there is a possibility that the axisymmetric equilibrium response possesses a snap-through point.

That $\omega(\rho)$ stays positive can be observed in a number of interesting cases. Take, for example, a circular plate ($\bar{\theta}(\rho) \equiv 0$). If $N_T \equiv$ const. ≤ 0 and $\lambda = 1$ (isotropic temperature field), then (4.52) will be satisfied for movable or immovable support, and the plate will never suffer snap-through in the space of axisymmetric solutions. Let the temperature field be distributed along the radius of the plate by the prescription $T = \alpha f(\rho)$. If, for example, $f = 1 - \rho^2$, then N_T is given by $N_T = \beta(1 - \rho^2)$, where α and β are constants with the same sign and $N_T' = -2\rho\beta$. Consequently, the inequality (4.52) will be violated if $\lambda = 1$ and $\alpha > 0$ (overall temperature rise), and snap-through is possible. The condition $\omega(\rho) \geq 0$ prevails only when $\alpha < 0$.

The opposite situation occurs when $f(\rho) = \rho^2$ and $N_T' = 2\rho\beta$. In this case, snap-through behavior can be expected when $\alpha < 0$ ($\lambda = 1$, as in the previous case), because (4.52) cannot be

$$q_R(\rho) = \begin{cases} -\lambda w(\rho) & \text{if } w > -\dfrac{æ}{\lambda} \\ 0 & \text{if } w < -\dfrac{æ}{\lambda} \end{cases}; \quad æ \geq 0 \tag{5.2}$$

in the case of a possible failure of the bond between plate and foundation. The foundation's reaction force is denoted here by $q_R(\rho)$.

The relation (5.1) represents the effect of the classical Winkler foundation, where λ stands for the coefficient of rigidity. The sign convention is such that q_R is taken as positive when w is negative; thus $q_R > 0$ when $w < 0$. In this case (5.1), the cohesion between plate and foundation cannot be undone.

In the second case (5.2), decohesion takes place if $w < -\frac{æ}{\lambda}$. The connection between the plate and its foundation is thus bounded by an ultimate tensile (failure) stress. When the reaction of the foundation is greater than this ultimate stress, decohesion takes place and a gap is formed between the plate and foundation. In the area of the gap, the reaction forces of the foundation are assumed to be zero. In regions where the bond between the plate and foundation exists (complete compatibility), the reaction forces of the foundation are of the classic Winkler type. The bound $æ > 0$ represents the ultimate (failure) tensile stress of the connection. Thus the first relation of (5.2) holds in bonded regions, while the second holds in the debonded regions. Possible interactions (tangential or normal) between the plate and the foundation in the debonded regions are ignored.

According to the model introduced above, there is either a complete bond between the plate and the Winkler foundation, or there is no bond. An intermediate condition is not considered. If $æ = 0$, we have a case where no adhesion between the plate and foundation exists.

This problem is governed by two different types of nonlinearities, one of which is the geometric nonlinearity of the plate itself. The second is due to the process of the separation of the plate from the foundation, where the regions of separation are *a priori* unknown. The first results of an analysis of this problem were described by Shilkrut [1983c, 1987]. A similar approach, applied to the problem of a delamination between two-layers in a composite laminate, was considered by Ascione and Bruno [1987]. The results that are discussed below were obtained by Shilkrut [1983c, 1987].

We shall use the basic formulation given in Section 4.3, appropriately adapted to the present model. These equations read:

$$L_k(\omega) = -\frac{k^2\theta^2}{2\rho}; \quad (0 \leq \rho \leq 1); \tag{5.3}$$

$$L_k(\theta) = -\frac{m_a}{\rho}\int_0^\rho (q + q_R)\rho d\rho + \frac{m_a \omega\theta}{\rho}; \tag{5.4}$$

where

$$L_k(\;)=(\;)''+(\;)'/\rho-k^2(\;)/\rho^2;\; k^2=E_\phi/E_r; m_a=12\left(1-\mu_r\mu_\phi\right)=12\left(1-\mu_r^2k^2\right)\;;$$

$$E_r\mu_\phi=E_\phi\mu_r.$$

$$w(\rho)=\xi-\int_0^\rho\theta(\rho)d\rho;\;\;\xi=w(0). \tag{5.5}$$

The boundary conditions are

$$\omega(0)=\theta(0)=0. \tag{5.6}$$

$$\alpha_1\omega'(1)+\beta_1\omega(1)=\gamma_1;\;\;\alpha_2\theta'(1)+\beta_2\theta(1)=\gamma_2. \tag{5.7}$$

Note that the integrand of the integral in (5.4) contains an additional term that represents the reaction force $q_R(\rho)$.

The numerical solution of this problem becomes easier if the right-hand side of (5.4) is written in another form that does not contain $w(\rho)$ explicitly ($w(\rho)$ is introduced by means of q_R). In order to do this, we shall use a transformation of the integral in (5.4) given by

$$\int_{\rho_i}^{\rho}\rho q_R d\rho=-\lambda\int_{\rho_i}^{\rho}\rho w d\rho=-\lambda\xi\int_{\rho_i}^{\rho}\rho d\rho+\lambda\int_{\rho_i}^{\rho}\rho\int_0^{\rho}\theta ds d\rho$$

$$=-\frac{\lambda\xi}{2}\left(\rho^2-\rho_i^2\right)+\frac{\lambda}{2}\left[\rho^2\int_0^{\rho}\theta d\rho-\rho_i^2\int_0^{\rho_i}\theta d\rho-\int_{\rho_i}^{\rho}\rho^2\theta d\rho\right]. \tag{5.8}$$

The entry ρ_i denotes the coordinate of the beginning of the ith bonded region, where the reaction q_R is acting. The last form in (5.8) is obtained by partial integration of the double integral in (5.8). Then, by means of (5.8), the right-hand side of (5.4) (labeled $Q(\rho)$) can be written for each specific distribution of debonded and bonded zones. For example, let there be full cohesion in the interval $(0\le\rho\le\rho_1<1)$; separation in the region $(\rho_1\le\rho\le\rho_2<1)$; while, in the interval $(\rho_2\le\rho\le 1)$, there is again full compatibility. We then have (see equation (5.4)):

$$Q(\rho)=\frac{m_a}{\rho}\left\{-\int_0^{\rho}\rho q d\rho+\frac{\lambda}{2}\left[\xi\rho^2-\rho^2\int_0^{\rho}\theta d\rho+\int_0^{\rho}\rho^2\theta d\rho\right]+\omega\theta\right\};\ \left(0\le\rho\le\rho_1<1\right). \tag{5.9}$$

$$Q(\rho)=\frac{m_a}{\rho}\left\{-\int_0^{\rho}\rho q d\rho+\frac{\lambda}{2}\left[\xi\rho_1^2-\rho_1^2\int_0^{\rho_1}\theta d\rho+\int_0^{\rho_1}\rho^2\theta d\rho\right]+\omega\theta\right\};\ \rho_1\le\rho\le\rho_2<1. \tag{5.10}$$

$$Q(\rho)=\frac{m_a}{\rho}\left\{-\int_0^{\rho}\rho q d\rho+\frac{\lambda}{2}\left[\xi\rho_1^2-\rho_1^2\int_0^{\rho_1}\theta d\rho+\int_0^{\rho_1}\rho^2\theta d\rho+\xi\left(\rho^2-\rho_2^2\right)-\rho^2\int_0^{\rho}\theta d\rho\right.\right.$$

$$\left.\left.+\rho_2^2\int_0^{\rho_2}\theta d\rho+\int_{\rho_2}^{\rho}\rho^2\theta d\rho\right]+\omega\theta\right\};\ \left(\rho_2\le\rho\le 1\right). \tag{5.11}$$

The governing equations (5.4) appear now in an integro-differential form, instead of the usual differential form, and explicitly contain the additional unknown ξ (i.e., the center displacement $w(0)$), which is not present in the simpler case of no foundation. The presence of this additional unknown ξ is due to the fact that, in the general case, the average displacement of the plate is an additional degree of freedom in the formulation, and this freedom is here replaced by the apex displacement. To determine ξ, the governing equation demands a supplementary boundary condition in addition to (5.7). Such a condition can always be formulated taking into account the constraints on w determined by the edge supports. For example, if the plate's edge is immovable in the Z direction, the additional boundary condition is $w(1) = 0$, and ξ is then directly determined by (5.5). If no constraints are imposed on the motion of the plate's edge in the Z direction, the additional boundary condition takes the form:

$$Q(1) = P, \tag{5.12}$$

where P is the external shear force per unit length along the plate's edge. The other two external forces that might act together (edge force N and edge-bending moment M) are taken into account by the right-hand sides of the two conditions (5.7).

The numerical solution of this problem and the presentation of its results are based on the deformation map technique that we described for the classical case in Sections 4.4, 4.5. The presence of the foundation does not introduce any fundamental changes in this technique. However, it is now necessary to take into account the existence of the additional variable ξ, which necessitates the construction of maps for varying values of ξ.

An example is given in Figure 5.1. This case is determined for the following conditions: $\xi = 1$; $q = 0$; the plate is elastic and isotropic ($k = 1$) and rests without additional support ($\alpha_1 =$

0, $\beta_1 = 1$, $\alpha_2 = 1$, $\beta_2 = \nu_r = 0.3$; (see 5.7)) on the foundation, with $\lambda = 20$ and $æ = \infty$ (thus, the classical Winkler's case). The parameters γ_1 and γ_2 are the dimensionless intensities of the edge force N and edge-bending moment M, respectively (see (5.7)).

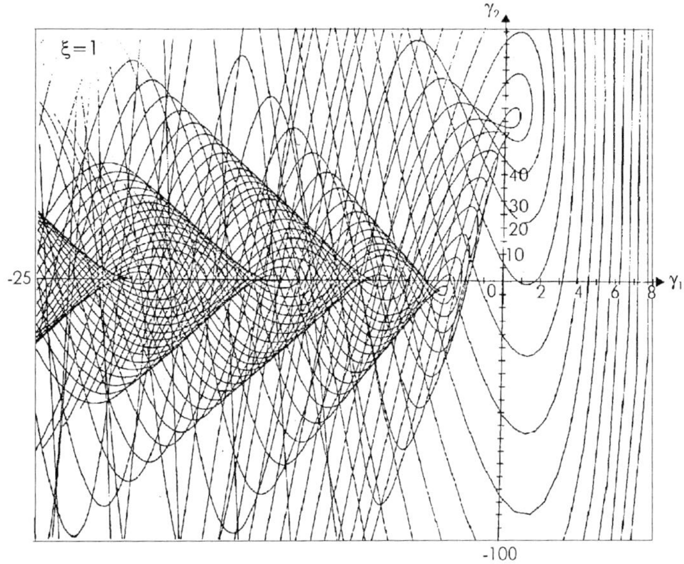

Figure 5.1 Deformation map for a circular plate on elastic foundation, γ_2 vs. γ_1; the case $\xi = 1$.

Some specific numerical results for the case that debonding does not occur are presented in Figures 5.2, 5.3, and 5.4. In Figure 5.2, a series of graphs, γ_2 versus $2Q(1)/m_a$, are given for $\gamma_1 = 0$ and different ξ. The intersections of these graphs with the γ_2-axis determine the behavior of the plate for "pure bending"; i.e., when only edge moments γ_2 are acting (see Fig. 5.3). The deformation of the plate induced by edge shear forces $Q(1)$ is described by the graphs given in Figure 5.4 for different values of $\gamma_2 \geq 0$, again for the case that $\gamma_1 = 0$ (γ_2 is considered positive when it causes compression in the upper layers of the plate). The graphs for $\gamma_2 < 0$ and $\gamma_2 > 0$ are symmetric with respect to the origin of the coordinate system, just as, for example, the graphs presented in Figure 5.2 for different values of ξ. It is interesting to note that, despite the possibility that more than one equilibrium state exists at a particular value of ξ (see Fig. 5.4), snapping phenomena do not occur, because $Q(1)$ is here the prescribed load and not the displacement ξ. The analysis of the case $\gamma_1 \neq 0$, can be obtained in the same way, but this will not be presented here.

In order to investigate the initiation and spread of a delamination in the way described, we shall consider an isotropic plate on an elastic foundation that is loaded by edge-bending moments γ_2 only. Other external forces are not considered (except for the reaction q_R in the regions where the connection plate-foundation is not destroyed). In this case, we deal with the following parameter settings: $k = 1$; $q = 0$; $\alpha_1 = 0$; $\beta_1 = 1$; $\gamma_1 = 0$; $\alpha_2 = 1$; $\beta_2 = \nu_r = \nu_\phi = 0.3$; $\gamma_2 = M_r(1)$; $P = 0$ (see (5.12)). This means that the plate is supported by the foundation itself; there are no other constraints. If one takes, for example, the case $\lambda = 20$ and $æ = 40$, according to (5.2), debonding occurs in the regions where $w(\rho) < -2$. (Note that the displacement component $w(\rho)$ is considered positive when it is pointing downwards.)

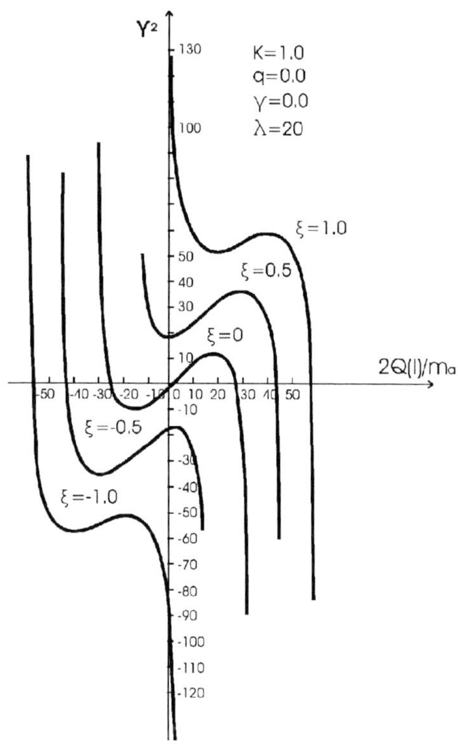

Figure 5.2 γ_2 versus $2Q(1)/m_a$ for $\gamma_1 = 0$ and various values of ξ.

Using the deformation maps that are constructed for various values of ξ, auxiliary graphs can be drawn that display γ_2 and $Q(1)$ versus $w(1)$ for different values of ξ. Two examples of such graphs are given in Figures 5.5. and 5.6. In the case $\xi = 1$, the graph $Q(1)$ versus $w(1)$ does not intersect the horizontal axis (see Fig. 5.6). This means that, in this case, there is no solution, because we cannot satisfy the additional boundary condition (5.12), $Q(1) = 0$.

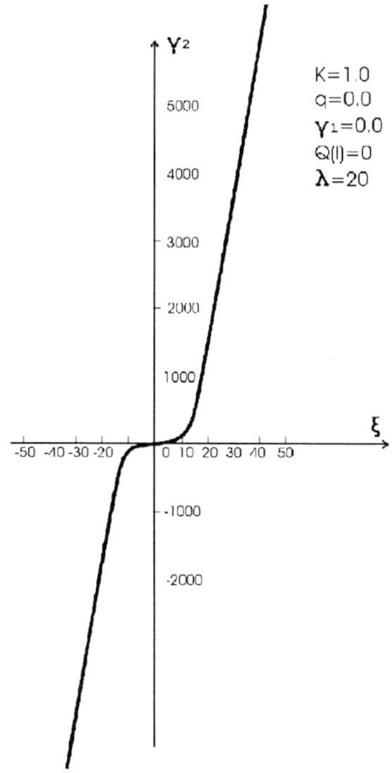

Figure 5.3 The influence of the edge moment, γ_2 vs. ξ.

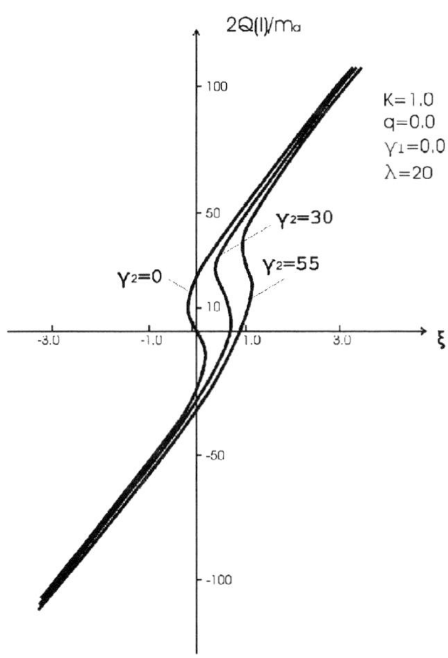

Figure 5.4 The influence of the edge shear force, $2Q(1)/m_a$ vs. ξ.

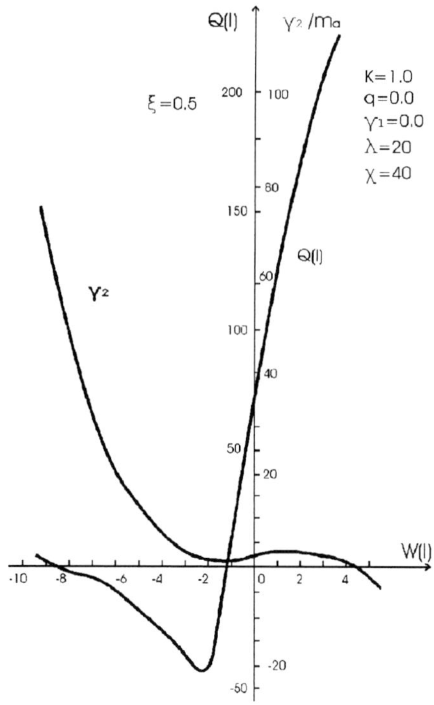

Figure 5.5 A case where debonding can occur: γ_2, Q(1) versus w(1); $\xi = 0.5$.

In contrast, when $\xi = 0.5$, the graph $Q(1)$ versus $w(1)$ intersects the axis $Q(1) = 0$ twice (see Fig. 5.5), and we see that two solutions exist: one for $w(1) > -2$, and another for $w(1) < -2$. Consequently, for $\xi = 0.5$, the plate and foundation must have separated in a region in the vicinity of the edge; while, in the previous case ($\xi = 1$), the condition for debonding is not reached in the domain of the plate, and there is thus no separation at all. For the specific case considered here, $\xi = 0.5$, only one region where debonding occurs can exist if $\xi > 0$, and this is located in the neighborhood of the plate's edge.

The results are summarized by the graph given in Figure 5.7, which is plotted in the following way. For each value of ξ, the corresponding map of the type shown in Figure 5.1 is constructed for the given values of k, q, λ, and $æ$. Sections of $\gamma_1 = 0$ of these maps make it possible (after the solution of the corresponding initial value problems) to compute the curves $Q(1)$ and γ_2/m_a versus $w(1)$, examples of which are presented in Figure 5.6. The intersections of the curve $Q(w)$ with the axis $w(1)$ determine the corresponding solutions. If such intersections are absent, a solution of the problem for the given ξ does not exist. If there are intersections, these points determine the graph $\gamma_2(\xi)/m_a$.

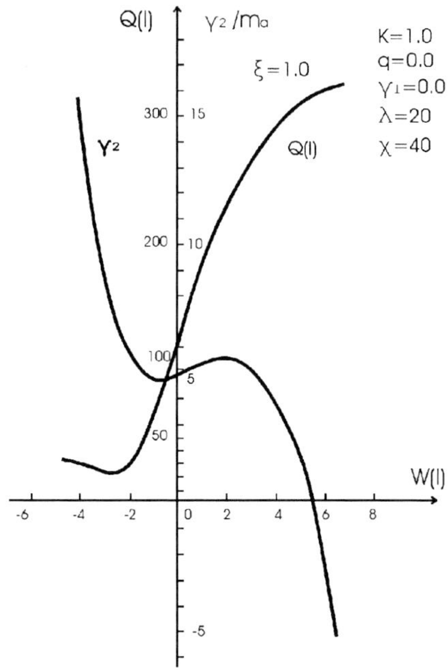

Figure 5.6 A case where debonding can occur: γ_2, Q(1) versus w(1); $\xi = 1$.

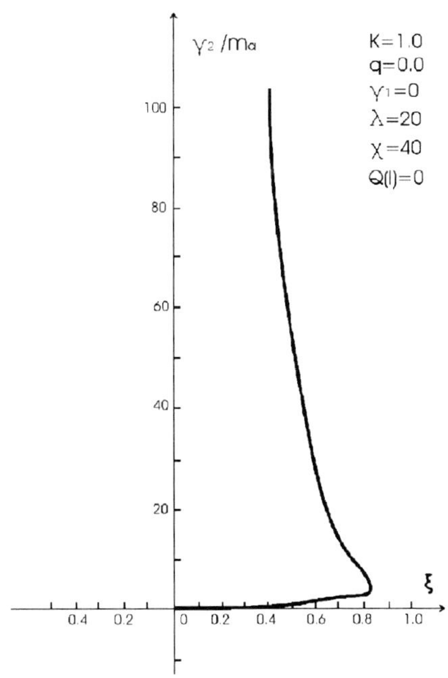

Figure 5.7 The deformation plot of the plate suffering debonding with the foundation.

The graphs in Figure 5.7, actually consist of two distinct parts. The first, where γ_2 is small, is connected with the origin of the coordinate system and the turning point of the curve where the derivative $d\xi/d\gamma_2 = 0$. For all the states of the plate that correspond to this part of the curve, there is a complete bond between the plate and the foundation, since $w(\rho) > -2$ everywhere. The second part of the curve drawn in Figure 5.7 starts from the turning point (at which $\gamma_2/m =$ 4.5). For $\gamma_2/m > 4.5$, each equilibrium state corresponds to the deformation state of a partly debonded plate, where the separation between plate and foundation is found in a circular, banded area that starts at the edge. A number of these deformation states are presented in Figure 5.8.

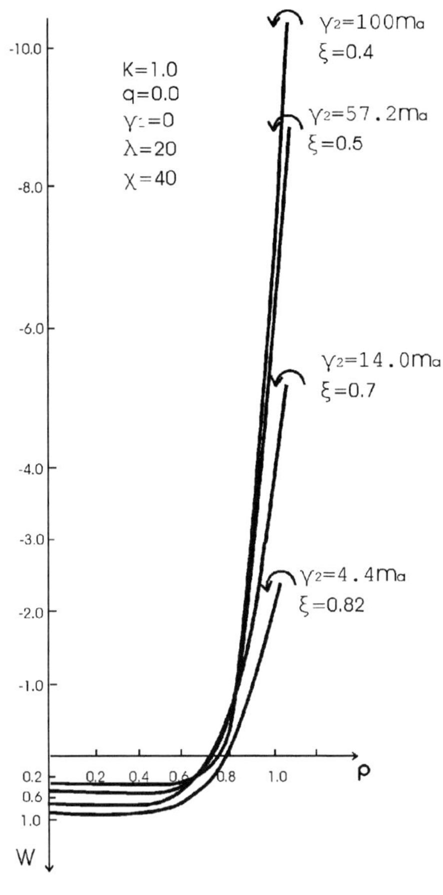

Figure 5.8 Some deformation states of the plate with debonded edge layer.

The dependence of the length ℓ of the debonded region versus the moment γ_2 has an asymptotic character (see Fig. 5.9). ℓ is related to the plate's radius; consequently, it is impossible to tear the plate completely from the foundation by means of the edge-bending moment only. This effect may be due to the restrictions posed by the theory of shallow shells. We expect that the more general theory of nonshallow shells will lead to another conclusion.

The graph presented in Figure 5.9 leads to an important observation. The process of the spread of the decohesion happens in terms of jumps (thus in a discontinuous fashion). At $\gamma_2 = 0$, the length of the crack $\ell = 0$, and the value of ℓ does not change until $\gamma_2 = \gamma_A$ (see Fig. 5.9). At this moment, the crack front proceeds, suddenly reaching distance $\ell = \ell_A$. After this, the length ℓ is a continuous function of γ_2. Thus, the progress of the front of the decohesion sometimes corresponds to a transient process that compares to a snap-through type of behavior.

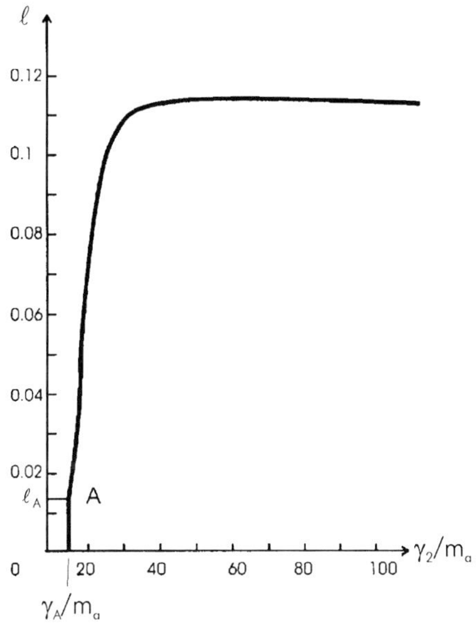

Figure 5.9 Depth of the debonded zone as a function of the edge moment.

We also present results for another case: $\lambda = 1$ and $æ = 1$. Figure 5.10 shows the graph γ_2/m versus ξ in two parts; i.e., one part for positive and one part for negative values of γ_2/m (the scale of the ξ axis of the two graphs in this figure differs by a factor of 10). For about $\gamma_2/m > 0.82$, separation between foundation and plate starts at the plate's edge. The graph of the length ℓ of the crack front versus ξ is given by the curve 2. Decohesion in the central part of the plate becomes possible when $\gamma_2 < 0$ at $\xi < -1$ (since $\lambda = æ = 1$ (see 5.2)). The corresponding graphs of γ_2/m and ℓ versus ξ for $\gamma_2 < 0$ and $\xi < 0$ are denoted by 3 and 4.

change in results, which also occurs in a contact problem with bars, as was already mentioned [Feodosjev, 1953, problem 45].)

The graphs q_r versus the angle $\Phi_0 = b/R$ (see Fig. 5.12), which changes in the interval ($0 \leq \Phi_0 \leq \alpha$), are presented in Figure 5.13 for three different values of $\alpha = 2^o, 5^o, 10^o$. These results were obtained by Updike and Kalnins [1972a] for a shell with $R/h = 30$ and $\nu = 0.5$, using a theory that takes shear deformation into account. It can be seen that, when $\alpha = 10^o$, q_r is very close to zero in the main part of the circular contact zone and increases sharply near the edge. This is a typical boundary layer effect, and it is close to the "classical" distribution shown in Figure 5.12. The graphs in Figure 5.13 are obtained by Updike and Kalnins [*ibid.*] on the basis of the extended equations of Reissner [1969] although, in this work, the flattened part of the shell is considered as a shallow shell.

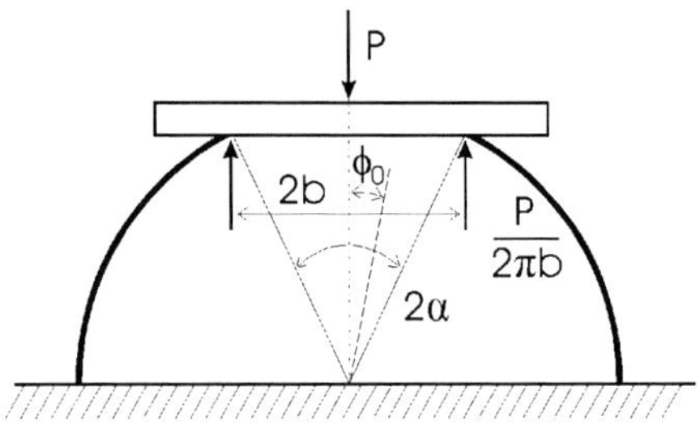

Figure 5.12 Conventional basic model.

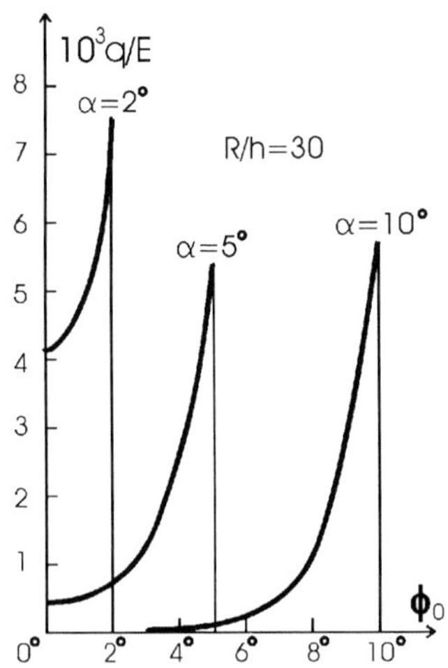

Figure 5.13 Results obtained by Updike and Kalnins [1972a].

Remaining in the frame of the usual theory (no shear strains in transversal sections of the shell), and following Updike and Kalnins [1970, 1972a], we can see that, at least qualitatively, the real distribution of q_r inside the contact zone does not affect the behavior and bifurcation phenomena of the shell. At some value of P, a buckling occurs when the initially flat "region" of radius b of the shell goes over into the axisymmetric form, qualitatively drawn in Figure 5.14. If P continues to increase, a new (secondary) bifurcation can be expected, when there will be a transfer from the axisymmetric state (Fig. 5.14) to a nonsymmetric state with lobes in the circumferential direction. However, the cap is always loaded axisymmetrically, because, during the entire process of loading, the rigid plate moves parallel to itself, and the contact remains uniform in circumferential direction.

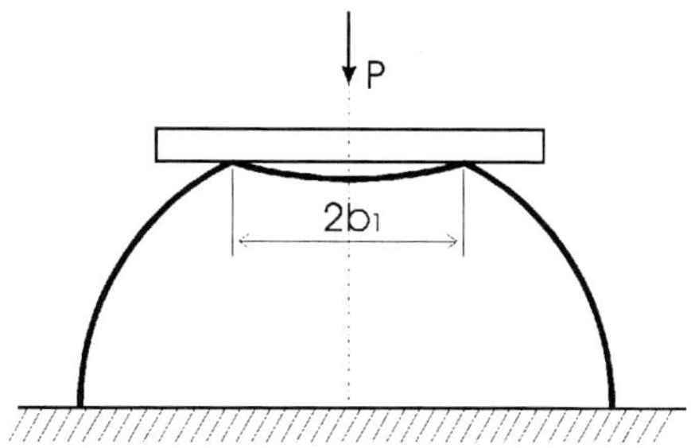

Figure 5.14 The sphere after an axisymmetric buckling mode has appeared.

Figure 5.15 shows the final deformation curve obtained for R/h = 100 and ν = 0.3, [Updike & Kalnins, 1970, 1972b]. The nondimensional parameter Δ/h represents the vertical displacement Δ of the loaded plate, and B denotes the first bifurcation point (a transfer from the flat state to an axisymmetric buckling mode in the contact zone). Two secondary bifurcation points with n = 3 and 4 are also shown along the loading curve. (The integer n denotes the number of circumferential waves of the buckling mode.) Updike and Kalnins [1972b] also presented experimental results which, in a qualitative sense, agreed very well with the numerical results. But, in spite of this, we are not completely sure that these findings are correct. In all the known cases of nonsymmetric bifurcation buckling of caps, the edges of the buckling zone remain circular during the development of the buckling state. The nonsymmetric buckling modes are located in a belt between the pole of the cap and its contour (see data in Chapters 2, 3, and 4). Because of this, it is natural that the contour that separates the contact zone and the free part of the semisphere will be circular, as in the case of caps.

In the case discussed by Updike and Kalnins, however, when the second bifurcation occurs, the boundary of the contact zone is no longer a plane circle but a spatial curve, which is the composition of a circle and a sine wave with an integral number of waves.

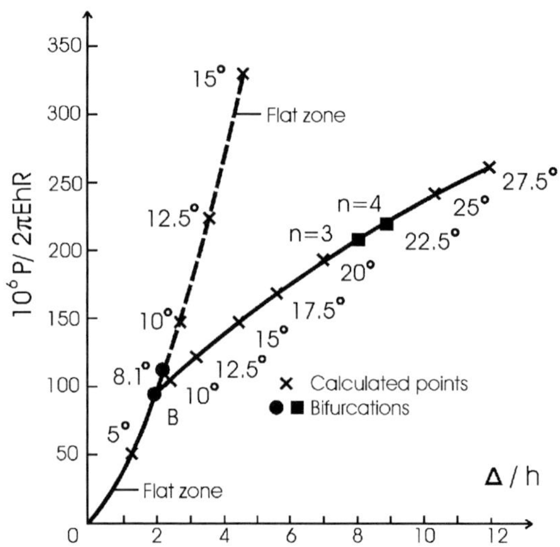

Figure 5.15 The response curve obtained by Updike & Kalnins [1970, 1972a].

Updike and Kalnins also claim [1972b] that the secondary bifurcation states are stable, because, there, $dP/d\Delta > 0$. But, as will be shown in the Chapter 6, $dP/d\Delta$ being positive can only be seen as a necessary condition for stability of the equilibrium state under consideration so that the latter issue is also still in doubt.

5.3 Bifurcation phenomena in a truncated spherical cap in tension.

The shell configuration is shown in Figure 5.16a. This truncated, hemispherical shell is (immovably) clamped at the equatorial edge. At the other edge, a uniformly distributed edge load is applied. We shall briefly look into this problem, because it is an interesting case, where bifurcation buckling occurs while the structure is subjected to pulling forces. The problem is investigated in detail by Wu and Cheng [1970], who based their investigations on Sanders's equations for shells of revolution [Sanders, 1963]. These equations are used to obtain the fundamental axisymmetric prebuckling equilibrium path, and to develop the corresponding linearized differential equations and boundary conditions that determine the bifurcation points along this path.

The principal results that were obtained by Wu and Cheng [1970] are presented in condensed form in Figure 5.17 in the case of an isotropic homogeneous shell wall material with $\nu = 0.3$. They are given for three different geometries of the problem in question. The basic geometric parameter here is $\kappa = a/h$, where a is the radius of the spherical surface of the shell and h is its thickness. P denotes the resultant of the load that is applied along the upper edge of the truncated shell. The plots represent the bifurcation buckling loads in dependence of the number of circumferential waves of the corresponding buckling modes.

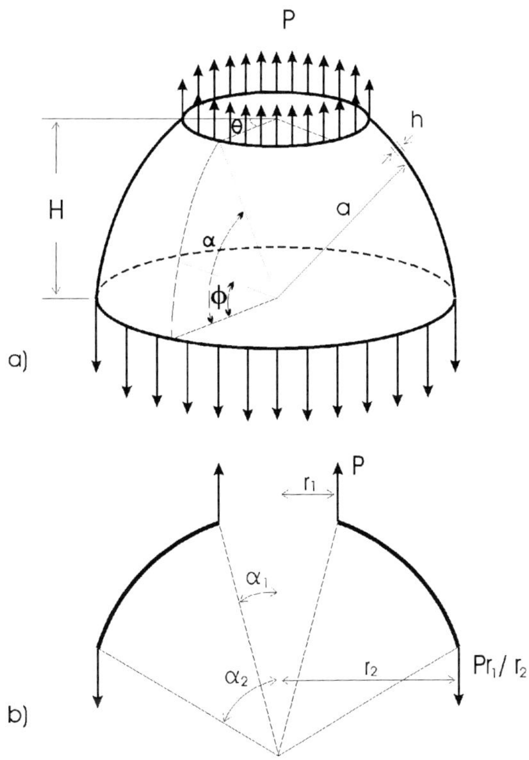

Figure 5.16 Truncated hemispherical shell loaded in tension.

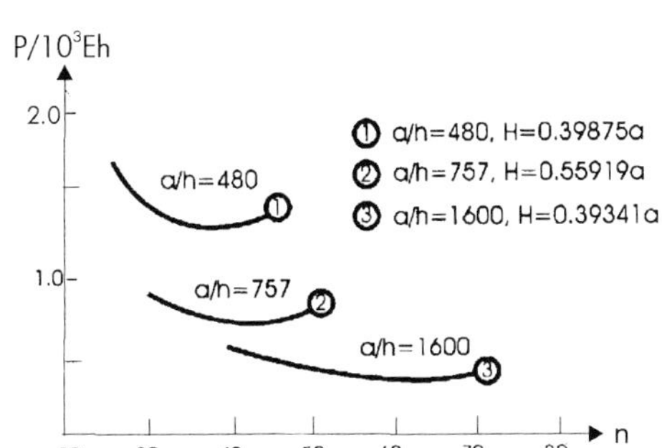

Figure 5.17 Buckling loads hemispherical shell in tension.

This problem was investigated earlier by Yao [1963], who introduced a number of simplifications; i.e., the prebuckling states were described by membrane theory, while the bifurcation equations were linearized along these states using Vlasov's formulation. Although the results of both of these investigations agree with each other to a large degree, we believe that in other cases this correspondence will cease to exist. Indeed, the prebuckling state that is approximated by a membrane state of stress leads to a (too) conservative estimate of the actual value of the bifurcation load [see Bushnell, 1967a, 1989]. This follows, because the hoop stress σ_ϕ, which is responsible for the occurrence of the nonsymmetric buckled states, is negative everywhere in the interior of the sphere in the (linear) membrane state, while for the actual nonlinear prebuckling state, σ_ϕ is a function of variable sign.

Bushnell [1967a, 1989] considered the more complicated case (shown in Fig. 5.16b), of a truncated spherical cap, the geometry of which is determined by two meridional angles, α_1 and α_2 (the previous case of a truncated hemispherical shell corresponds to $\alpha_2 = \pi/2$). Bushnell's analysis (which was based on the simplified equations of E. Reissner) shows that the difference between the two analyses — the simplified membrane prebuckling state versus the full nonlinear prebuckling state — is very small for the truncated hemisphere, but increases when $\alpha_2 < \pi/2$. For example, when $\alpha_2 = \pi/2$ and $\alpha_1 = 1.16$, the difference is about 1%; but when $\alpha_2 = 0.6$, $\alpha_1 = 0.1$, and $a/h = 455$ or 1600, this difference is more than 100%. The conclusions that we can draw from these investigations can now be summarized as follows.

Increasing α_1 leads to an increase of the buckling load and an increase in the number of circumferential waves of the corresponding buckling mode. Thus, when the height of the truncated shell decreases, its rigidity increases, so that the bifurcation occurs at a higher load.

When α_1 tends to zero, the problem and its solution can be compared to that of a hemisphere with a movably clamped edge subject to a concentrated central force. When α_1 is very small and α_2 is close to $\pi/2$, the edge effect is minimal. Many other aspects of this problem and additional numerical results can be found in Bushnell [1989].

5.4 Deformation and stability of deep spherical caps.

The results described above are based on shallow shell theory. In fact, a large majority of the investigations that deal with shell stability problems use this theory. The reason for this preference can easily be understood. It concerns the relative simplicity of the basic equations of the theory of shallow shells, due to the basic assumption that the metrics of the undeformed midsurface of the shell can be replaced by the metrics of its base plane. This approximation is, of course, not always allowed, and, in that case, equations that are more generally applicable should be used.

There are several formulations available that surpass the restrictions posed by the shallow shell equations to a degree that differs from case to case (see Section 1.1). This is due to the tendency to introduce simplifications in the formulation with the aim to facilitate the solution of the final governing equations.

The present state of affairs in the area of computer-installed numerical solution techniques makes it possible to solve the problems of the theory of nonlinear shells without the help of simplifying assumptions. In other words, numerical solution methods do not require the equations to be simplified, and this means that we can use the most exacting version of the theory. We shall start therefore with a survey of solutions that are obtained from the general equations of axisymmetrically deformed spherical caps [see Shilkrut, 1969b, 1974; Gerlaku, Morar & Shilkrut, 1970] (see Section 1.1).

a) ***Caps loaded by external bending moments (pure bending) uniformly distributed along the edge.***

In the general case, the behavior of the spherical cap is governed by two parameters instead of one, as in the shallow shell case. The non-dimensional apex-rise $\overline{\xi}$ is now defined by the exact relation:

$$\overline{\xi} = \frac{W^*}{h} = \frac{R}{h}(1 - \cos\theta_0), \tag{5.13}$$

where h is the shell thickness, R is the radius of the cap's midsurface, θ_0 is half of the central angle of the undeformed cap, and W^* is physical value of the apex rise. When θ_0 is small (i.e., $\theta_0 << 1$), we can write:

$$\overline{\xi} \approx \frac{R}{h}\frac{\theta_0^2}{2}. \tag{5.14}$$

This expression is used in the shallow shell formulation.

The symmetry and boundary conditions for pure bending and an immovable, simply supported edge are (see Section 1.1):

$$\Phi(0) = u(0) = 0 \tag{5.15}$$

$$u(\theta_0) = 0; \quad M_\xi(\theta_0) = M. \tag{5.16}$$

For movable supports, the conditions (5.16) must be replaced by:

$$N_\xi(\theta_0) = N, \text{ or } H(\theta_0) = H \tag{5.17}$$

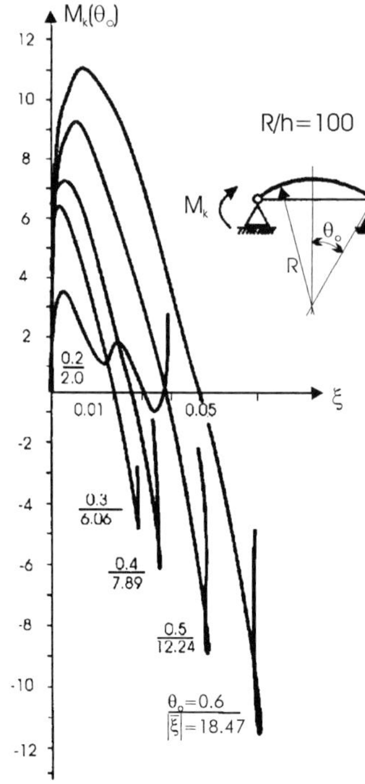

Figure 5.21 Results for the deep shell, immobile hinged, R/h = 100.

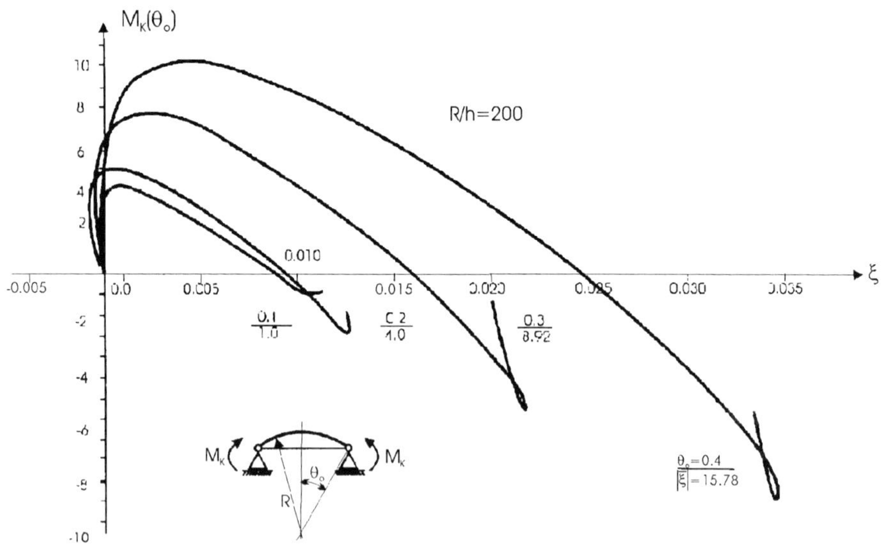

Figure 5.22 Results for the deep shell, immobile hinged, R/h = 200.

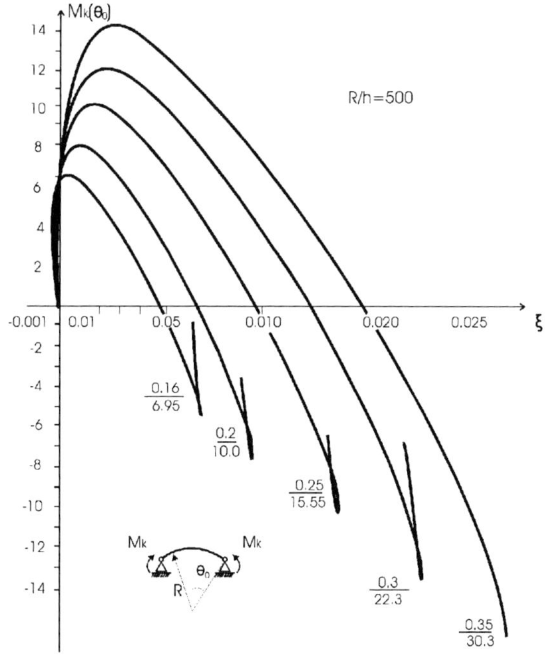

Figure 5.23 Results for the deep shell, mobile hinged, $R/h = 500$.

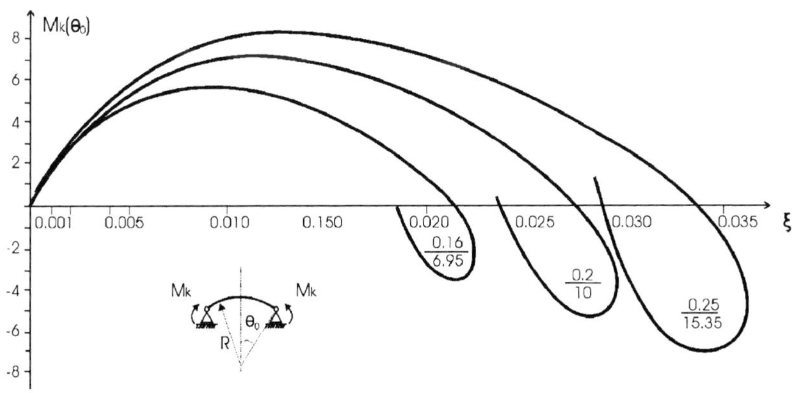

Figure 5.24 Solution curves, mobile hinged, $R/h = 500$.

The dependence of the limit load M_k^+ versus θ_0 for different values of R/h is shown in Figure 5.25. The graphs in this figure all emerge from a fixed value, $M_k^+ = 1.3$, which turns out to be the minimum value (for the considered class of caps) at which limit points occur for the first time. Like the minimum value $M_k^+ = 1.3$, it appears that the magnitude of the apex rise $\overline{\xi}$, at which limit points occur for the first time, is a constant given by $\overline{\xi} \cong 0.7$. The values of θ_0 that correspond to this special point are, as can be expected, dependent on the radius to thickness ratio R/h (see Fig. 5.25). Indeed, when $R/h = 20$, then $\overline{\theta}_0 = 0.26$, for example; if $R/h = 100$, then $\overline{\theta}_0 = 0.12$; if $R/h = 1000$, $\overline{\theta}_0 = 0.037$; and so on. In all these cases, $\overline{\xi} = \overline{\xi}_0 \cong 0.7$. The same characteristics occur with an immovable hinged shallow cap (see Section 3.7).

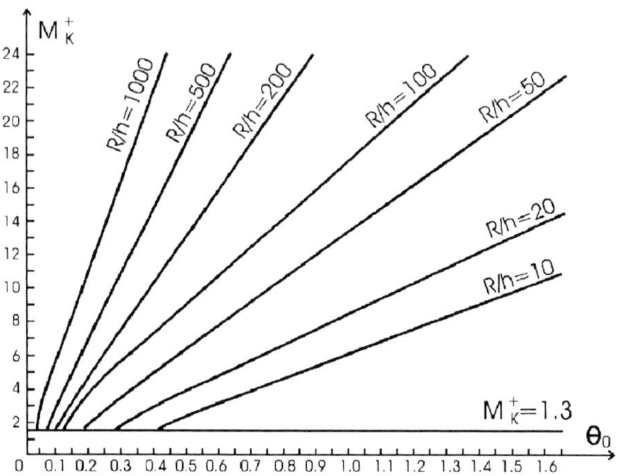

Figure 5.25 Limit loads for the deep shell vs. θ_0 for various R/h values.

At $\overline{\xi} = \overline{\xi}_0 \cong 0.7$, the corresponding value of M_k^+ and M_k^- is given by $M_k^+ = M_k^- = 1.3$, in agreement with the symmetry property of nonshallow shells delineated in Section 2.9, where the relation (5.18) between upper M^+ and lower M^- limit loads was given by

$$M^+ + M^- = 2(1 + \nu) \tag{5.18}$$

It follows directly from (5.18) that the smallest value of $M^+ = M^- = 1.3$ belongs to a cap with $\overline{\xi} = \overline{\xi}_0 = 0.7$, which corresponds to the center of symmetry C of the corresponding deformation curve (see Section 2.9). The graph $M_k^- = M_k^-(\theta_0)$ (for fixed R/h) is the mirror image of the graph $M_k^+ = M_k^+(\theta_0)$ with respect to the ordinate $M_k = 1.3$ (see Fig. 5.25). This effect is a consequence of relation (5.18). Figure 5.25 shows that the curves $M_k^+ = M_k^+(\theta_0)$, and thus also $M_k^- = M_k^-(\theta_0)$ (R/h fixed), are almost straight except at the region close to $M_k^+ = 1.3$.

The family of curves M_k^+ (R/h) for θ_0 = const. (immovable hinges) is presented in Figure 5.26. Figure 5.27 presents the equilibrium curve for the case $R/h = 500$, $\theta_0 = 0.3$, and immovable edges (only half of the total response curve — up to the center of its symmetry C

— is shown). The first part of this curve was already displayed in Figure 5.23. It is noted that, from the qualitative point of view, the behavior is similar to that of a shallow shell (see, e.g., Figs. 2.7 and 2.8).

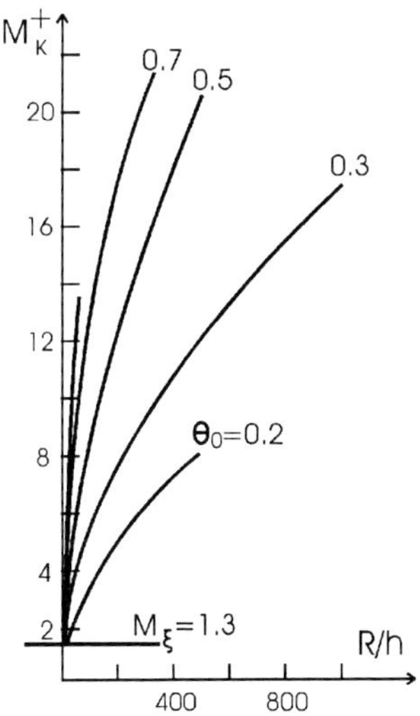

Figure 5.26 Limit loads vs. R/h for $\theta_0 = 0.2$.

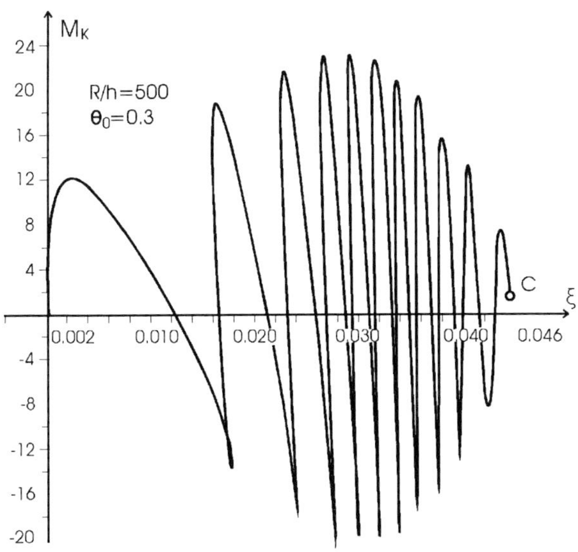

Figure 5.27 The response curve for the deep shell, the particular case $R/h = 500$, $\theta_0 = 0.3$.

As a matter of interest, we also show some results for plates with movable and immovable supports in Figures 5.28 and 5.29, respectively.

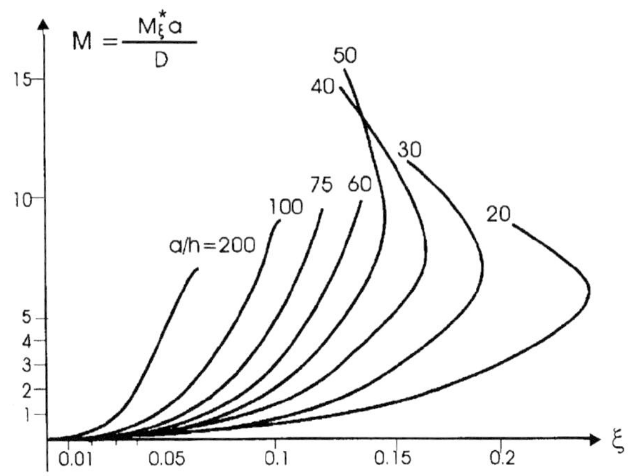

Figure 5.28 Results for plates with movable supports.

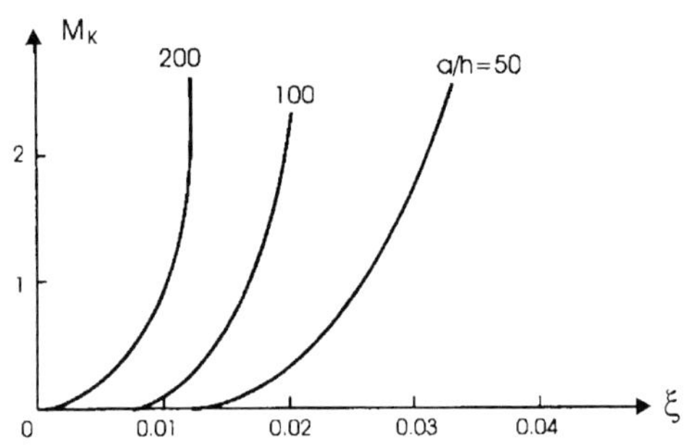

Figure 5.29 Results for plates with immovable supports.

As can be seen, there are essential differences between the behavior of circular plates described by shallow shell theory and the behavior described by general nonlinear shell theory. In the latter case, the deformation curves, contrary to the former, depend on the radius-to-thickness ratio a/h, where a is the initial radius of the plate. For example, for fixed M_k, the displacement of the plate's center decreases when a/h increases.

It is necessary to emphasize here that the circular edge of the plate can change its magnitude considerably during deformation, because the theory of nonshallow shells does not restrict the magnitudes of the displacements or rotations. This means that the nonlinear behavior of the plate can be compared to that of the Euler beam in Euler's famous elastica problem [see e.g., Timoshenko & Gere, 1963]. Observe, for example, the turn in the deformation curve in the case of movable edges shown in Figure 5.28.

Thus, if the deformation curve $M_k = M_k(\xi)$ of the plate is monotonic according to the theory of shallow shells, it ceases to be monotonic when the general theory is applied. This effect is not very well known. It is probable that it is a consequence of the circumstance that, for large values of M_k, the deformed shape of the plate is of such a type that some circles lying in the midsurface in the undeformed state are deformed to circles with a larger radius. The deformed state, in such a case, takes a form that is somewhat similar to a "chef's hat," which is wider near the top than at its base. Note that this is a shape that can never be described by shallow shell theory.

Comparison of numerical results obtained by the two different theories for plates shows that, in order to reach the same central deflection ξ, "shallow" plates need a 20%-40% higher edge moment than "nonshallow" plates. This means that "shallow" plates behave as if they are stiffer than "nonshallow" plates.

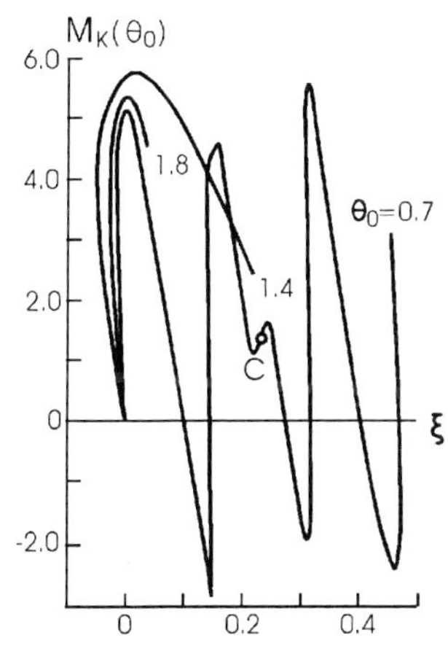

Figure 5.30 Some equilibrium curves for shells with different depths.

Returning to the cap problem, it can be said that, generally speaking, it is difficult to characterize the difference between the solutions produced by both theories. This can, for example, be seen in Figure 5.30, where three equilibrium curves ($M_k(\theta_0)$ versus ξ are shown for a cap with an apex rise of $\bar{\xi} = 4$ for three different magnitudes of θ_0: $\theta_0 = 1.8$, 1.4, and 0.7 (the cap is immovably hinged). The numerical solutions show that the curves seem to converge to a result that we expect for the shallow shell at about $\theta_0 < 1/3$. As can be concluded from the above presented matter, the problem of deformation of shells described by the more exacting general theory, which covers deep (nonshallow) shells, requires further investigation.

sis was based on Sander's [1963] equations. It can categorically be stated that, for shallow spherical caps, these differences in load application are not very significant.

The problem of the deformation of spherical caps under hydrostatic pressure was also considered by Gotsuljak, Guliajev and Melnitchenko [1974], Gospodarikov and Terentjev [1977], and Kobrits and Terentjev [1977] with results that are in good agreement with the data in Table 5.1. In Figure 5.31, we present some of the force displacement graphs obtained by Gotsuljak, Guliajev and Melnitchenko [1974], which are valid for caps with hinged supports that are movable in the meridional direction. These results are obtained for the following parameter values: $R/h = 250$; $E = 2.1\times10^6$ *kgf*/cm^2; $\nu = 0.3$. The pressure is plotted along the vertical axis in terms of q^*, the true pressure in *kgf*/cm^2. For $\theta_0 = 10^{\mathrm{o}} \approx 0.17$ rad. $\left(\left|\overline{\xi}\right| \approx 3.8;\ \lambda = 2.69\right)$, the response curve practically coincides with the one obtained by shallow shell theory [see Valishvili, 1976; Shilkrut, Shevandronov, Morar & Maksimov, 1969], and also Figure 2.10 for the case $\overline{\xi} = -4$ and $M = 0$.

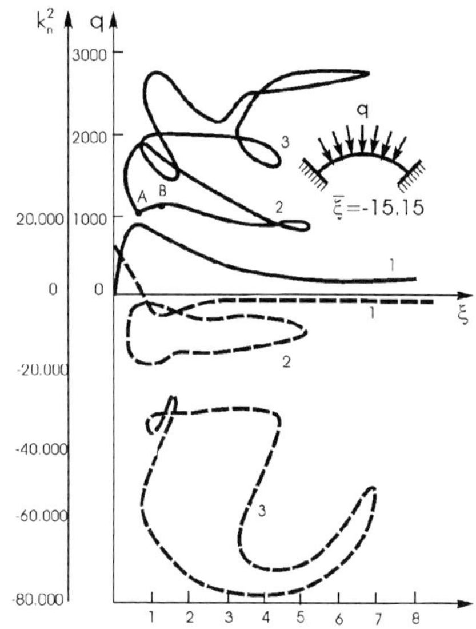

Figure 5.31 Results obtained by Gotsuljak, Guliajev & Melnitchenko [1974].

For $\theta_0 = 20^{\mathrm{o}} \approx 0.35$ rad. $(\lambda = 5.34)$, as can be seen, the lower limit load is still negative, and the difference between the upper and lower limit loads is markedly larger than in the previous case. But for $\theta_0 = 30^{\mathrm{o}} \approx 0.52$ rad. $(\lambda \approx 15)$, this difference begins to decrease, and when $\theta_0 = 40^{\mathrm{o}} \approx 0.7$ rad. $(\lambda \approx 20)$, the limit points disappear. Thus, in the last case, snapping in the subspace of the axisymmetric class of deformations is not possible. This is a phenomenon that is also described by the theory of shallow shells.

Investigations of the same sort were also carried out by Vorovitch and Minakova [1968a,b], who used governing equations that can be placed between the shallow shell equations and the most general equations. The results of these investigations present further confirmation of the validity of the shallow shell equations when used in the range for which they are derived.

It should also be mentioned that the simplified equations of E. Reissner (see Section 1.1) also produce results that compare to those produced by the shallow shell equations.

c) Axisymmetric deformations of deep caps under concentrated force at the apex.

Melnitchenko [1974], considered the problem of the axisymmetric deformation of a spherical cap subjected to an axial force applied at the apex for the particular case: $R/h = 250$; $E = 2.1\times10^6$ *kgf*/cm^2; $\nu = 0.3$. The results are summarized in Figures 5.32 and 5.33 for fully clamped edges, and also for hinged edges, movable in the meridional direction. The load plotted along the vertical axis is the true load P^* given in *kgf.*

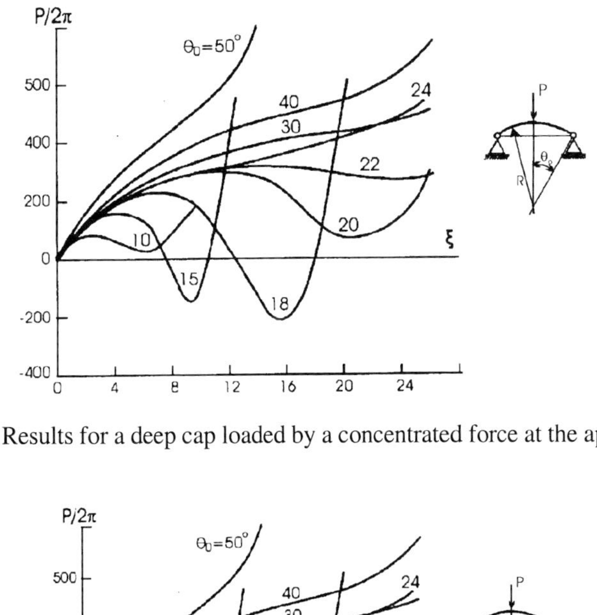

Figure 5.32 Results for a deep cap loaded by a concentrated force at the apex, detail 1.

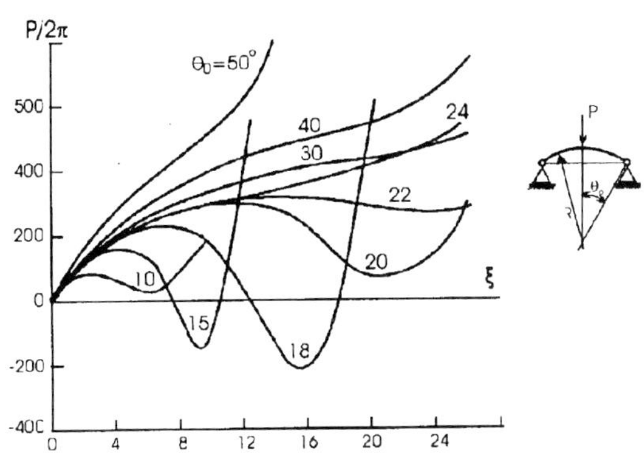

Figure 5.33 Results for a deep cap loaded by a concentrated force at the apex, detail 2.

An analysis of these data show that the solutions exhibit the same qualitative features as those of the previous problem (see Fig. 5.31). Again there is a value of θ_0, beyond which snap-through can no longer occur. This rather significant property has not been observed for the case of "pure bending" (see part a) of the present section). It deserves a detailed investigation, because it shows that the removal of the restrictions posed by the shallow shell equations reveals a behavior that has not been detected before.

To the best of the writer's knowledge, no results are available that deal with bifurcation buckling of shells of revolution based on the exact thermo-elastic equations. The stability of the axisymmetric equilibrium states of deep caps is, however, investigated and discussed in Section 6.6.

5.5 Axisymmetric deformation of deep hemispheric shells.

a) Uniformly distributed external loading.

Apart from its practical significance, the problem in question is important from the theoretical point of view. It is also a good starting point for the analysis of the stability and deformation of complete spherical shells under external pressure.

We shall consider this problem for the case of external pressure in combination with two different boundary conditions [see Bauer, Reiss & Keller, 1970, 1973]. The boundary conditions are: (i) clamped edges with mobility in the base plane of the cap; and, (ii) fully clamped edges. The formulation of these conditions along the edge is given by:

(i) partially clamped: $$\beta\big|_c = H\big|_c = w\big|_c = 0\,; \tag{5.19}$$

(ii) fully clamped: $$\beta\big|_c = w\big|_c = u\big|_c = 0\,. \tag{5.20}$$

β, H, w, and u are the rotation of the shell's normal to the midsurface, the stress resultant in the horizontal direction, and the displacement components of the midsurface in vertical and horizontal directions, respectively.

These two problems are essentially different in character, despite the fact that, at first sight, it seems that they should be nearly the same. The first problem has a closed form solution of the form:

$$N_r = N_\theta = \frac{qh}{2}; \quad w = (1-\nu)\frac{qR^2}{2Eh}, \tag{5.21}$$

where N_r, N_θ are the membrane forces in meridional and circumferential directions, respectively, while w is the radial deflection. This prebuckling solution corresponds to a uniform contraction of the shell. It is the same solution that holds for a complete spherical shell, because the latter can be reduced to the present problem if symmetry is taken into account around one of the great circles (see conditions (5.19)).

The second problem does not permit a pure axisymmetric membrane solution because of the restriction $u|_c = 0$. This basic difference in the axisymmetric prebuckling states leads to two different kinds of stability loss. In the first case, loss of stability happens at a bifurcation point (in compression), while, in the second, loss of stability always occurs at a limit point.

The first case was investigated by Bauer, Reiss and Keller [1970]. Taking the membrane state as the fundamental (prebuckling) state (see (5.21)), the corresponding linear boundary value problem for bifurcation buckling can be formulated. The eigenvalues of this problem yield a spectrum of bifurcation pressures with corresponding (nonsymmetric) modes. This spectrum is a part of that for the complete spherical shell. The spectrum of buckling loads for the complete sphere was obtained by several researchers [e.g., Zoelly, 1915; Leibenzon, 1917; Schwerin, 1922; van der Neut, 1932; Flügge, 1962] and is given by:

$$P_n(s) = \frac{s}{1-\nu^2}\left[n(n+1)-1-\nu\right] + \left[n(n+1)-1-\nu\right]^{-1}; \quad (n = 2,4,6,\ldots), \tag{5.22}$$

where

$$P = \frac{q^* R}{2Eh}; \qquad s = \frac{1}{12}\left(\frac{h}{R}\right)^2 \tag{5.23}$$

[see Bauer, Reiss & Keller, 1973]; n is the wave number of the corresponding buckling mode in the meridional direction (and not in the circumferential direction) because the modes are axisymmetric. That n is even follows from the symmetry conditions (5.16). For the complete shell, the spectrum (5.22) must be considered for odd, as well as even, wave numbers n.

The fourth and fifth column of Table 5.2 presents the first bifurcation load $\underline{P}(s)$ and corresponding wave number n for different magnitudes of the geometric parameter s. The meaning of the other entries, P_U and P_L, will be discussed later.

From each bifurcation point determined by P_n, a branch of equilibrium states emerges with a wave pattern characterized by n. Figure 5.34 shows a series of such branches computed for $s = 0.001$. In this diagram, the load P is plotted along the vertical axis, while the average buckling displacement A, defined by

$$A = \left(\frac{1}{2}\int_0^{\pi} w_b^2 \, d\theta\right)^{1/2}, \tag{5.24}$$

is plotted in the horizontal direction. In this expression, w_b, the buckling mode, is measured with respect to the prebuckling state.

Table 5.2

$\nu = 0.32$

s	$P_U \times 10^2$	$P_L \times 10^2$	$\underline{P} \times 10^2$	n	$P_U/\underline{P}$
1×10^{-3}	6.57144	1.40161	7.061597	6	0.931
5×10^{-4}	3.91113	--	4.759904	6	0.822
4×10^{-4}	3.52130	0.688578	4.299565	6	0.819
2×10^{-4}	2.96804	0.410734	3.003954	8	0.988
1×10^{-4}	1.99919	--	2.138047	8	0.935
4×10^{-5}	1.32751	0.128967	1.338653	12	0.992
2×10^{-5}	0.881715	--	0.945602	14	0.932
1×10^{-5}	0.669125	0.047213	0.671713	16	0.996
7.5×10^{-6}	0.550495	--	0.578724	18	0.951
1×10^{-6}	0.206175	--	0.211214	30	0.976

In this case, the lowest bifurcation load is P_6 (see Fig. 5.34), which is the origin of branch $B_{6,1}$. From P_6, another branch, $B_{6,2}$ originates, which, forming a loop, turns to point P_4, from which other branches emerge, and so on. (It is necessary to note that these branches correspond to the numerical solution of the governing equations for deep shells [Bauer, Reiss & Keller, 1970] — equations that are similar to the Reissner equations (1.23) and (1.24), which are used in this book.) As can be seen, in the last figure, all the postbuckling branches are closed, locked onto bifurcation points along the prebuckling state; i.e., the *P*-axis. Only one branch $B_{6,1}$ (the lowest) has a parabolic character, as can be seen in Figure 5.35. Bauer, Reiss and Keller did not investigate the stability of all these postbifurcation states. Because they failed to do this, and, instead, relied on ill-conceived stability concepts, their speculations about the possible ways of loss of stability turned out to be of no value

The rising branch of the deformation curve from P^L, which is more completely represented in Figure 5.35, is the only stable branch of the set of axisymmetric equilibrium states. All other rising branches that emanate from the other bifurcation points P_n ($n \neq 6$) represent unstable states, as will be shown in the next chapter. The loss of stability of the cap will happen by snap-through at the critical point P_6 to a new axisymmetric state that belongs to the stable branch just mentioned or to a new nonsymmetric equilibrium state (these were not determined in this investigation). It is of interest to note that Bauer, Reiss and Keller did reveal the existence of isolated (separate) branches, but these were not further analyzed. It seems that these branches are of the same type as the ones that were found for clamped caps (see Fig. 2.6).

The presented results are probably valid for hydrostatic pressure; in their paper, Bauer and his colleagues [1970] did not mention which type was applied. This also applies to the second case (edge conditions (5.20)).

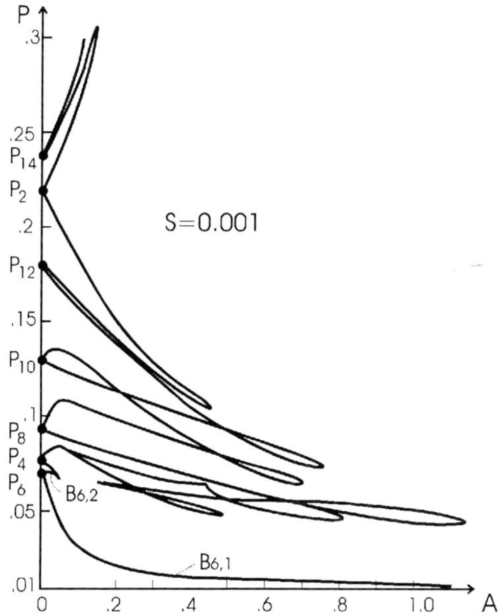

Figure 5.34 Results for the partially clamped case.

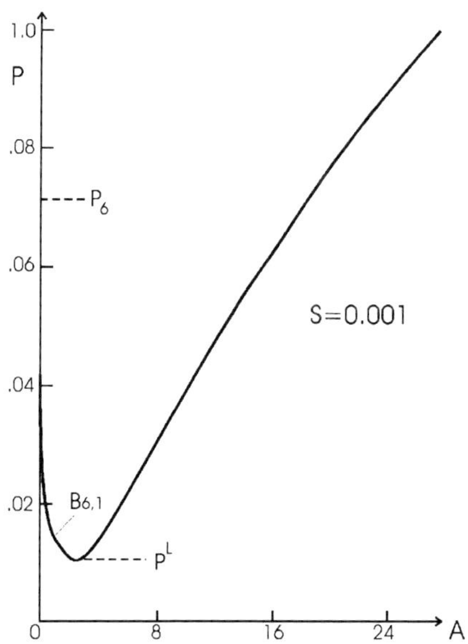

Figure 5.35 Detail of Figure 5.34: The stable (axisymmetric) postbuckling branch.

For the second case — the fully clamped cap — numerical solutions obtained by Bauer, Reiss and Keller [1973] are presented in Table 5.2 (above) and in Figures 5.36 and 5.37.

Chapter 6
The Stability of Equilibrium States of Geometrically Nonlinear Shells

One of the main complications in the analysis of the behavior of geometrically nonlinear shells is the existence of so many solutions; i.e., the multitude of solutions that exist at the same value of the load. This is not only the case for a multi-parameter loading system; it also holds for a single parameter system although it can be said that, in the former case, the difficulties multiply (see, e.g., Section 4.4).

The existence of so many solutions raises the question in what way the behavior of the actual structure can be deduced from these solutions, and it is this question and the need to answer it that leads us to the notion of stability, or more precisely, the distinction between unstable and stable solutions. Failure to understand these concepts properly may lead to a misinterpretation of the actual behavior that can have grave consequences in applications.

The analysis of the stability of the whole multitude of equilibrium states is the principal subject of the present chapter. An analysis of this kind permits the unfolding of new phenomena. We shall focus on the notion of stability and its interpretation from both the mechanical and formal points of view.

6.1 Some remarks on the stability of equilibrium states of shells and geometrically nonlinear conservative structures in general.

a) General Remarks.

It is not necessary to emphasize the importance of the subject that deals with the stability of structures from both theoretical and practical points of view. There is a great variety of these problems, and their treatment is often based on ill-founded notions concerning the meaning of the word "stability." Bellman once illustrated this by stating that, according to his perception, "Stability, is a heavily loaded concept with unstable definition." In our opinion, this statement still reflects the state of affairs to this day. As in no other area of mechanics, the analysis of stability problems is often marred by misunderstandings or mistakes. A number of such cases will be discussed below. Some interesting historical examples were collected and described by Panovko and Gubanova [1965].

The principal source of these misunderstandings must be sought in the formal transposition of features associated with the loss of stability of simple (single degree of freedom) models to the more general case without an appropriate and careful consideration of the differences that exist between them. In this section, we will discuss some of the concepts of the theory of elastic stability, with the aim to provide a basis for interpretation that may help to prevent the occurrence of at least a part of these misunderstandings.

We are dealing here with the stability of equilibrium states of linearly elastic, geometrically nonlinear structures subjected to loadings that do not change their direction during deformation of the structure (dead loadings). Thus, so-called "follower" forces, which are nonconservative in the majority of cases, are excluded from our examination here. We shall also not consider either of the following two special cases:

1) Cases when the magnitude of external forces depends on the deformed shape of the structure, as, for example, in some problems of aeroelasticity.

2) Cases when some basic features of the structure can change during its deformation. In particular, contact problems where the contact zones depend on the deformed state of the structure.

An intuitive physical concept of stability can be formulated in the following manner. An equilibrium state is said to be stable if small disturbances of this state do not lead to a motion that transforms this state to a new state far away from its initial state. Before considering a more strict formulation of this concept, it should be noted that, although the theory of stability is considered to have been founded by Euler in connection with the nonlinear behavior of a beam, Euler, did, in fact, not examine a stability problem. Euler's theory deals first of all with bifurcation phenomena, where two basic equilibrium curves cross in one point (the bifurcation point). In the classical terminology of elastic stability, the investigation of such a point is based on the concept of new, additional, equilibrium states that emerge adjacent to the basic state. Bifurcation and loss of stability are two concepts that should not be confused with each other: the first is a geometrical concept, while the second is a physical concept. They do happen to have a close connection with each other, however, in some special, but rather important, cases.

Lagrange approached the notion of stability of equilibrium by considering the motion from a perturbed state to the initial state for small but arbitrary perturbations. The time and also the character of the motion are immaterial in this consideration. The pervading requirement for stability is that, beginning from the instant the perturbation is applied, the ensuing motion remains in the small neighborhood of the equilibrium state under investigation. This concept has thus an intrinsic dynamic character and is focused on a behavior that takes place in a small locality of the equilibrium state in question. At about the same time, Lagrange also formulated his famous energy principle of stability, which, at later date, was put on a firm foundation by

Legeune and Dirichlet.

A configuration of a conservative system of n-degrees of freedom is in a state of stable equilibrium if the total potential energy in this configuration is a local minimum with respect to the generated energy of all neighboring states that are compatible with the kinematic constraints. These virtual variations of the potential energy, which are introduced by virtual displacements, are determined while the external forces, which generate the equilibrium state under investigation, are kept fixed [see, e.g., Lanczos, 1949; Langhaar, 1962; Gantmacher, 1960]. This criterion is thus also of a local static type. Liapunov later showed that equilibrium is unstable if the local minimum of the total potential energy does not exist [see, e.g., Gantmacher, 1960].

Liapunov's kinetic criterion is more general than the energy criterion, since it also applies to structures that are non conservative. According to Liapunov's definition, the equilibrium of a given mechanical system can be declared stable, if and only if it is possible to find a finite bound (no matter how small) for the size of otherwise arbitrary initial disturbances, which, when applied to the system, will result in a motion that always remains in a finite region around the equilibrium state. By this definition, "arbitrary" means any possible initial displacement and/or velocity distribution that is compatible with the constraints imposed on the structure. If the test described above fails, the equilibrium state is declared unstable.

Because the criterion calls for a *finite* bound on the size of the initial disturbances — *no matter how small this bound is chosen to be* — it is possible to restrict the investigation to the response of the system for arbitrarily small disturbances. The test for stability is then carried out by inspection of the solutions of the linearized version of the equations of motion of the system around the equilibrium state under investigation [Langhaar, 1962; Gantmacher, 1960; Bolotin, 1965].

The basic ideas of these approaches to the testing of the stability of elastic structures have the following common features:

1) Only local phenomena in an infinitesimal neighborhood of the tested equilibrium state are considered. This means that the stability is investigated in the small.

2) If the examined equilibrium state is unstable, the structure will not remain in this state, but will undergo a motion that ends in another equilibrium state; however, the way this happens is not part of the analysis and thus remains unknown.

From this point of view, the above given criteria can be tested using either a linearized or a nonlinear analysis. In order to obtain the postbuckling behavior of the structure, however, the problem must be examined in a nonlinear context.

3) The basic approaches of the stability test demand a strict definition of the class of the perturbed or virtual states. The equilibrium state can be stable in the small for a specific class of perturbations and unstable for another class. For example, Figure 6.1 depicts a typical deformation curve for circular plates. Branch OAB describes axisymmetric deformations of the plate, while branch AC describes the nonsymmetric postbuckling equilibrium state. The equilibrium state of branch OA is stable in both classes of axisymmetric and nonsymmetric perturbations, while the equilibrium state of branch AB is stable relative to symmetric perturbations, but unstable in the general case of nonsymmetric forms. The equilibrium state of branch AC, however, turns out to be stable initially, both in the symmetric and nonsymmetric classes of perturbations. Note that each nonsymmetric state of the branch AC corresponds to a specific number n of circumferential waves. When the loading increases, it is possible that a secondary bifurcation occurs in point C, where the form with n waves becomes unstable and the shell moves to a new shape with another number of waves j. In such a manner, higher bifurcations, and, consequently, new postbifurcation branches, can occur.

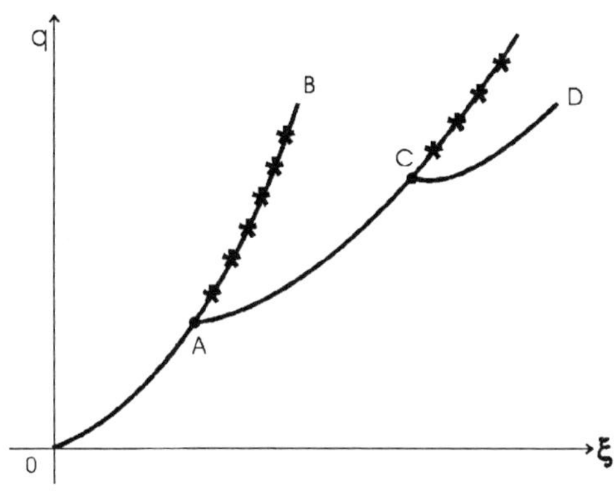

Figure 6.1 Qualitative behavior of circular plates in compression.

4) The local character of stability in the small justifies the linearization of equations of motion by using infinitesimally small perturbations from an equilibrium state. As a result, a set of linear equations of motion (so-called "equations in variations," according to the terminology of Poincaré) around the tested equilibrium state is obtained. If all members of the spectrum of proper frequencies (in terms of squares) of these oscillations (in absence of the dissipation factors) are positive, then the considered equilibrium state is stable in the small. If only one member of this spectrum is negative, then the corresponding equilibrium state is unstable. When the first, or the first (smallest) member of the spectrum becomes equal to zero, then the corresponding equilibrium state

is called critical. (Corresponding examples will be presented in Section 6.3.). The method is valid for both conservative and nonconservative problems.

5) The theory of these approaches is developed for systems with a limited number of degrees of freedom. Their extension to continuous systems necessitates additional mathematical analysis and justification, and the problem is not solved as yet, although there have been some serious attempts in this direction [Koiter, 1963a, 1965, 1969a, 1971; Knops & Wilkes, 1973].

6) It is generally accepted that the unstable equilibrium states cannot be maintained in the physical situation [Bolotin, 1965]. This is true in a strict sense; however, in practice, we can, often by small changes, restabilize a give situation. For example, by changing the mode of loading, the former unstable states can be stabilized (forced displacements instead of dead weight loads) [e.g., Panovko & Gubanova, 1965; Croll & Walker, 1972]. These measures thus correspond to a small change in the conditions, and thus to a change of the problem.

As is well-known, loss of stability manifests itself in two different ways. There is one type whereby the transfer from the critical fundamental (prebuckling) equilibrium state to another state takes place in a gradual way. The second type corresponds to a loss of stability in the "large," which occurs by a sudden transition to another, stable, postbuckling or post-snapping state that is not necessarily close to the previously held fundamental state. Well-known examples of these types of loss of stability are presented in Figures 6.2 and 6.3a.

In Figure 6.2, the branching diagram shows the behavior of the Euler beam. The vertical axis corresponds to the initially straight beam. At $P = P_{cr}$ (point A), bifurcation buckling takes place where the straight form of the beam goes over into a bent form. The slowly increasing load from this point is accompanied by a slowly growing buckling mode. This is possible because the new bifurcated state is stable in the small, as is the fundamental state for $P \leq P_{cr}$.

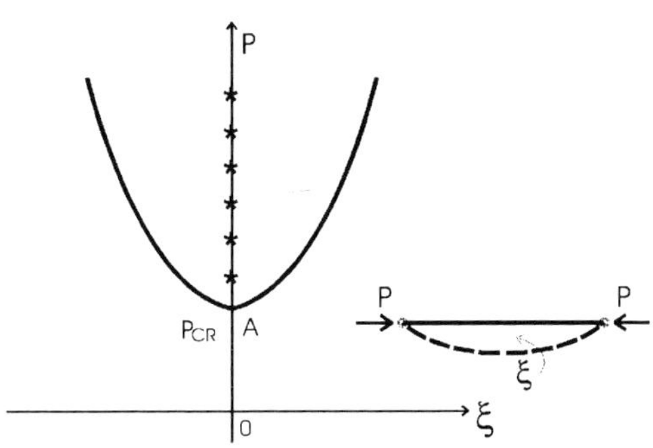

Figure 6.2 The behavior of the Euler strut.

librium state at K is axisymmetric, while the bifurcation branch KL correspond to nonsymmetric deformation states.

Another well-known illustration of loss of stability in the large at a bifurcation point is the axially compressed cylindrical shell or cylindrical panel (point A in Fig. 6.4) [Hutchinson & Koiter, 1970]. At the bifurcation point, where the fundamental equilibrium state represents a simple compressed state, the adjacent bifurcation states are unstable and descending nonsymmetric buckled states. Point B represents the deep post-snapping (stable) state of the shell. The basis of the modern engineering theory of bifurcation was worked out by W. T. Koiter [1963b, 1967]. Some aspects of this theory will be discussed below.

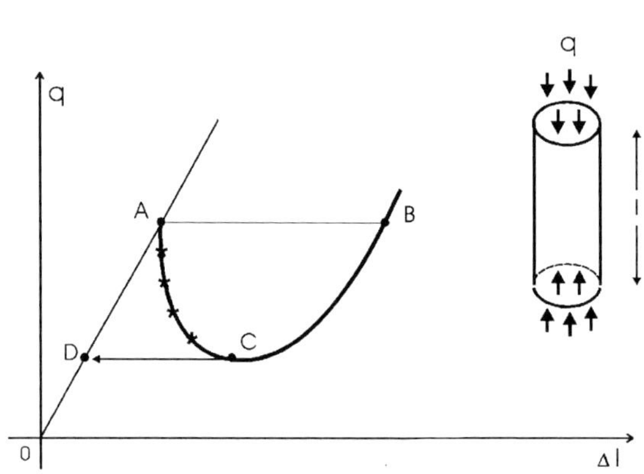

Figure 6.4 Typical behavior of a cylindrical shell in compression.

The process of transition from one equilibrium state to another, whereby both states are stable in the small, is here called loss of stability in the large. But in this case, the process of stability loss is not initiated at a limit or bifurcation point, but at a regular stable equilibrium state by means of an externally applied disturbance. The pair of points E and F in Figure 6.3a are intended to illustrate this process. The transition from E to F can be realized by an impulsive charge which supplies energy necessary to leave the stable state E. Thus, the equilibrium state marked by E or F is stable in the small, but unstable in the large (relative to a large perturbation). We shall not examine these scenarios, but will restrict our attention to snap-through processes that are triggered at limit or bifurcation points.

The phenomena of loss of stability in the small is connected with the concept of what is called "change of stability." It takes place at bifurcation and limit points. Let us consider, for example, the branching diagram shown in Figure 6.2, where the unstable part of the branches are marked by crosses "†." Up to the bifurcation point A, for $P < P_{cr}$ the fundamental state is stable, but after the bifurcation point, this state has become unstable. In the terminology of Poincaré, a "change of stability" has occurred. Moreover, the stability of the fundamental state has been transferred to the two bifurcating branches; this phenomenon is called "exchange" of stability.

A similar case is shown in Figure 6.1, where the prebuckling branch OB changes its stability in the bifurcation point A. The process is repeated along the bifurcating branch in the second bifurcation point C. The change of stability that takes place at a limit point is shown in Figure 6.3a if we discard the possible existence of a bifurcation point K before the limit point A. Pictures of the basic forms of structural behavior shown in Figures 6.2, 6.3a, or 6.4 can be found in almost all textbooks, monographs, and papers that deal with elastic stability problems.

We shall conclude this part of the present section with the following remarks.

Until now, we were mainly concerned with the question of loss of stability in the case where the external influence parameter is the intensity of the (live) load. However, the loading system can also consist of a set of prescribed displacements. This does not give rise to any changes in the way the stability of these systems should be analyzed. Only, the behavior of a structure that is loaded by live forces is in principle different from that loaded by prescribed displacements, even in cases where it appears that the two cases are identical. We will illustrate this by an example.

Let the intensity parameter of the load system be denoted by q. The energy U_L that is supplied to the structure during loading varies with $dU_L = q(\xi)d\xi$ where ξ is some generalized displacement (in the case of an axially compressed plate or cylindrical shell, ξ is called the end-shortening). In the particular case that the load system consists of a single point force, the displacement ξ is the displacement of the structure at the point where the force is applied and is measured in the direction of the force. For convenience, we will now focus our attention on this case of a structure that is loaded by one point force only.

Instead of using the force as a loading device in this particular case, we could also use the displacement ξ as a means to load the structure. This is, for example, the way buckling tests are often carried out. When such a switch is effectuated, the displacement ξ becomes the external influence parameter and the load q becomes the unknown. It turns out that the solution to this new problem, in terms of the static equilibrium states, is identical to the previous case, where the load q was active. However, the same does not hold true for the stability of these solutions.

We shall now compare the loss of stability in the above-mentioned two cases of control parameters (the load intensity q and its conjugate displacement ξ) in the case of a cap, which is loaded by a point force at the apex. Let us consider the response of this case, which is qualitatively illustrated in Figure 6.3a. We will assume that our problem exhibits along the fundamental state a bifurcation at K from this state. If q is the external parameter (q is applied), the bifurcated branch is unstable (denoted by crosses † in Fig. 6.3a).

Because of this, Poincaré [1885, p. 274] affirms that, if *BS* is the branch of the unstable, and *B′S* of the stable, equilibrium states when $\lambda < \lambda_0$, then the change of stability of these branches will take place for $\lambda > \lambda_0$. We also refer to the observations of Appel [1932, p. 281], who discusses some particular examples. It is of interest to note that Poincaré, himself, encountered these phenomena during an investigation of a specific problem related to the stability of the steady state of rotating fluid masses [1902, pp. 162-164, and 170-171]. The observations are also applicable to other singular points, such as limit points (point *B* in Fig. 6.5).

The considerations presented above can be summarized as follows.

1) Change of stability can only take place along an equilibrium curve that is stable initially. This occurs at a singular point, called the critical equilibrium state, which is either a bifurcation point or a limit point. Bifurcation points and limit points also exist along branches that are unstable, but, in this case, change of stability is not a meaningful concept. In other words, change of stability along equilibrium branches is a rare phenomenon in the case of a multi-dimensional system.

2) If the system is one dimensional, change of stability always occurs at singular points. The cyclic changes from stable to unstable and back to stable, etc., is one of the main features of a one-degree of freedom system, which ceases to exist for $n > 1$ (see examples below).

d) The procedure of investigation of equilibrium state stability by reducing the case of n > 1 to that of n = 1.

This procedure, briefly and formally described in part b of this section, needs some additional discussion to elucidate the ways in which it can be applied correctly. There are many examples of an inadequate use of this procedure, which is valid only under certain restrictions. When these restrictions are not properly taken into account, the formal use of the method can lead to incorrect results.

Before we proceed to a clarification of the validity of the reduction technique (discussed in the previous part), one should realize that the test of stability of the solutions must take into account all virtual perturbations from these solutions [see e.g., Thompson and Hunt, 1973]. Consequently, the argument $x_1, x_2,...,x_n$ of the energy function *E* also defines the set of virtual displacements that must be taken into account.

Another interesting view on this procedure is given by Appel [1932, p. 278]. If energy *E*, as a function of $\{x_i\}$, possesses a minimum at an equilibrium state, then the reduced function $E^* = E(x_1,\phi_i(x_1,\lambda),\lambda)$ also possesses a minimum (relative to the variable x_1). On the other hand, the minima of E^* do not in general correspond to minima of $E(x_1, x_2,...,x_n,\lambda)$. Consequently, the test for stability or instability established for a system with $n = 1$ is not applicable to a system of $n > 1$ degrees of freedom.

Procedures specifically developed for the analysis bifurcation phenomena, which are based on the elimination of the so-called "passive" coordinates [see, e.g., Thompson & Hunt, 1973], are now known as the Liapunov-Schmidt reduction and the Koiter method. But the general idea of such reduction was first conceived and formulated by Poincaré [1885, 1902].

To conclude this subsection, we can say that change of stability always takes place if the prebuckling (pre-bifurcation) states are stable. In order to illustrate this fact, we shall consider a classical problem — the case of a doubly hinged rod of uniform stiffness in compression; i.e., the problem of the Elastica. The equilibrium curves that include postbuckling deformations are plotted in Figure 6.8 in terms of the load intensity q vs. the nondimensional displacement $\xi = f\ell$ measured at the middle of the rod (f is the displacement component normal to the axis of the undeformed rod, and ℓ is its length). These curves correspond to some of the exact solutions of the nonlinear equation for the bending of the rod. The bifurcations that are displayed correspond to buckling states with an uneven number of half waves in an axial direction. The nonsymmetric states are absent because $\xi = 0$ for these forms.

The bifurcation points A, B, C, etc., in Figure 6.8 are symmetric with respect to the prebuckling state of the rod, the axis $\xi = 0$.

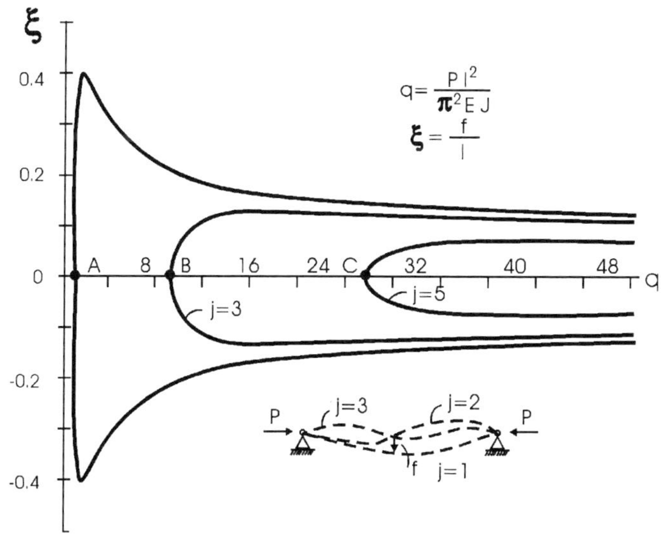

Figure 6.8 Equilibrium branching diagram for the Euler beam.

Thus, in accordance with Poincaré's theory for the one-dimensional case $n = 1$, the change of stability must take place along this prebuckling state at points A, B, C, etc., in an alternating fashion; i.e., OA stable, AB unstable, BC stable, etc. But this does not refer to the actual situation, because the part of the prebuckling state beyond A is unstable, as can be proved easily by any method that tests the stability in the small. Thus, the change of stability takes place only at the bifurcation point A (as in Fig. 6.3); at other bifurcation points $B,C,...$, no changes occur.

If we consider the equilibrium states that correspond to a fixed value of q, then it can be seen that only two of the whole set of possible equilibrium states are stable, and all the others are unstable when $q > q_A$. Before q_A, there is only one stable state. Consequently, the effect of alternating stable and unstable states, characteristic for systems with $n = 1$, does not take place in the case of a system defined as a beam or continuum.

Another example where the inadequacy of one dimensional theory can be demonstrated is the geometrically nonlinear, axisymmetrically loaded spherical cap considered in detail in the next section.

e) ***The applicability of the approximating model of n degrees of freedom for investigation of continuum systems.***

So far, the discussion concerned systems of n degrees of freedom. The question arises whether the theory can be extended to systems with an infinite number of degrees of freedom; i.e., continua, and, the related (reverse) question under what conditions a continuum can be replaced by an n-dimensional system. We will try to deal with the latter question by using the axisymmetric deformations of a geometrically nonlinear shell of revolution as an example.

In this case the deformed state of the shell can be characterized by two field quantities, for example, the displacements $u(\rho)$ and $w(\rho)$, where ρ is the radial shell midsurface coordinate. We now try to approximate this continuum description by a finite number of degrees of freedom. We take n values of function $u(\rho)$, breaking the interval of variation of ρ into $n - 1$ parts, and approximate the function $w(\rho)$ by m values (where m is not necessarily equal to n). In this way we come to a system of $(n + m)$ degrees of freedom.

The following important questions arise naturally. How must the process of the transition from an $n + m$ degrees system to the original continuum be realized? Which limit process must be used for such a transition? For example, the transition can be materialized by letting n tend to ∞, keeping m constant, and after that we can let $m \to \infty$. But we can also do this in the opposite way, or we can let n an m go to infinity simultaneously without varying n, or do the same in the opposite order; or we can let m and n tend to infinity simultaneously, assuming some kind of relationship between them. Only if the result that is obtained is the same in all these cases can we can state that the investigation of the model with a finite number of degrees of freedom will yield a valid result for the continuum under consideration.

Even if the condition stated above is satisfied, however, it is still necessary to know the number of degrees of freedom of the model that we minimally require in order to be sure that the model represents the principal features of the original continuum. We believe that this important problem can only be solved in a local sense; that is, in the neighborhood of a specific point of the loading path. Unfortunately, these problems, to the best of our knowledge, have not yet been addressed and solved in the general case.

The question we have just posed, implies that the approximation of a continuum by a system of a finite number of degrees of freedom cannot always ensure the preservation of the actual behavior of the continuum. The one-dimensional representation of a cap under external pressure (discussed in Chapter 2, Section 2.8) is a striking example of such a failure.

Thus, any approximate treatment of the stability of a continuum cannot provide, as was noted above, a guaranteed correct result, and conclusions drawn from any such approximate analysis must be treated with great caution. The best option available is to investigate the problem along different pathways using different numerical techniques, and if these all yield the same result, then one might expect that the result can be trusted. On the basis of such an approach, we constructed our own investigations [Shilkrut, 1980; Shilkrut & Vyrlan, 1977; Shilkrut, Shevandronov, Morar & Maksimov, 1969]. Some of these results are discussed in this book.

f) ***The significance of the equilibrium curve. Energy barriers and paths of realization of the snap-through process.***

The equilibrium curve of a given structure represents, to some extent, its behavior. This curve is constructed in an $(n + 1)$ dimensional space if the system is of finite dimensions; otherwise, for continuum systems (such as beams, plates, shells, and deformable bodies in general), the corresponding space is infinite. Hence, all the plane curves that plot the loading parameter versus the displacement of some point of the body are only projections of the multi-dimensional deformation curve onto some plane spanned by the load parameter q and the displacement component ξ that we select to draw the plot. Thus, such planar curves of deformation contain only partial, incomplete information about the behavior of the system. One should be aware of these limitations, because, otherwise it is possible to come to serious mis-calculations. For example, if a planar equilibrium curve of a multi-degree of freedom system intersects itself, this does not imply that at this point a bifurcation occurs. Only in the case of a system of one degree of freedom is the deformation curve complete in the plane (q, ξ), and only in this case does it contain all the information about the response of the system.

We shall now consider this issue in more detail. In the case of one degree of freedom, there exists (as was shown in part b of the present section) a cyclic change between stable and unstable equilibrium states at a fixed value of the load parameter. Transition from one stable equilibrium state to another can then only be accomplished by going, in between, through the unstable equilibrium states (see for example the curve in Fig. 6.5, where, at $\lambda = \lambda_0$, transition from x_1 to x_3 can only be realized by going through the unstable equilibrium state x_2). This intermediate unstable equilibrium state, however, possesses a higher energy level than the neighboring stable equilibrium state. Because of this difference, a transition cannot take place that does not correspond to bifurcation from one stable equilibrium state to the next stable state without overcoming an energy barrier determined by the difference between the energies of the stable state and the in-between unstable state. It should be emphasized that this type of transition, from one stable equilibrium state to another, can only take place in the way described when we deal with a system of one degree of freedom. For a multi-degree of freedom system or continuum, the situation is quite different, as we will see below.

The stable and unstable solutions that are found for a one-degree-of-freedom system at the same value of the load alternate along the axis x, in conformity with the theory of Poincaré (see part b). Let us assume that the deformation curve of this one dimensional system is of a type depicted in Figure 6.9 and that the load is slowly increasing. (This figure is tailored to the behavior of spherical caps; that is, a two-dimensional continuum. But we are using this response curve now as an example that is more complicated than the case shown in Fig. 6.5). When the loading is applied, slowly starting from zero, the system will eventually reach the first limit point at *A* in Figure 6.9. At this point *A*, a snap-through process takes place and the

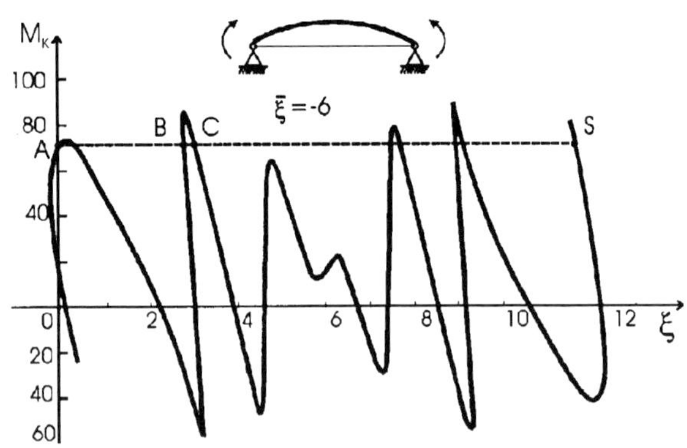

Figure 6.9 Branching behavior of a simply supported cap under edge moments.

system will “arrive” at a new stable equilibrium state (point *B*) (again, as before, we assume the presence of some damping). Naturally, this event, which is transient, takes place rapidly at the value M_{KA} of the loading, which does not really change during the snapping motion (the loading is varied infinitely slowly). After the snap, the system will follow the stable branch *BC*, and, at the second limit point *C*, a second snap-through event will occur, and so on. Thus, a series of successive snaps will occur until the system finally reaches a stable equilibrium state located on the last branch of the curve emanating from the last minimum.

A system of finite number (more than one) of degrees of freedom or a continuum is not constrained to behave in this manner and shows a behavior that is quite different from that of a one-dimensional system. Consider, for example, the shell, the behavior of which is again represented by the equilibrium curve in Figure 6.9. The planar curve depicted in this figure is only a projection of the actual, infinite-dimensional, solution curve onto the plane (ξ, M_K). In reality (as will be shown in detail in Section 6.2), the snap-through loss of stability of the shell happens by transition from the first limit point A directly to the last stable equilibrium state, point S, which is located on the same level as A. The intermediate unstable states between A and S at the load level M_{KA} are simply bypassed in this process. That this is possible follows because the system has $n > 1$ degrees of freedom, so that the transition from A to S can be realized in an infinite number of ways. An illustration of such a snap-through trajectory is plotted in a qualitative way in Figure 6.10, where ξ_1, ξ_2 are the coordinates that correspond to the degrees of freedom and M is the loading parameter. P is the plane $M = M_A$ that is perpendicular to the load axis M at the load level of the first limit point A.

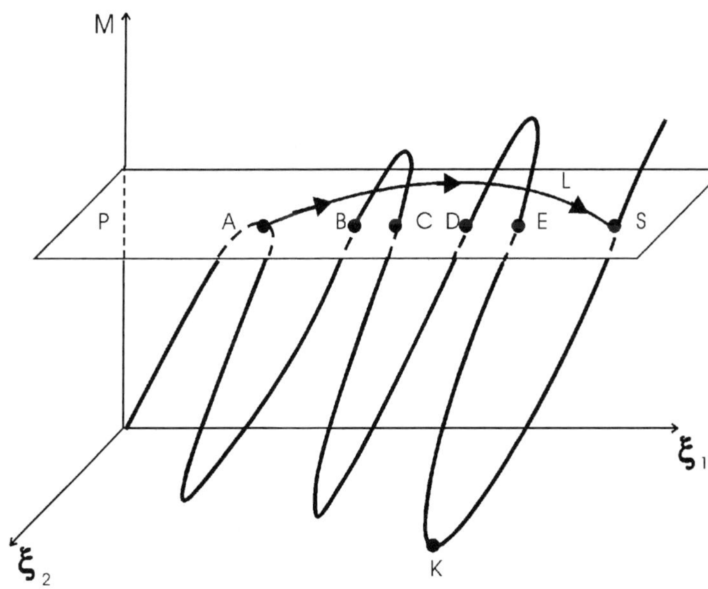

Figure 6.10 Diagram of an actual snap-through process in the n dimensional case.

It is clearly seen that the snap-through process, which moves the deformation state from A to S, can be realized in many ways by a trajectory that lies in the plane P; for example, the curve L in the Figure 6.10. Each of these (transient trajectories), like L, do not need to pass through the unstable equilibrium states.

Transitions described above actually happen, as will be shown below, and they characterize the actual phenomenon of loss of stability in terms of a jumping process when we deal with a system of n ($n > 1$) degrees of freedom. In the general case, there are no successive snap-through effects and there is no change of stability at the intermediate limit points except for those at the limit points A and K (see Fig. 6.10). It should be noted that, in light of the description of the behavior of complex systems given above, it must be clear that Thompson's [1979] investigation of a succession of snap-through processes can only be related to a one degree of freedom system, but not to any other system.

Special cases exists where a succession of snap-throughs can take place (see, e.g., the deformation curves in Figs. 3.18, 3.20, and 3.21). In these cases, the second minimum of the deformation curve is higher than the first snap-through point. Successive snaps can also occur in certain plate and panel buckling problems [e.g., Riks, Rankin & Brogan, 1997; Riks, 1998].

g) The influence of imperfections on the loss of stability of equilibrium states.

The theory of W. T. Koiter [1963b, 1967] describes the essential influence of geometric imperfections on the process of loss of stability in the neighborhood of bifurcation points. By "imperfections" are meant small deviations from the idealized model of the structure. There are many different types of imperfections, such as: a) imperfections of the structure's geometry; b) deviations of the directions of the forces and/or eccentricities in the points of application; c) imperfections in the mechanical properties of the material (e.g., deviations from the homogeneity or isotropy); d) imperfections in the support conditions; and so on. We shall restrict ourselves mainly to the question of the influence of geometric imperfections, which belong to the type of imperfections most frequently considered and which are known to be of some importance.

Figures 6.11, 6.12, and 6.13 portray the most common types of bifurcation behavior that we can encounter in elasticity for perfect, as well as slightly imperfect, structures. In these figures, the load λ is plotted against the characteristic displacement a, which can be identified with the amplitude of the buckling mode. The dashed lines denote unstable branches, and the bifurcation points are denoted by A. Branches marked by the additional symbol *Im* are related to the system with small geometrical imperfections, while the curves without the symbol *Im* represent the bifurcating branches of the idealized system. Bifurcation point A in Figure 6.11 is called nonsymmetric, while the two other cases are called symmetric.

Koiter found that, for simple bifurcation points, a falling branch of an initially stable pre-buckling state implies that critical state A is unstable. On the other hand, if all bifurcation branches are rising, the equilibrium at the critical point is stable. This means that the bifurcation points in Figures 6.11 and 6.13 are unstable while the bifurcation point in Figure 6.12 is stable. Koiter's analyses lead to the following conclusions:

1) The branches of the bifurcation point A, which characterize the behavior of the ideal structure, can be seen as the asymptotes of the behavior of the imperfect systems; i.e., the branches of the imperfect systems approach the branches of the perfect system infinitely closely when the amplitude of the imperfections becomes smaller and smaller. In the case of falling (unstable) bifurcation branches, the corresponding imperfect branches become unstable at limit points (points L in Figs. 6.11 and 6.13). In the case of rising branches of the bifurcation point, the corresponding imperfect branches are also transformed into smooth curves, but this time they contain no limit points.

2) In the case of unstable bifurcation points, imperfections can reduce (in some cases, to a very large degree) the load at which snap-through occurs. (Compare values of λ in points A and L in Figs. 6.12 and 6.13.)

We note here that these conclusions are often presented without much comment and without the warning that they only hold under certain restrictive conditions. Failure to recognize this can lead to errors in the prediction of the expected behavior of a given problem. We will illustrate this with an example below.

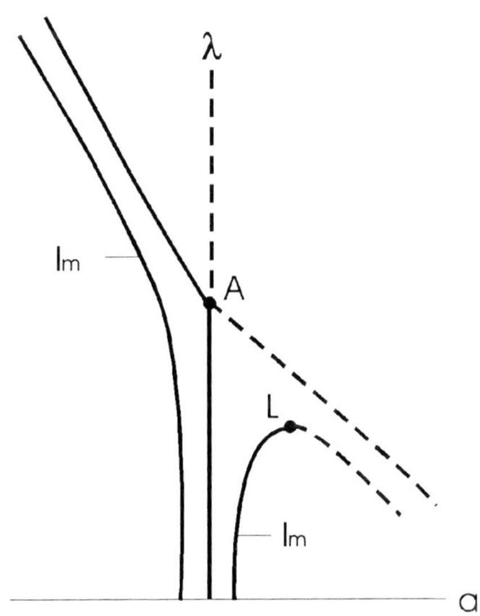

Figure 6.11 Perfect and imperfect behavior related to nonsymmetric bifurcation.

system can produce spurious results.

The principal result of these investigations is the conclusion that all axisymmetric forms of equilibrium of geometrically nonlinear spherical shells between their first maximum and the last minimum (moving along the curve from the undeformed state) are unstable in general, and that only the equilibrium state up to the first maximum and beyond the last minimum are stable in the small. Consequently, the "curse of ambiguity" [Feodosjev, 1969] assigned to the multiplicity of equilibrium states does not hold: no more than two stable axisymmetric states of equilibrium can exist at the same value of the externally applied load if the considerations are restricted to axisymmetric deformations. The practical significance of this conclusion lies in the fact that, combined with results available for the determination of the bifurcation points at which nonaxisymmetric forms of equilibrium make their appearance, it enables us to establish the theoretical values of the buckling loads of geometrically nonlinear shells of revolution loaded by axisymmetrically applied loads, and it provides an exact method for the determination of the upper and lower critical states along the basic solution curve.

This important characteristic is established for single- and, in some cases, two-parameter, external loading (excitations). In the case of multi-parameter external loads, the phenomena are usually much more complicated (see Section 6.8).

b) ***Problem statement.***

Stability in the small of all possible axisymmetric equilibrium states is discussed for geometrically nonlinear, isotropic, elastic shells of revolution. (There are no obstacles to extending these investigations to the other axisymmetric cases, as, for example, circular orthotropic shells of revolution.)

Consider a shell subjected to a specified external load which is applied statically. The problem of the stability in the small of the equilibrium states will be resolved independently, with no regard to other equilibrium states that exist at the same value of the load parameter. The first way to investigate the stability of the state of equilibrium, is the universal method of small oscillations [Ziegler, 1968], which considers small perturbations imposed on the investigated state of equilibrium. The results obtained by this method can be (Section 6.4) verified by other methods.

If the shell undergoes small oscillations around the state of equilibrium in question, the equilibrium state is stable in the small for the class of perturbations under consideration. But if the small perturbations that are imposed on the considered equilibrium state, give rise to an a-periodic motion of the shell, the tested equilibrium state is declared unstable, both in the small and in the large. The stability criterion thus demands that all eigenfrequencies of the linearized equations of motion of the shell around the equilibrium state of interest are real. If one (or more) of the frequencies is imaginary, then the equilibrium state under investigation is unstable. The investigation of stability by the method of small oscillations is thus reduced to the determination of the spectrum of eigenfrequencies of the linearized equations of motion around the equilibrium state.

The practical execution of the method goes as follows. The eigenfrequencies are obtained numerically using the well-known method of separation of variables, while, for the determination of the equilibrium state themselves, the shooting method is used [see, e.g., Keller & Wolfe, 1967; Roberts & Shipman, 1972; Shilkrut, 1980; Shilkrut, Shevandronov, Morar & Maksimov, 1969; Shilkrut & Vyrlan, 1975a, 1979; Vyrlan & Shilkrut, 1978].

c) Equations of small oscillations of shallow caps.

To start this overview, we first recapitulate the equations of motion for shells of revolution (see property 18, Chapter 2).

$$L[\omega(\rho,\tau)] = -\frac{1}{2\rho}\left[\theta^2(\rho,\tau) + 2\theta(\rho,\tau)\bar{\theta}(\rho)\right]. \qquad (6.24)$$

$$L[\theta(\rho,\tau)] = -\frac{m}{\rho}\left\{\int_0^\rho \left[q(\rho,\tau) - s\frac{\partial^2 w(\rho,\tau)}{\partial\tau^2}\right]d\rho + P(\tau) + P_1(c,\tau)I(\rho - c) - \omega(\rho,\tau)\left[\theta(\rho,\tau) + \bar{\theta}(\rho)\right]\right\};$$

$$(0 \le \rho \le 1); \quad (0 \le \tau \le \infty). \qquad (6.25)$$

The boundary conditions are in this case dependent on the dimensionless time $\tau = t\sqrt{g/h}$:

$$\theta(0,\tau) \equiv \omega(0,\tau) \equiv 0; \quad \alpha_1\omega'(1,\tau) + \beta_1\omega(1,\tau) = \gamma_1(\tau);$$

$$\alpha_2\theta'(1,\tau) + \beta_2\theta(1,\tau) = \gamma_2(\tau). \qquad (6.26)$$

Some of the symbols that did not appear earlier in these equations, have the following significance: $s = \gamma a^4/Eh^3$ is the dimensionless parameter of the specific mass of the material related to a unit surface area of the middle shell's surface; γ is the specific weight of the material of the shell; and, P and P_1 are the concentrated apex force and intensity of the uniformly distributed load along a circle of constant radius c, respectively. The symbol I stands for the Heaviside step function:

$$I(\rho - c) = \begin{cases} 1 & \text{when} \quad \rho > c \\ 0 & \text{when} \quad \rho < c \end{cases}. \qquad (6.27)$$

The corresponding initial conditions are as follows:

$$\theta(\rho,0) = f_1(\rho); \quad \frac{\partial\theta(\rho,0)}{\partial\tau} = f_2(\rho), \tag{6.28}$$

where $f_i(\rho)$; $(i = 1,2)$ are given functions that satisfy the above-presented boundary conditions for $\theta(\rho,\tau)$.

As is customary, in the equations presented above, only inertial forces due to the deflections $w(\rho,\tau)$ are taken into account, because these are the ones that are dominant in these types of problems. The initial conditions for $\omega(\rho,\tau)$ are not specified here, since they result from (6.24), $\bar{\theta}(\rho)$ and $f_i(\rho)$.

The given integro-differential equations with boundary and initial conditions yield a complete set of relations for the axisymmetric transient deformations of shallow, geometrically nonlinear, linear elastic, isotropic shells. It is convenient to write the above given equations (6.25) in the form given below, (6.29), which does not explicitly contain the displacement $w(\rho,\tau)$, and which is obtained by integration by parts of the right-hand side of (6.25).

$$L(\theta) = \frac{m}{\rho}$$

$$\left\{-\int_0^\rho q\rho d\rho + \frac{1}{2}s\frac{\partial^2}{\partial\tau^2}\left[\rho^2\int_0^1 \theta d\rho - \rho^2\int_0^\rho \theta d\rho + \int_0^\rho \rho^2\theta d\rho\right] + \omega\left(\theta+\bar{\theta}\right) - P - P_1 I(\rho - c)\right\}. \tag{6.29}$$

We should note that this equation is obtained using the condition $w(1,\tau) = 0$, which means that the edge of the cap is immovable in the z-direction.

To obtain the perturbation equations we write:

$$\omega(\rho,\tau) = \omega_s(\rho) + \delta\omega(\rho,\tau); \quad \theta(\rho,\tau) = \theta_s(\rho) + \delta\theta(\rho,\tau); \quad 0 \le \rho \le 1; \quad 0 \le \tau < \infty, \tag{6.30}$$

where the pair ω_s and θ_s is the static solution that describes the equilibrium state under consideration. The functions $\delta\omega$ and $\delta\theta$ are the small perturbations superimposed on the static solution. It is clear that both the transients (ω and θ) and static solutions (ω_s and θ_s) must satisfy the same boundary conditions.

Upon substituting expressions (6.30) into equations (6.24) and (6.29), recalling the meaning of ω_s, θ_s, and discarding quantities of the order $(\delta\theta)^2$, $\delta\omega \cdot \delta\theta$, the following system of linearized equations are obtained for $\delta\omega(\rho,\tau)$ and $\delta\theta(\rho,\tau)$:

$$L(\delta\omega) = -\frac{\delta\theta}{\rho}\left(\theta_s + \bar{\theta}\right); \tag{6.31}$$

$$L(\delta\theta) = \frac{m}{\rho}\left\{\delta\omega\left(\theta_s + \overline{\theta}\right) + \omega_s\delta\theta + \frac{s}{2}\frac{\partial^2}{\partial\tau^2}\left[\rho^2\int_0^1 \delta\theta d\rho - \rho^2\int_0^\rho \delta\theta d\rho + \int_0^\rho \rho^2\delta\theta d\rho\right]\right\}. \tag{6.32}$$

The corresponding boundary and initial conditions take the form (6.33) and (6.34) if one takes into account the relations (6.26), (6.28), and, naturally, (6.30).

$$\delta\omega(0,\tau) \equiv \delta\theta(0,\tau) \equiv 0; \quad \alpha_1\delta\omega'(1,\tau) + \beta_1\delta\omega(1,\tau) \equiv 0; \quad \alpha_2\delta\theta'(1,\tau) + \beta_2\delta\theta(1,\tau) \equiv 0 \tag{6.33}$$

$$\delta\theta(\rho,0) = f_1(\rho) - \theta_s(\rho); \quad \frac{\partial\theta(\rho,0)}{\partial\tau} = f_2(\rho). \tag{6.34}$$

Note that boundary conditions are homogeneous.

Equations (6.31) and (6.32) are linear equations in partial derivatives, and to solve them we resort to the method of separation of variables, presenting $\delta\omega(\rho,\tau)$ and $\delta\theta(\rho,\tau)$ in the form of a product of two functions, each of which is solely dependent upon one variable.

$$\delta\omega(\rho,\tau) = \Omega(\rho)\varphi(\tau); \quad \delta\theta(\rho,\tau) = \Theta(\rho)\varphi(\tau). \tag{6.35}$$

The separation of the variables brings us to the following equations for the unknown functions:

$$L(\Omega) = -\frac{\Theta}{\rho}\left(\theta_s + \overline{\theta}\right); \tag{6.36}$$

$$L(\Theta) = \frac{m}{\rho}\left[\omega_s\Theta + \Omega\left(\theta_s + \overline{\theta}\right)\right] - \frac{k^2}{2\rho}\left[\rho^2\left(A - \int_0^\rho \Theta d\rho\right) + \int_0^\rho \rho^2\Theta d\rho\right]; \tag{6.37}$$

$$\ddot{\varphi}(\tau) + p^2\varphi(\tau) = 0, \tag{6.38}$$

where the dots denote the derivative with respect to the time τ, and A is given by:

$$A = \int_0^1 \Theta(\rho)d\rho. \tag{6.39}$$

The other symbols are $k^2 = msp^2$, where p^2 is the square of the dimensionless, circular frequency of the small oscillations about the given equilibrium state. The physical value of this frequency is $p^* = p\sqrt{g/h}$. The corresponding boundary conditions are:

$$\Theta(0) = \Omega(0) = 0; \quad \alpha_1\Omega'(1) + \beta_1\Omega(1) = 0; \quad \alpha_2\Theta'(1) + \beta_2\Theta(1) = 0. \tag{6.40}$$

The general solution of (6.38) is given by:

$$\varphi(\tau) = c_1 \sin p\tau + c_2 \cos p\tau, \tag{6.41}$$

where the integration constants c_i ($i = 1,2$) are determined by the initial conditions. Their values do not influence the stability of the investigated equilibrium state, however; this is completely determined by the sign of p^2 and does not depend on c_i. When $p^2 > 0$, the shell executes small oscillations about the equilibrium state $[\omega_s(\rho), \theta_s(\rho)]$ in question. But as soon as there is one $p^2 < 0$, the function $\varphi(\tau)$ — and with it $\delta\omega$, $\delta\theta$ — will eventually increase beyond bounds (according to the linearization applied here). This means that the shell will not oscillate around the equilibrium state under investigation, but that the motion will take the shell to another stable state. In that case, the equilibrium state is unstable.

We now consider the boundary problem (6.33) and (6.34) for the system (6.31 and (6.32), the solution of which will enable us to determine p^2 in terms of k^2. The non-vanishing (nontrivial) solutions of this homogeneous problem only exist at specific values of the parameter k^2 (the eigenvalues). Since the discrete spectrum of eigenvalues for this problem is infinite [Collatz, 1963], we label each of them and their corresponding eigenfunctions by the subscript n; that is, we label the eigenvalues as k_n^2 together with the eigenfunctions $\Theta_n(p)$ and $\Omega_n(p)$, where n = 1,2,3,... and $k_1^2 \le k_2^2 \le k_3^2 \le \ldots$. The equality sign holds for multiple eigenvalues.

We are interested first and foremost in the eigenvalues k_n^2, since they determine the stability in the small of the equilibrium state. The numerical methods of computing k_n^2 will be considered in Section 6.3. Before doing this, however, we first discuss some basic features of the spectrum k_n^2 and corresponding eigenfunctions.

d) Some basic features of eigenfrequencies and eigenfunctions.

First of all, we shall consider some features of the solutions that are connected with reciprocal systems, which were discussed in Sections 2.2-2.6. In the theory of reciprocal systems, two equilibrium states of the shell are called symmetric if their deformed meridians are mutually symmetric about the base plane of the cap. In Sections 2.2-2.5, it was shown that hinged spherical caps are reciprocal systems when they are subjected to edge bending moments M_K plus other arbitrary transverse load systems that generate axisymmetric deformations.

Calculating the eigenfrequencies of geometrically nonlinear shells is a fairly laborious task, sometimes requiring a large volume of computation; thus, the reduction of this expense is of some importance. Symmetry properties of shells enable us to demonstrate that the eigenvalues of two symmetric states of a shell are equal, and in this way the computational operations can be cut in half. They also facilitate the control and verification of the computations. We will show this in the following consideration.

Let ω_{s1} and θ_{s1} correspond to the first equilibrium state and ω_{s2} and θ_{s2} to the second state, which is symmetric to the first. For these two symmetric states of equilibrium, the equations

$$\theta_{s1} + \theta_{s2} = -2\bar{\theta}; \quad \omega_{s1} \equiv \omega_{s2} \tag{6.42}$$

hold (see Section 2.2).Then, taking into account (6.42) and (6.35), one can rewrite equations (6.36) and (6.37) for the two pairs of eigenfunctions (Θ_1, Ω_1) and (Θ_2, Ω_2), which correspond to a pair of symmetric equilibrium states in the following manner:

$$L(\Omega_{n1}) = -\frac{\Theta_{n1}}{\rho}\left(\theta_{s1} + \bar{\theta}\right),$$
$$L(\Theta_{n1}) = -\frac{m}{\rho}\left\{\omega_{s1}\Theta_{n1} + \Omega_{n1}\left(\theta_{s1} + \bar{\theta}\right) - \frac{k_{n1}^2}{2m}\left[\rho^2\int_\rho^1 \Theta_{n1}d\rho + \int_0^\rho \rho^2\Theta_{n1}d\rho\right]\right\}; \tag{6.43}$$

$$L(\Omega_{n2}) = \frac{\Theta_{n2}}{\rho}\left(\theta_{s1} + \bar{\theta}\right),$$
$$L(\Theta_{n2}) = \frac{m}{\rho}\left\{\omega_{s1}\Theta_{n2} - \Omega_{n2}\left(\theta_{s1} + \bar{\theta}\right) - \frac{k_{n2}^2}{2m}\left[\rho^2\int_\rho^1 \Theta_{n2}d\rho + \int_0^\rho \rho^2\Theta_{n2}d\rho\right]\right\}. \tag{6.44}$$

From these relations, the following results are obtained:

$$k_{n1}^2 = k_{n2}^2; \quad \Omega_{n1} \equiv \Omega_{n2}; \quad \Theta_{n1} \equiv -\Theta_{n2}. \tag{6.45}$$

Other features of the considered functions and parameters can be obtained by means of the inequality (6.46) (see Section 2.8)

$$\int_0^1 \eta L(\eta)d\rho \le 0 \tag{6.46}$$

if the function $\eta(\rho)$ satisfies one of the following two-boundary conditions at $\rho = 1$:

$$\eta(1) = 0; \text{ or } \eta'(1) \pm \nu\eta(1) = 0, \tag{6.47}$$

(ν is Poisson's ratio and, of course, $\eta(0) = 0$).

Taking for η the function Ω, and using equations (6.46) and (6.36), the following inequality results:

$$\int_0^1 \frac{1}{\rho}\Omega_n T_n\left(\theta_s+\overline{\theta}\right)d\rho \geq 0. \tag{6.48}$$

Applying (6.46) for equation (6.37) and taking into account (6.48), we obtain

$$m\int_0^1 \frac{\Theta_n^2\omega_s}{\rho}d\rho - \frac{k_n^2}{2}\int_0^1 \frac{\Theta_n}{\rho}\left\{\int_0^\rho \lambda^2\Theta_n(\lambda)d\lambda + \rho^2\int_\rho^1 \Theta_n(\lambda)d\lambda\right\}d\rho \leq 0. \tag{6.49}$$

From (6.49), we directly derive the important corollary that, when $\omega_s \geq 0$ everywhere, $k_n^2 \neq 0$. As is well-known, $k_n^2 = 0$ at bifurcation points and at limit points. When $\omega_s \geq 0$, therefore, the system cannot escape the equilibrium state spontaneously and move to an other stable state; i.e., it is stable both in the small and in the large. As demonstrated in Section 2.8, if a force N is applied to the edge of the shell, satisfying the constraint

$$N \geq \frac{1}{4}\int_0^1 \left(1-\lambda^2\right)\frac{\overline{\theta}^2(\lambda)}{\lambda}d\lambda, \tag{6.50}$$

then $\omega_s \geq 0$. Accordingly, by applying the corresponding edge force, N, we can completely eliminate loss of stability with respect to axisymmetric forms both in the small and in the large. Using N, we can also come to the situation when $\sigma_\phi(\rho) = \omega'(p) \geq 0$ (see Section 2.8), which should exclude the occurrence of bifurcation into nonaxisymmetric forms as well. Consequently, by applying the corresponding force N to the edge, we can control the stability of the shell all the way until complete stability is assured.

Let us now consider some of the properties of the eigenfunctions by comparing any two. To do this, we resort to Green's formula (6.51) for a self-adjoint operator $\rho L()$:

$$\int_{\rho_0}^{\rho_1} \left[v\rho L(u) - u\rho L(v)\right]d\rho = \left[\rho(u'v - uv')\right]\Big|_{\rho_0}^{\rho_1}. \tag{6.51}$$

$u(p)$ and $v(p)$ are arbitrary functions that admit the operations carried out here.

Applying this formula to any pair of eigenfunctions Ω_j and Ω_m, we obtain (taking the equation (6.36) and boundary conditions (6.40) into account) the following relation:

$$\int_0^1 \Omega_j\Theta_m\left(\theta_s+\overline{\theta}\right)d\rho = \int_0^1 \Omega_m\Theta_j\left(\theta_s+\overline{\theta}\right)d\rho; \tag{6.52}$$

and similarly for Θ_m and Θ_j,

$$\int_0^1 \left\{ \int_0^\rho \lambda^2 \left[\Theta_j(\rho)\Theta_m(\lambda) + \Theta_m(\rho)\Theta_j(\lambda) \right] d\lambda \right\} d\rho = 0 \quad . \tag{6.53}$$

In deriving this last formula, we made use of (6.52) and assumed that $k_j^2 \neq k_m^2$.

The relationships ((6.52) and (6.53), which define the orthogonality of the eigenfunctions, enable us to follow the procedure put forth by Collatz [1963], by which it follows that all of the eigenvalues k_n^2 are real. We omit the proof here, because there is nothing new to add to Collatz's demonstration. We also infer from (6.53) that only one out of all the eigenfunctions Θ_n can hold its sign in the integration interval; the remaining ones must be of an alternating sign. If even as much as two of the eigenfunctions held their sign in the interval, it would contradict (6.53).

There is also another interesting question connected with the stability of equilibrium states. Up to this point, we have been dealing with oscillations and ignoring possible dissipative forces. In the case of systems with a finite number of degrees of freedom, the influence of the viscous damping on the stability of equilibrium states is well known [see, e.g., Appel, 1932]. Using the example of shallow shells, we shall prove below that a state of stable equilibrium, when acted upon by conservative forces alone, remains stable when dissipative forces such as viscous damping are also taken into account. Damping can be defined by

$$R = -\mu \frac{\partial w}{\partial \tau}, \tag{6.54}$$

where the parameter $\mu > 0$ is the damping factor.

The equations of small vibrations, similar to eqs. (6.32), with a term that accounts for effects of dissipative forces, can be written in the form.

$$L(\delta\theta) = \frac{m}{\rho} \left\{ \delta\omega\left(\theta_s + \bar{\theta}\right) + \omega_s \delta\theta + \frac{1}{2}\left(s\frac{\partial^2}{\partial\tau^2} + \mu\frac{\partial}{\partial\tau} \right) \left[\rho^2 \int_\rho^1 \delta\theta d\rho + \int_0^\rho \rho^2 \delta\theta d\rho \right] \right\}. \tag{6.55}$$

Equation (6.38) will then take the form

$$\ddot{\varphi}(\tau) + \frac{\mu}{s}\dot{\varphi}(\tau) + p^2\varphi(\tau) = 0. \tag{6.56}$$

The behavior of $\varphi(\tau)$ is determined by the two roots r_i of the characteristic equation of (6.56), which are given by:

$$r_{1,2} = \left[-\mu \pm \sqrt{\mu^2 - 45k_n^2 / m} \right] / 2s. \tag{6.57}$$

The analysis of these roots gives the following results:

1) If $k_n^2 < 0$, then $r_1 > 0$, $r_2 > 0$ and $\varphi(\tau) = c_1 e^{r_1 \tau} + c_2 e^{-|r_2|\tau}$. (6.58)

Consequently,, when at least one eigenvalue k_n^2 satisfies $k_n^2 < 0$, the equilibrium state is unstable [Malkin, 1966], since, in that case, $\varphi(\tau)$ grows monotonically with the passage of time τ, and the functions $\delta\theta(\rho,\tau)$ and $\delta\omega(\rho,\tau)$ will therefore become unbounded.

2) When $k_n^2 > 0$, two variants are possible:

a) $\mu < 45\, k_n^2/m$, then $r_{1,2} = -a \pm ib$ (where a and $b > 0$). The corresponding expression for $\varphi(\tau)$ is of the type:

$$\varphi(\tau) = e^{-\alpha\tau}\,(c_1 \text{ sine } b\tau + c_2 \cos b\tau), \tag{6.59}$$

and the shell executes damped oscillations about the equilibrium state.

b) $\mu > 45\; k_n^2/m$, then both $r_i < 0$, and the shell moves aperiodically to the equilibrium state.

Consequently, the tested equilibrium state is asymptotically stable.

When $k_n^2 = 0$, we have $r_1 = 0$; $r_2 < 0$, and the equilibrium state is indifferent or, in the terminology that we used before, critical.

From the above discussion, we arrive at the conclusion [Appel, 1932; Timoshenko, 1955], that a state of stable equilibrium, when acted upon by conservative forces alone, will remain stable when dissipative internal forces are added to the system. This conclusion follows from a theorem proved by Tait and Thomson [see Koiter, 1965], according to which dissipative forces that are added to a conservative system cannot alter the nature of the equilibrium state. It thus follows that the sign of the eigenvalue k_n^2 can fully serve as a criterion of stability in the small.

e) Equations for small vibrations around equilibrium states of nonshallow caps.

By the same procedure as for shallow shells (see the previous parts of the present section), the linearized equations of small oscillations around the equilibrium state of non-shallow caps can easily be obtained. These equations have, of course, a more complicated form than their counterparts based on shallow shell assumptions. We shall take the Reissner equations (1.25) and (1.26) (derived in Section 1.1; see also (1.44) and (1.45)). The basic unknowns are the horizontal component of the displacement $u(\xi, \tau)$, and the rotation $\Phi(\xi, \tau)$, which determines the orientation of the normal to the midsurface of the cap. In order to obtain the equations of motion, it is necessary to add the d'Alembert forces as an externally distributed pressure $q^* = -s^* \dfrac{\partial^2 w_n^*}{\partial t^2}$, where s^* is the specific mass per unit area of the midsurface, and w_n^* is the displacement of the midsurface in the normal direction of the surface in the undeformed state. w_n^* is dependent on w^* and u^* in the following way:

$$w_n^* = w^* \cos \Phi - u^* \sin \Phi, \tag{6.60}$$

where w^* is the vertical component of the displacement of the cap. Asterisks denote here the physical values of the variables. The d'Alembert forces due to motions tangential to the shell midsurface and rotations of the cross-sections of the cap are neglected in comparison to those that are due to the acceleration terms $\ddot{w}_n$. This simplification is fairly common in the theory of shells.

The development of the small vibration equations are again started from the decompositions:

$$u(\xi,\tau) = u_s(\xi) + \delta u(\xi,\tau); \quad \Phi(\xi,\tau) = \Phi_s(\xi) + \delta\Phi(\xi,\tau), \tag{6.61}$$

where u_s and Φ_s denote the basic static equilibrium state. Using the method of separation of variables for solving the linearized boundary and initial value problem, we can obtain equations and relations similar to those valid for shallow shells, as discussed earlier. Thus, no new features emerge when the equations are extended to general shells; therefore, we do not present this extension here. This also counts for the resulting solutions. We refer to Shilkrut and Vyrlan [1977] for details.

6.3 Investigating stability of equilibrium states of shallow spherical shells by the method of small oscillations.

The numerical solution of this problem concerns the determination of the spectrum $\left\{k_n^2\right\}$ and corresponding eigenfunctions. It can be carried out by means of an approach described in [Shilkrut, 1984; Shilkrut & Vyrlan, 1976c, 1977]. First of all, the functions $\omega_s(\rho)$, and $\theta_s(\rho)$ are obtained by a shooting method that solves the boundary value problem of the equations of equilibrium. After this, it is necessary to solve the linearized set of equations ((6.36) and 6.37) with (6.39), with boundary conditions 6.40).

a). Consider again the states that exist at $M_K = 7.5$. Of these seven states, only two, 1 and 7, are stable. At the other states, at least one eigenvalue is negative. At state 2 and 6, the first eigenvalue k_1^2 is negative, at state 3 and 5, the first two (k_1^2 and k_2^2) eigenvalues are negative, while in state 4 (the center of symmetry), the first three eigenvalues are negative. Consequently, the conclusions that we reached in the previous case also apply for this case. Note that, whenever an eigenvalue vanishes, an extremal point of the deformation curve is reached.

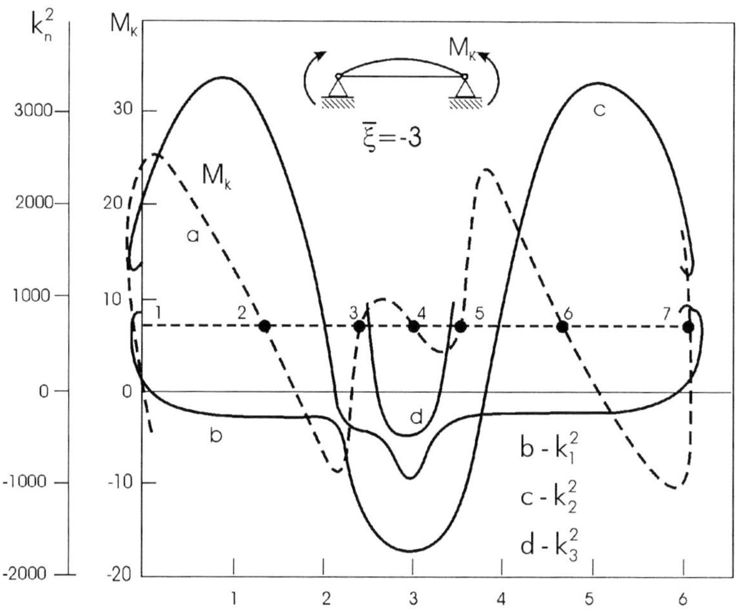

Figure 6.23 Solutions for the cap with $|\xi| = 3$.

We can also draw one additional and, in our view, crucially important conclusion from the results of Figure 6.23. As the load parameter varies, some of the eigenvalues tend to come close to each other; i.e., the spectrum of natural frequencies becomes denser at some places. This phenomenon differs from the effect observed by Bolotin [1966] of the nonuniform, banded structure of the spectrum of eigenfrequencies. Bolotin's conclusions concerned linear problems and therefore relate here only to a distribution of eigenfrequencies for $M_K = 0$. In the case of a nonlinear problem, such as the cap under consideration, the eigenfrequencies depend on the intensity of the load. Their distribution therefore varies with the load, as is clearly evident from the graphs plotted in Figure 6.24.

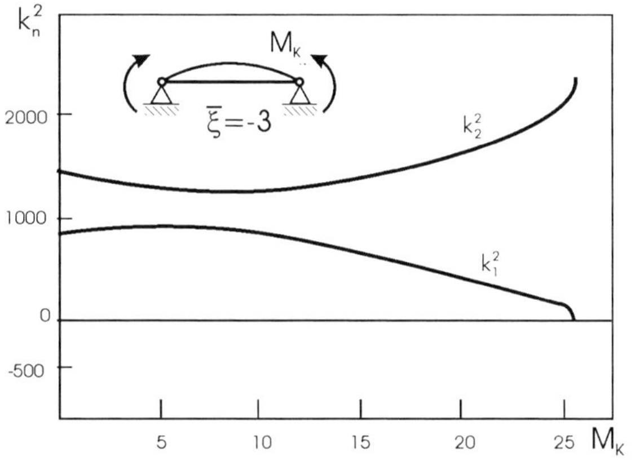

Figure 6.24 The behavior of the first two eigenfrequencies as a function of the load.

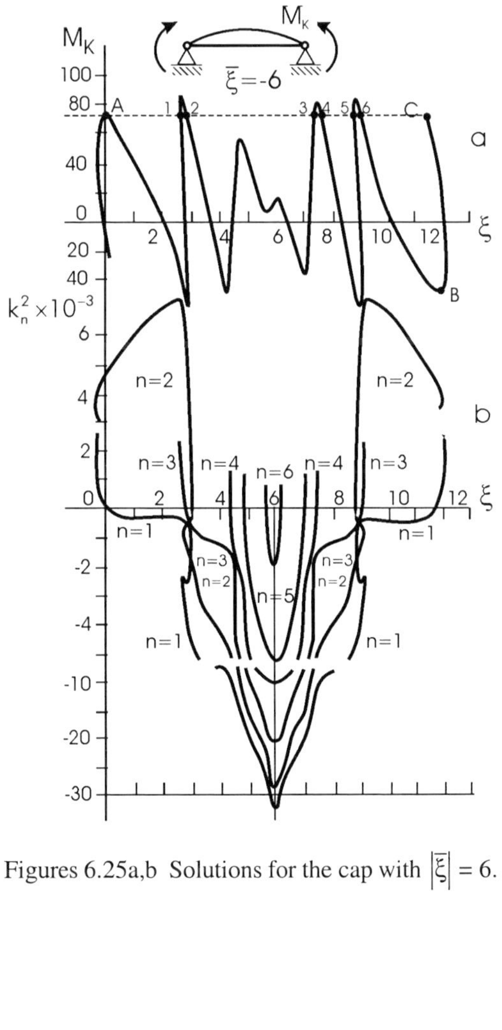

Figures 6.25a,b Solutions for the cap with $|\bar{\xi}| = 6$.

Table 6.1

Forms			$d\theta_1$					$d\theta_2$				
Fig.	q	ξ	A					A				
6.29			-0.10	-0.05	0.00	0.05	0.10	-0.10	-0.05	0.00	0.05	0.10
a	240	0.239821	0.0509	0.0129	0.0000	0.0122	0.0493	0.0403	0.0111	0.0000	0.0102	0.0380
b	240	0.396854	0.0336	0.0087	0.0000	0.0072	0.0302	-0.0300	-0.0095	0.0000	-0.0016	-0.0142
c	240	1.34670	0.0322	0.0081	0.0000	0.0077	0.0391	0.0263	-0.0055	0.0000	-0.0092	-0.0326
d	240	1.49731	0.0421	0.0105	0.0000	0.0100	0.0410	-0.0291	-0.0047	0.0000	-0.0117	0.0409
1	270	1.63368	0.0276	0.0071	0.0000	0.0062	0.0254	-0.0452	-0.0106	0.0000	-0.0128	-0.0484
2	270	3.63238	0.0100	0.0036	0.0000	-0.0009	0.0008	-0.1027	-0.0303	0.0000	-0.0116	-0.0649
3	270	3.92826	0.0117	0.0040	0.0000	-0.0005	0.0024	-0.1049	-0.0307	0.0000	-0.0128	-0.0692
4	270	1.88003	0.0374	0.0095	0.0000	0.0086	0.0350	-0.0503	0.0118	0.0000	-0.0147	-0.0557

energies for only the first two forms of deviations ((6.70) and (6.71)). Actually, it is clear from Table 6.1 that the energy Π has a minimum with respect to $\delta\theta_1(p)$ for all forms of equilibrium (a, b, c, d) at level $q_1 = 240$, while the energy for these states of equilibrium (except state a) have a maximum for the second type of deviations (6.71). Accordingly, the states of equilibrium labeled by points b, c, d in Figure 6.29 are unstable. We also realize from the table that all the equilibrium configurations (1, 2, 3, 4) that exist at $q_2 = 270$ are unstable, since the energy at all these states has a maximum of the potential energy with respect to the second type of deviations $\delta\theta_2$.

Finally, we note that, the correspondence between the Trefftz criterion and the small vibration criterion only exists if the structure is conservative. If this is not the case, the small vibration analysis is still valid as a test for stability, but this is no longer the case with the Trefftz method. The small vibration criterion is therefore the more general of the two.

b) Description of the jumping (snap-through) process.

When the considerations are restricted to axisymmetric deformations, we arrive at the conclusion that snapping, or jumping, occurs at the first limit point of the response curve and ends at that part of this curve that emerges after the last minimum (snap-back point). Jumping is a transient process that corresponds to a motion that makes its way around all the unstable equilibrium states that may be encountered. The jumping process is described by the equations of motion of the shell.

The nonlinear equations of motion for axisymmetric deformations of a cap of revolution are given by (6.24) and (6.25), where it is noted that the equation for $\omega(\rho,\tau)$ is formally the same as in the static case. As a specific example, we select again the rigidly clamped spherical cap with $|\xi| = 7$ subjected to external pressure. (The static deformation curve of this cap is given in Fig. 6.29.) In order to investigate the snap-through process, we shall load the investigated cap in accordance with the recipe:

$$q(\tau) = q_0[1 - \exp(-\alpha\tau)], \tag{6.73}$$

where q_0 is the maximum value of q attained in the limit $\tau \to \infty$, and τ is the dimensionless time.

This concept of loading allows us to vary its rate over a broad range, from quasi-static loading to impulsive loading, simply by manipulating the coefficient α. In order to have the transient process close to the static process along the stable part of the static response curve, the coefficient α is taken to be very small; i.e., $\alpha = 0.001$ to 0.01. The value of $q_0 = 300$ was assigned such that it slightly exceeded the first limit load for that shell ($q_{cr} = 245$), so that the snap-through process could be triggered. We solve the spatial part of the problem in two different ways: by Galerkin's method, with quite a few trial functions, and a finite difference method.

It was found that, in comparison with the static analysis, the transient procedure yields practically the same results up to the first limit point of the static response curve (there are very small oscillations about the equilibrium state in the transient case when α is small). When the limit load is reached (i.e., at the start of the jump), the imbalance between the internal and external forces increases drastically, even for very small α, so that the motion gains momentum. The transient part of the response curve (Fig. 6.35, dashed curve) shows that intermediate equilibrium states (points b and c on the $q(\xi)$ curve) are bypassed in the jumping process. At the new stable static equilibrium state, the shell finally executes oscillatory motions (point D in Fig. 6.35). (Incidentally, these computations were carried out for a model that possessed a small amount of damping.) The outcome thus confirms the correctness of our conclusions with regard to the instability of all the intermediate equilibrium states.

What we describe here is a transient method that is used to solve a static problem (in part, of course). This approach is very universal, and the solution process follows, to some extent, the path the actual structure would follow if it were loaded in the same manner (see (6.73)). It is of interest to note that Feodosjev [1963] was one of the first to consider the transient approach for the solution of static problems in nonlinear shell theory.

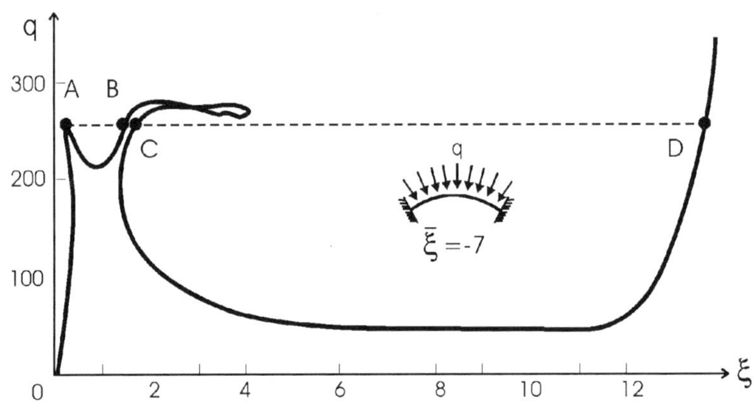

Figure 6.35 Transient response of a clamped cap under quasi-statically applied external pressure.

Due to its nature, the transient approach is only able to determine stable equilibrium states. Consequently, during the solution process, the stability of equilibrium states, if they are encountered, are automatically tested. With reference to Figure 6.35, one should observe the following. The value of the upper critical loading obtained by the transient approach is somewhat higher than the one determined statically (see the dashed curve in the region of point A). At first, it was thought that this effect, which can always be observed, was the result of unavoidable numerical errors. It appears, however, that there is a delay of the snap-through phenomena that is due to the fact that snap-through can only take place in some "preferred" direction (the escape direction), and the gain in kinetic energy at the very early stage of the snap turns out to be very slow. This effect has the appearance of the structure having some "inertia." The load will therefore exceed the static limit point by some finite amount before the snapping motion reaches full momentum. The effect of the delay in the passage through a bifurcation point for a slowly changing parameter has been discussed by Arnold [1988; see also, Riks 1998].

6.5 The total potential energy of equilibrium states and a physical explanation of the snap-through process.

In order to come to an understanding and explanation of the phenomenon of snap-through buckling, we will consider the energy levels that belong to the equilibrium states that are involved in this process. The results of such computations are shown in Figure 6.36 for a shell that exhibits no more than three solutions at some range of the load parameter. This particular problem corresponds to the case of pure bending of a cap with a geometry determined by $\left|\xi\right| = 1$ and an immovably hinged edge. It is thus one of the simplest cases that we have considered so far.

At values $M_K < 1.8$ (below the A-B level, which corresponds to the lower limit point; see the dashed curve in Fig. 6.36), the equilibrium states are unique; they are stable in the small and in the large. At $M_K = 1.8$, two equilibrium states exist; i.e., the two that are indicated by points A and B. The equilibrium state A is stable, but the one corresponding to point B is unstable.*

The value of the total potential energy along the response curve, which is also shown in Figure 6.36, clearly indicates that the energy of the equilibrium state at B is higher than that at A. This explains how the potential energy can be converted to kinetic energy during the snap-back motion. The same can be said of the jump that will take place from the limit point E toward the point G. In this case also the potential energy belonging to E is larger than the energy that belongs to G.

* Editor's Note: The point B is a limit point — in this case, a snap-back point — so that, at this state, the quadratic form of the potential energy Π_2 is semi-positive definite. It can be shown, however, that, in this critical case, the potential Π does not possess a minimum at this point.

We shall now consider another load level — say, $C\Gamma D$ ($M_K = 2.5$), where there are three equilibrium states: C, Γ, and D. The corresponding values of the potential energies are denoted by points C', Γ', and D'. Equilibrium states C and D are stable, while state Γ is unstable. If the shell somehow comes to occupy state Γ, which exhibits a higher potential energy than states C and D, then the shell will move to one of these last states. Just which one of these locations will be occupied by the shell probably depends on the direction of the disturbance. One fact stands out clearly: the shell will leave state Γ because it is unstable. On the other hand, in order for the shell to leave the stable state C and reach the stable state D, it would have to overcome an energy barrier that is at least equal to the difference between the levels that correspond to states D and C; i.e., the transition from C to D requires a supply of additional external energy.

In Figure 6.36, it can be seen that, when the load approaches the limit load at E, the difference (surplus) of the energy in the pre-snapping state with respect to the post-snapping state changes from negative to positive. In order to leave a stable equilibrium state before the limit point E, however, a local barrier around this state must be overcome, and this can only be accomplished by an appropriate externally applied impulse.

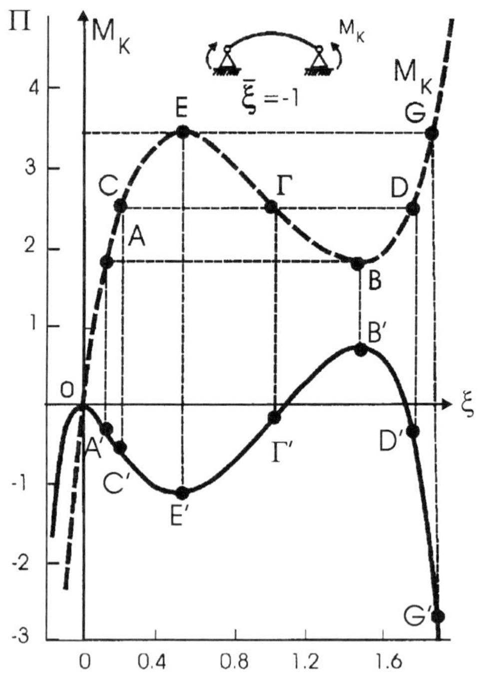

Figure 6.36 Shallow cap in bending; the energy levels.

In another example (shown in Fig. 6.37), the energy computed along the response curve of the shell is plotted as a function of the load parameter for the same problem discussed above, but this time with a geometry given by $|\bar{\xi}| = 2.16$. (The arrows in this diagram indicate the direction of the loading.) The deformation curve and graph of $k_1^2(\xi)$ that correspond to this case were presented in Figure 6.22. Let us now focus on the five solutions at the load level $M_K = 8$. It is observed that the intermediate equilibrium states (2, 3, and 4, which are unstable) possess a higher energy level than the stable forms 1 and 5.

When the load parameter attains the value $M_K = 15.3$, only two states of equilibrium exist. The energy levels that correspond to these states are indicated in Figure 6.37. The state corresponding to the critical point A has a higher potential energy, so that the shell makes its transition to state B by shedding energy. The shell does not have to overcome any energy barrier in this case and is free to transform to the stable state without resistance. Figure 6.37 also shows the jump to K from the snap-back point D.

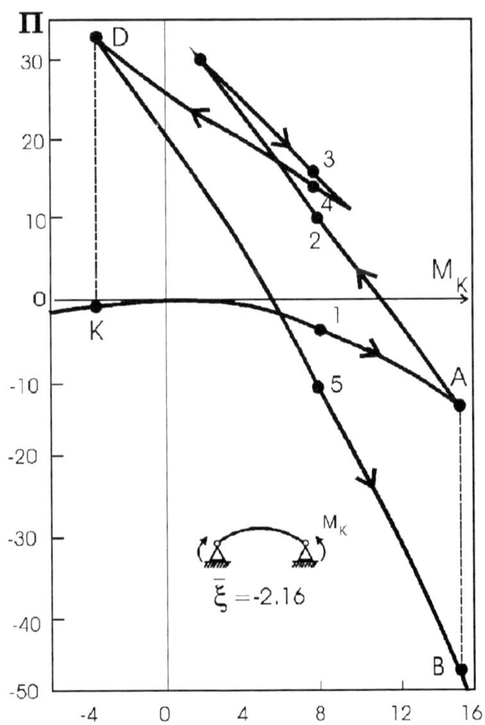

Figure 6.37 Energy versus load for the case $|\bar{\xi}| = 2.16$.

Chapter 4

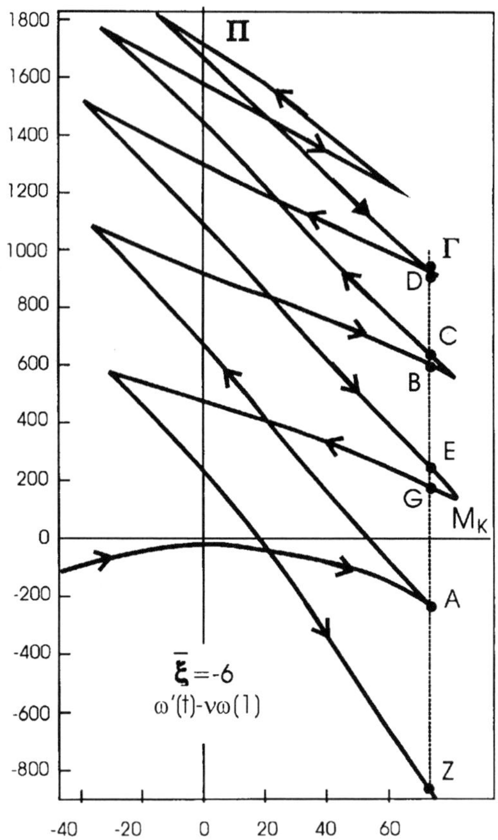

Figure 6.38 Energy versus load for the case $|\bar{\xi}| = 6$.

A contrasting picture is observed in the case of $|\bar{\xi}| = 6$. A graph of the energy Π along the response curve for this shell is shown in Figure 6.38. The same solution is also shown in Figure 6.39, where the energy variation is given by the dashed curve and the response curve is drawn as a heavy line. It is clear from this diagram that, at the level of the first limit point at A', eight equilibrium states exist beyond the state A'. The potential energy levels that correspond to these states are indicated by the same letter, but unprimed (see also Fig. 6.25).

The equilibrium state of the shell, becomes unstable at state A' (Fig. 6.39) and then moves to the stable state Z'. It would appear that, in this process, several energy barriers must be overcome that are in the way of the jumping motion. The explanation is, however, quite trivial.

The graphs that are shown in Figure 6.39, just as in the other Figures, are projections of trajectories that occur in a multidimensional space. These trajectories are paved in such a way that the energy barriers are avoided (see discussion in Section 6.1; also Fig. 6.10).

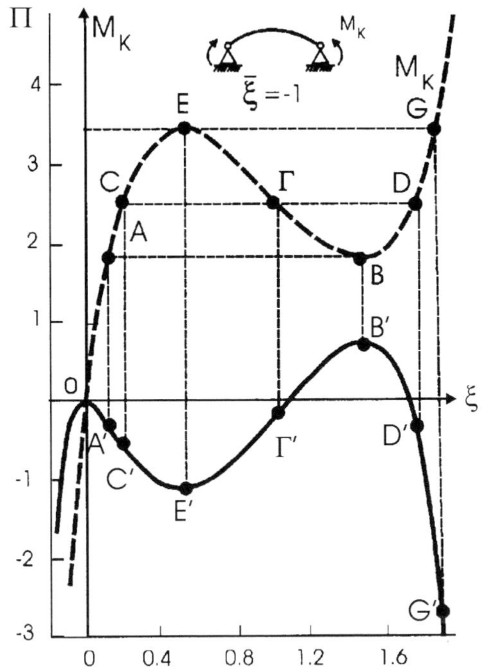

Figure 6.39 Energy and equilibrium diagram for the case $|\bar{\xi}| = 6$.

We emphasize here that it is important to recognize this aspect of the multi-dimensionality of the snap-through phenomenon, because, in many publications of repute, incorrect explanations about energy barriers have been given, either directly or indirectly [e.g., Keller & Wolfe, 1967; Thompson, 1979; Thompson & Hunt, 1984; Valishvili, 1970]. Thus, the question about these barriers has been clarified, at least in principle. But then the following, rather important, question has not yet been answered. How can the transition from one stable equilibrium state to another, which is also stable, be realized when the energetic levels are different in the sense that the energy in the other stable state is lower than the original?

Consider, for example, the transition from stable state 1 to stable equilibrium state 5 at the level $M_K = 8$ in Figure 6.22. The "intermediate" equilibrium states are labeled by 2, 3, and 4.

The graph of the energy $\Pi(M_K)$ in this case is given in Figure 6.37, where it is shown that the energy levels in positions 2, 3, and 4 are higher than in 1 and 5, while the energy of the cap in position 1 is higher than in position 5.

We can now ask the question why there is no (spontaneous) transition from state 1 to 5 while the difference between the levels of energies in 1 and 5 is finite and positive.

In answering this question, we describe the following experiment. Keeping the load constant at $M_K = 8$ in state 1, a small disturbance (in terms of initial conditions) is applied to the cap. This will start an oscillation with small amplitudes around this equilibrium state. The magnitude of these oscillations naturally depends on the value of the initial perturbations. If the latter exceed some "critical" level, then a jumping phenomenon can take place, and, as a result, oscillations around the new stable state 5, or oscillations of very large amplitudes around both states 1 and 5, can occur. Thus, in the infinite dimensional phase space of the system, one must expect the existence of areas of attraction for stable (in the small) states, similar to the case of autonomous, dynamic, conservative systems of one degree of freedom. If the initial perturbation of equilibrium state 1 is in the domain of attraction of this state, then the corresponding oscillations will occur around this state and remain within this domain.

In the opposite case, the oscillations will "escape" from this domain of attraction. Such a phenomenon is characteristic in particular for conservative dynamic systems of one degree of freedom. We shall now consider such a system here.

The stable and unstable equilibrium states of the system are represented in the phase plane as "singular" points of a center or saddle type that alternate with each other (see, e.g., the simplest case shown in Fig. 6.40). (In this terminology, a singular point is an equilibrium state, a center is a stable equilibrium state, and a saddle point is an unstable state).

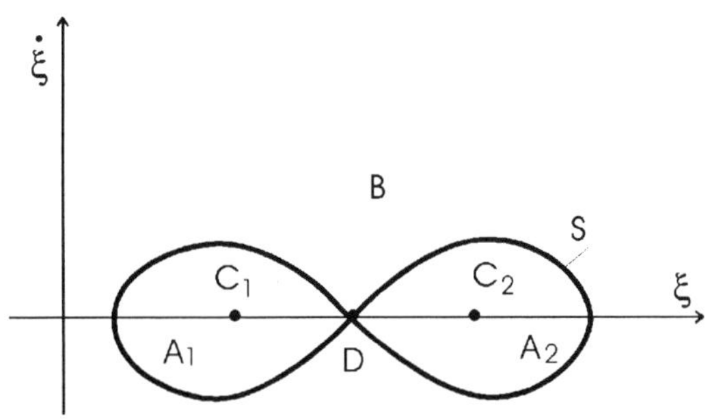

Figure 6.40 One dimensional harmonic oscillator in the phase plane.

The two centers, C_1 and C_2 (stable equilibrium states), between which there is a saddle D (unstable equilibrium state), are enclosed by a curve S (the so-called "separatrix," which is similar to a number 8. Curve S has no singular points except at D, where it intersects itself.

Fields A_1 and A_2 within the loops of the separatrix are the areas of attraction for centers C_1 and C_2, respectively. The trajectories inside $A_1(A_2)$ correspond to oscillations around $C_1(C_2)$. Field B outside separatrix S (Fig. 6.40) is the area of attraction that determines the oscillations around three equilibrium states together: C_1, C_2, and D. Thus, the separatrix is the boundary of attraction for the areas A_i. Special degenerate cases can occur when, for example, C_1 and D coalesce. The new compound singular point (say K) will then represent an unstable equilibrium state, and only one area of attraction will remain around equilibrium state of center C_2. If we want to transit the system from stable state C_1 in area A_1 to C_2 in area A_2, it is necessary to intersect the separatrix S, which represents an energy barrier, since singular point D is located on the way. The transition can only be realized by introducing into the system the corresponding portion of energy that will make overcoming the barrier possible. These are the main lines of behavior of conservative, autonomous, dynamic systems of one degree of freedom, which are described by ordinary differential equations.

Our case of the dynamics of a spherical cap must be based on a qualitative theory of a system of nonlinear partial differential equations, but such theory does not yet exist. It is not known which type of orbits are the analog to the specific set of solutions in the phase plane of systems of one degree of freedom, but it seems that they must be, to some extent, similar and determined by the boundaries of the corresponding areas of attraction in the multi-dimensional phase space. These areas of attractions — or, more precisely, their boundaries — erect the energy barriers that prevent the spontaneous transition from one stable state to another state (with less energy). Such a transition can only be effectuated by the addition of a certain amount of kinetic energy in order to overcome these obstacles (barriers).

The situation described above changes when, by variation of the load, the topology of the potential energy distribution changes and the energy barrier around the equilibrium state in question becomes impaired — usually in only one particular direction — so that transition to another stable state can occur spontaneously; i.e., with only an infinitesimally small amount of additional energy to induce the escape motion. This situation arises first at unstable critical states of equilibrium. At an ordinary unstable state, the energy barrier is definitely breached in at least two opposite directions. Departure from these states is spontaneous and its course has a snap-through character [see, e.g., the analysis by Riks, 1997].

Thus, the basic characteristic, discovered here, that all intermediate equilibrium states of our axisymmetrically deforming caps are unstable, and that, for any value of the load parameter, only one or, at most, two stable equilibrium states exist, can be considered proved and explained from all points of view. It is interesting to note that this basic feature is also present in the simplest possible approximation of the geometrically nonlinear shell as a system of one degree of freedom when the deformation curve is a simple cubic parabola. Consequently, this feature seems to be an invariant that does not depend upon the type of shell. It could therefore well be a general property that is valid for all types of elastic deformations of shells, and not only for the axisymmetric deformations of elastic spherical shallow shells considered here. Of course, these remarks are meant for single-parameter load systems only. For multi-parameter load systems, the behavior is much more complicated (see Section 6.8).

6.6 Stability of axisymmetric equilibrium states of deep spherical caps subjected to one-parametric systems of loading.

The two equations that describe the equilibrium state of deep caps of revolution undergoing axisymmetric deformations are given by (1.44) and (1.45). We shall consider the case of a shell of uniform rigidity and ignore thermal effects, and make use of the notation introduced in Section 1.1. The cap's opening angle is 2θ. The initial slope $\overline{\Phi}$ along the meridian varies in the interval $0 \leq \overline{\Phi} \leq \theta$, where, naturally, the value $\overline{\Phi} = 0$ is attained at the apex of the cap. The cap's rise H, in nondimensional form $|\xi|$, is given by the formula:

$$H = R(1 - \cos\theta); \; |\xi| = H/R. \tag{6.74}$$

Thus, the initial geometry of the nonshallow cap is determined by two parameters (e.g., the initial radius R and the opening angle θ), and this circumstance complicates the comparison of results, because shallow shells are only characterized by one parameter $|\xi|$.

For nonshallow caps, the dimensionless parameter λ_n^2 used as an eigenvalue is connected with the corresponding parameter k_n^2 (used above) for shallow caps (see Section 6.2) by the formula:

$$k_n^2 = \frac{a^4}{R^2 h^2} \lambda_n^2. \tag{6.75}$$

This relation results from the expressions for k_n^2 and λ_n^2;

$$\lambda_n^2 = m \frac{\gamma R^2}{Eh} p_n^2; \quad k_n^2 = m \frac{\gamma a^4}{Eh^3} p_n^2; \quad m = 12(1 - \nu^2).$$

Here γ is the specific weight of the cap's material, and p_n is the dimensionless proper frequency. If the approximate relation $|\xi| \cong a^2/2Rh$ is used (see(3.2)), the relation (6.75) gains the following approximate form:

$$k_n^2 \cong 4\xi^2 \lambda_n^2. \tag{6.76}$$

We shall consider some illustrative examples that will lead us to the main qualitative conclusions. In all these cases, we consider a shell with $R/h = 100$; $\nu = 0.3$, an immovable, hinged edge, and a load that consists of an edge moment M_K. When the rise parameter is given by $|\xi| = 2$ — which, according to (6.74) corresponds to the angle $\theta \approx 0.2$ — no more than five forms of equilibrium exist for the same M_K value (see the dotted curve $M_K(\xi)$, Fig. 6.41). In this figure, it is also shown that the dependence of the first eigenvalue λ_1^2 on the deflection ξ (dimensionless displacement of the apex of the shell) is similar to the behavior of $k_1^2(\xi)$ in Figure 6.22 (curve *b*). At the beginning, λ_1^2 is positive and equal to 25.9 in the unloaded state; but it immediately decreases as M_K increases, and vanishes when M_K attains the critical value at point *A* on the $M_K(\xi)$ curve in Figure 6.41. After that point, the eigenvalue λ_1^2 becomes negative and remains so until M_K reaches its lower critical value (point *D*), after which λ_1^2 becomes positive again.

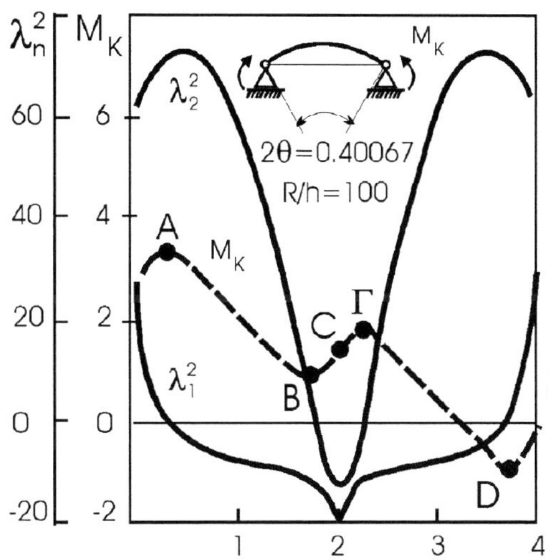

Figure 6.41 Results for a deep shell, case $2\theta = 0.4$, $R/h = 100$.

At the center of symmetry, λ_1^2 attains its minimum value. Inflection points of the curve for λ_1^2 occur at the intermediate extremal points on the deformation curve (points *B* and Γ). The second eigenvalue λ_2^2 decreases slightly with increasing M_K at the outset, then increases in the neighborhood of the critical value (point *A*), but decreases again as M_K changes along the deformation curve, and it becomes zero when M_K attains the value at the second limit point of the deformation curve (point *B*). For all forms of equilibrium corresponding to points lying between *B* and Γ, both eigenvalues, λ_1^2 and λ_2^2, are negative. All the equilibrium states between *A* and *D* then have at least one negative eigenvalue, which provides evidence of the instability of those states.

Figure 6.42 presents results for the shell $\theta = 0.283794$; $R/h = 100$, corresponding to $|\xi| = 4$, which admits as many as nine solutions at some range of the load parameter. These results also show that the parameter λ_3^2 takes on a positive value along the part of the equilibrium response curve up to the third extremum, and also along the part of the equilibrium branch that is skew symmetric with respect to this part. For all the remaining intermediate states of equilibrium, $\lambda_3^2 < 0$. The eigenvalue λ_4^2 vanishes at the fourth extremal point, remains negative for all the forms of equilibrium denoted by points between the fourth and fifth extrema, and then later becomes positive again. The first and second eigenvalues, λ_1^2 and λ_2^2, of this shell obey the same rules, as do the corresponding eigenvalues in the preceding case. For eigenvalues higher then the fourth (λ_n^2; $n > 4$) is positive everywhere; but only the first, λ_1^2 determines the stability of the equilibrium states, as discussed before. In the case of nonshallow shells, all the forms of equilibrium up to the first limit point (upper critical point) and beyond the last extremum of the deformation curve (the snap-back point) are stable, at least in the small, but the remaining, intermediate states are unstable.

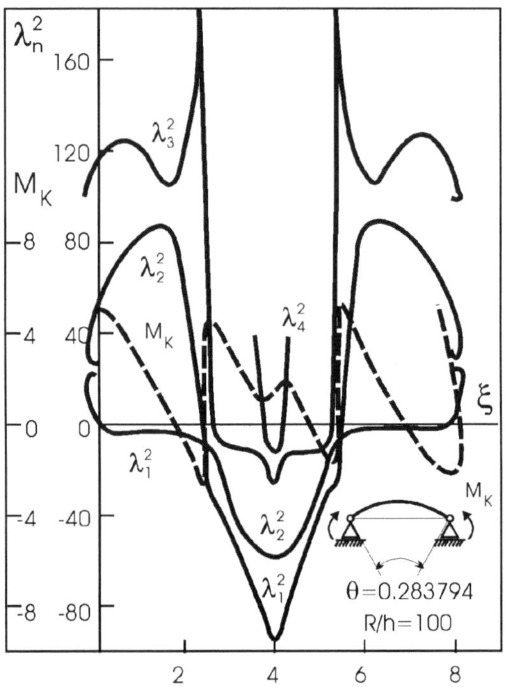

Figure 6.42 Results for a deep shell; the case $2\theta = 0.283$, $R/h = 100$.

It is worth comparing the magnitudes of eigenvalues for shallow and nonshallow caps for the same value of $|\xi|$. Let us do this for $|\xi| = 6$ for the shallow shell in Figure 6.25 and for the deep shell in Figure 6.43. In this case $\theta = 0.348166$ and $R/h = 100$. A part of the dependence of the first three eigenvalues λ_n^2 upon the load is shown in Figure 6.43.

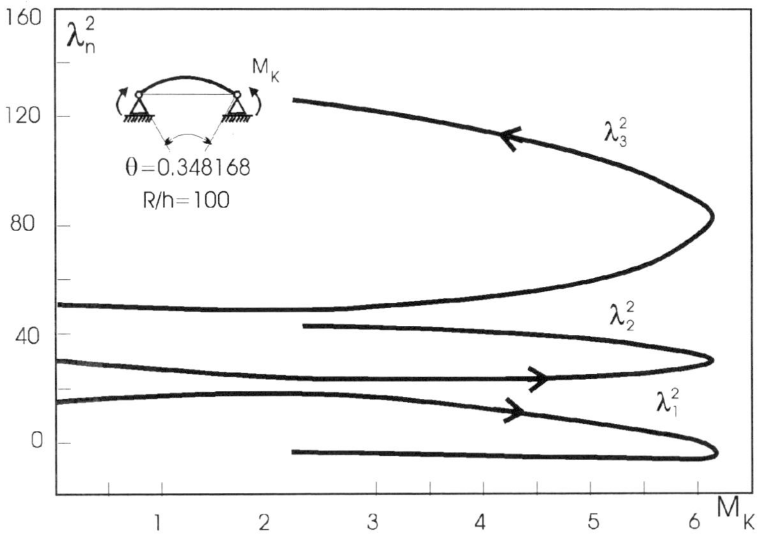

Figure 6.43 The variation of λ_1^2 with M_K.

In the case of a shallow shell, the first three eigenvalues have the following value (see Fig. 6.25) in the undeformed state:

$$k_1^2 = 2279, \quad k_2^2 = 4096, \quad k_3^2 = 5700. \tag{6.77}$$

In the case of the nonshallow shell, the corresponding results are:

$$\lambda_1^2 = 15.61, \quad \lambda_2^2 = 28.2, \quad \lambda_3^2 = 50.24, \tag{6.78}$$

which can be converted by (6.75) to:

$$k_1^2 = 2250, \quad k_2^2 = 4081, \quad k_3^2 = 7250. \tag{6.79}$$

Comparing these values, we note that the first two k_n^2 almost coincide. The discrepancy is within 1%, while the third eigenvalue for the nonshallow shell is higher than that for the shallow shell by a factor of 1.27. In order to show that the first two eigenvalues of shallow and nonshallow shells virtually coincide over the entire range of the load considered here, we computed them at a relatively large value of $M_K = 60.95$. The transition from the nondimensional moment in the case of shallow cap M_k^s to the one used in the case of nonshallow shell M_k^d is given by the formula

$$M_k^s \cong 2|\xi|M_k^d. \tag{6.80}$$

The latter results from the following expressions:

$$M_k^s = \frac{M_k^* a}{D}; \quad M_k^d = \frac{M_k^* R}{D}, \tag{6.81}$$

where M^* is the real (physical) moment; consequently, the dimensionless edge moment M_k^s = 60.95 for a shallow shell corresponds to the value of M_k^d = 5.08 for the deeper shell. It turns out that the first eigenvalues practically coincide, while the second eigenvalues differ only by 0.25%. The difference between the two third eigenvalues, however, is more than 20%. The same picture emerges for other values of M_K.

The examples given above show that the character of the solutions (the multiplicity of the equilibrium states and their stability) found for shallow shells does not change if the shallow shell model formulation is expanded to the more general (nonshallow) model forumultation. In other words, the use of general nonlinear shell equations will not lead to the discovery of new phenomena that are not already described by the shallow shell equations.

6.7 Comparison of reported numerical results and data of other authors about eigenfrequencies of nonlinear shells.

Little attention has been given to the calculation of eigenfrequencies of geometrically nonlinear shells. There are only a few sources we can refer to [Evensen & Evan-Iwanowski, 1967; Margolias & Weingarten, 1971; Okubo & Whittier, 1967; Valishvili, 1970]. The influence of the frequency on the amplitude of the oscillations of shells, in the case of forced oscillations, has been discussed in several papers [e.g., Grigolyuk, 1955; Grossman, Koplik & Yu, 1969; Rabinowitch, 1968]. Rayleigh-Ritz, Galerkin, and other direct variational techniques were used here, which resulted in the reduction of the governing equations to a system of one degree of freedom. Because the type of oscillations must be specified *a priori* when these methods are applied, sometimes substantial errors occur; in particular, in the case of thin spherical shells. Moreover, the direct variational approach (in the classical sense; i.e., with the help of global analytical functions) is difficult to apply and does not always provide satisfactory accuracy, especially when the frequency spectrum is dense [Korenev, 1972]. Of course, at the present time, there are finite element programs available by means of which one can calculate the spectrum of small oscillations around a nonlinear equilibrium state of a shell of general shape under very general conditions, Their description is, however, outside the framework of this book.

In the evaluation of the results of some of the researches where frequencies were computed for shells modeled as a continuum [Evensen & Evan-Iwanowski, 1967; Margolias & Weingarten, 1971; Okubo & Whittier, 1967; Valishvili, 1970], we will begin by comparing the eigenfrequencies of unloaded, geometrically nonlinear shells. In this case, the computed eigenfrequencies must be identical to the frequencies computed by the linear theory, as we argued before, and these frequencies depend solely on the shell geometry and mode of support. Figure 6.44 shows curves depicting the dependence of the "reduced" lower frequency ω (ω = the ratio of the lowest natural frequency of the shell with respect to the corresponding frequency of a circular plate with the same boundary conditions and with the same radius as the basis of the shell. The reduced frequency is plotted as a function of the rise parameter $|\bar{\xi}|$ for shallow shells with a rigidly clamped edge.

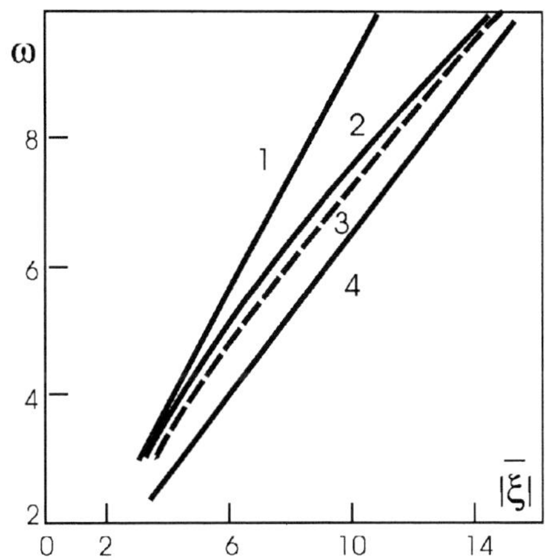

Figure 6.44 Frequency ratio ω as function of $|\bar{\xi}|$.

Curve 2 in Figure 6.44 is plotted on the basis of a frequency formula given in the reference book by Gontkevitch [1964], and curve 4 is based on a formula from Oniashvili [1957]. Curve 3 was plotted on the basis of our own numerical computations [see Shilkrut & Vyrlan 1975a, 1976c, 1977; Vyrlan & Shilkrut, 1978]. Clearly, the results that are the closest to ours (curve 3) are those based on the formula from Gontkevitch [1964] (curve 2). Some small discrepancies between the last two results are the consequence of the approximate transposition of the initial shape of nonshallow shells to the corresponding geometric parameters for shallow shells. (Note that Gontkevitch's paper contains some inaccuracies in the section on spherical shells.) Curve 4 tends asymptotically to curves 2 and 3 when $|\bar{\xi}|$ increases. Deviations between curve 4 with our data and with Gontkevitch's data is due to the fact that the frequency formula shown in Oniashvili [1957] is a crude approximation and therefore inaccurate. But as $|\xi|$ increases, this error goes to zero.

In order to compare the results to those obtained by a first approximation, whereby the shell is replaced by a system of one degree of freedom, curve 1 belonging to data presented by Grigolyuk [1955] is plotted in Figure 6.44. As might be expected, these results are less accurate. Frequencies computed by Grigolyuk for unloaded shells are generally too high, since approximation of continuum systems by a system with one degree of freedom is equivalent to the imposition of additional constraints, which cause an increase in the natural frequencies.

The effect of initial stresses on the frequencies of natural oscillations has previously been shown by Kalnins [1979] and Yang and Kim [1973]. The initial stresses are here specified arbitrarily in the form of a function of the coordinates, and they are thus not related to an equilibrium state of the structure.

We now proceed to analyze the results reported in Evensen and Evan-Iwanowski [1967], Margolias and Weingarten [1971], Okubo and Whittier [1967], and Valishvili [1970], where frequencies were calculated on the basis of a nonlinear continuum model. Evensen and Evan-Iwanowski [1967] considered the responses of a geometrically nonlinear, shallow, spherical cap rigidly clamped and subjected to uniformly distributed loads that varied with time by the law $P(t) = P_0 + P_1(t)$, where the transient component $P_1(t)$ was a periodic function of time, while the component P_0 was constant. The results were plotted in the form of the variation of the natural frequencies with the load parameter, this within the range from zero to the critical value of the load. Figure 6.45 shows the variation of $k_1^2(q)$ obtained by these authors (curve a) and by our investigation (curve b), in the case of a spherical segment ($|\xi| = 8$) with rigidly clamped edges loaded by uniform pressure. (Curve b is obtained by interpolation of our results for shells with $|\xi| = 5$, 7, and 9.7 [Shilkrut & Vyrlan, 1975a, 1977; Vyrlan & Shilkrut, 1978].)

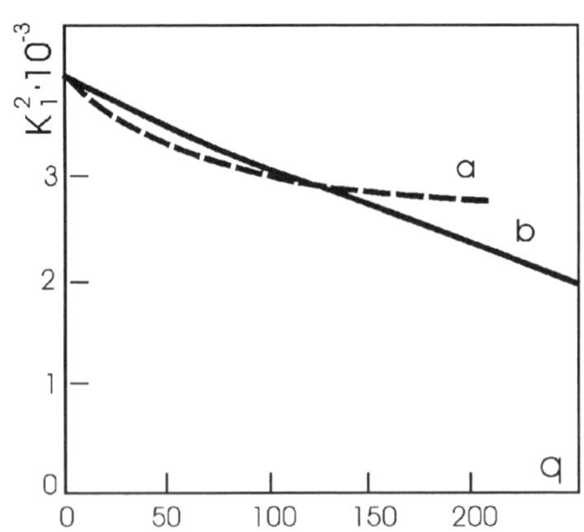

Figure 6.45 A comparison of our results with those of Evensen and Evan-Iwanowski [1967].

Okubo and Whittier [1967] tested six, nominally identical, shallow spherical shells to determine the effect of pressure on resonance frequencies. Extrapolation of the measured frequency vs. pressure load relation allows one to determine the critical pressure of these shells in a nondestructive way. The static external pressure was applied pneumatically, and small, forced, nonsymmetric oscillations with three nodal wavelengths were excited by a system of electromagnets. The lowest resonance frequency for each of the six shells tested was selected as the determining parameter in the experiment. The pressure at which this frequency vanishes corresponds to the first critical bifurcation or limit load. A graph of the normalized frequency plotted as a function of normalized pressure was presented in this paper [Okubo & Whittier, 1967]. (The normalized frequency here is the actual frequency divided by its peak value, while the normalized pressure is the actual pressure divided by its critical value). The dashed curve in Figure 6.46 is a reproduction of results from Okubo and Whittier for a spherical shell $|\xi| = 9.7$, clamped rigidly along the edges and loaded by a pressure q.

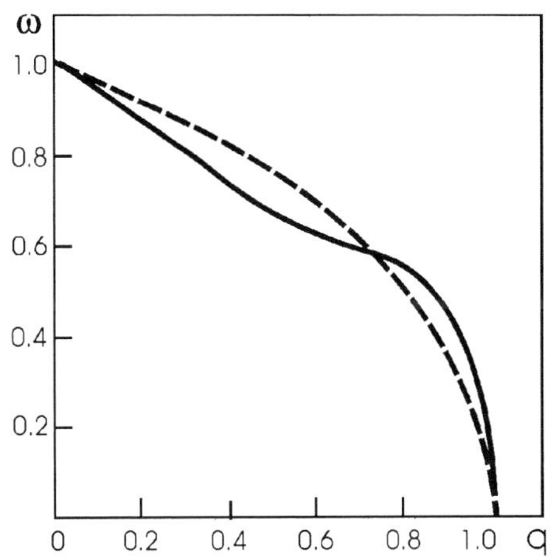

Figure 6.46 Comparison of our results with those obtained by Okubo and Whittier [1967].

Okubo and Whittier assumed that the lowest frequency belongs to a mode with three nodal waves. This assumption accounts for the discrepancy between our results (the heavy curve in (Fig. 6.46)) and the Okubo and Whittier results (the dashed curve). The discrepancy provides evidence that the eigenfrequency of the nonsymmetric vibration modes analyzed by Okubo and Whittier in the range of normalized pressures from 0.0 up to 0.73 is not the lowest frequency. In this interval, the lowest frequency is related to an axisymmetric mode. Only when the normalized pressure reaches 0.73 does the frequency belonging to a vibration mode with three waves become less than the frequency of oscillations for the axisymmetric mode. This means that the point where the curves intersect (pressure ≈ 0.73) is the change-over point of the solutions, after which the axisymmetric mode goes over into a nonsymmetric mode with three waves. However, this point does not correspond to the lowest load at which bifurcation of the solutions occurs. The transition of the axisymmetric basic state of equilibrium to a nonaxisymmetric state takes place for this shell in a mode with four waves at the value of the load parameter of 0.65 [Shilkrut & Gerlaku, 1969; Gerlaku & Shilkrut, 1971].

Margolias and Weingarten [1971], whose published results were the closest to ours, analyzed the influence of the load parameter on the lowest eigenfrequency of vibrations around the axisymmetric of equilibrium states. In Figure 6.47, the lowest eigenvalue k_1^2 is plotted as a function of pressure q for shells clamped along their edges. The top pair of curves is obtained for $|\xi| = 15.15$, and the bottom curves for $|\xi| = 9.7$. Curves b and d were plotted on the basis of results obtained by Margolias and Weingarten, while curves a and c were plotted on the basis of our own data. For $|\xi| = 15.15$, the results coincide, whereas significant discrepancies occur in the case of the $|\xi| = 9.7$ shell. We believe that some kind of error crept into Margolias and Weingarten's work, since their curve for $|\xi| = 9.7$ has a rather unusual shape.

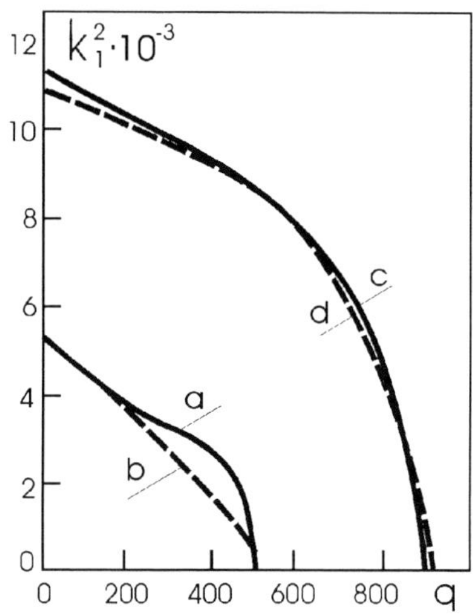

Figure 6.47 Comparison with the results obtained by Margolias and Weingarten [1971].

It is clear from this short survey that, so far, little attention has been given either to the computation of eigenfrequencies, or to the frequency dependence of the amplitude of the oscillations (in the case of forced nonlinear oscillations). Nowhere has use been made of the eigenfrequencies for a direct study of the stability of the forms of equilibrium of the shells. At best, the critical load was determined as the load at which the lowest frequency vanishes [Archer

& Lange, 1965]. Moreover, the frequencies that were calculated in all these cases only belonged to the first part of the basic state of the shell (i.e., for load parameter values below the upper critical value), and no effort was made to determine the spectrum of frequencies in postcritical states of equilibrium of the shell [see, e.g., Archer & Lange, 1965]. Given the scarcity of these data in the literature, it is not possible to compare small vibration analyses over the entire range of equilibrium solutions, the post-critical ones included.

In general, the determination of the eigenfrequencies of small oscillations around stable equilibrium states of nonlinear shells and plates is a separate, but important, problem, which, as is mentioned above, is often omitted. Some results for hinged circular plates that undergo pure bending are presented in Figure 6.48 in terms of the relation $k_1^2 = k_1^2(M_K)$. The plates are circular orthotropic with the coefficient of orthotropy $k = \sqrt{E_\varphi / E_r}$, where E_φ and E_r are the Young's moduli in circumferential and radial directions, respectively. Considered were the cases $k = 0.5$, 1, and 1.2.

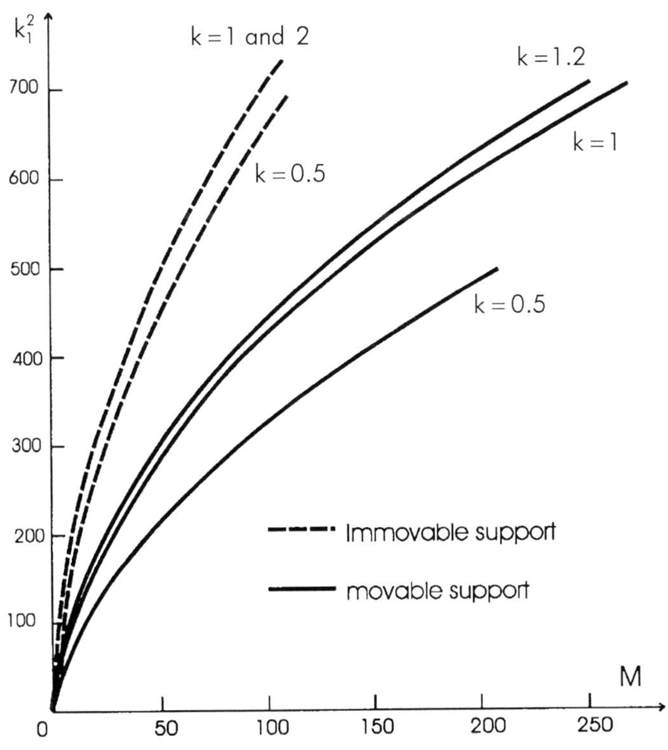

Figure 6.48 Some results for hinged circular plates in pure bending.

6.8 Concluding remarks on the stability of single parametric loading systems.

This brings to an end the presentation of the author's main results of investigations of the stability of all the possible axisymmetric equilibrium states of isotropic, spherical caps of uniform rigidity subjected to a single-parameter loading. These investigations are for the present the only efforts (to the best of our knowledge) that consider the whole set of equilibrium solutions of an important class of nonlinear shells. Despite the fact that these results are obtained for a specific group of problems, they are qualitatively of a general character. The conclusion that there are one or two (and no more) stable (in the small) solutions at a given value of the external load, is, in our opinion, valid in the most general class of conservative geometrically nonlinear structures — in particular, the class of geometrically nonlinear shells.

This particular property is already present in the first, rather crude approximation of the behavior of a cap, where the deformation curve is a cubic parabola. But it remains an essential aspect of the solutions, independent of the degree of approximation and the number of possible solutions the equations of equilibrium admit. In the case of a multi-parameter load system, however, the phenomena are more complicated, and these are considered in the next section.

a) *Stability of equilibrium systems of shallow spherical caps loaded by multi-parameter load systems.*

Up to this point, the results discussed were related to the simplest case of loading. We shall now consider the stability of equilibrium states that are induced by multi-parameter load systems. Some of these results were already discussed in other parts of the book, but here we shall look at them again in an attempt to provide additional and new information that will facilitate the understanding of the mechanical behavior of these systems.

Figure 2.10 presents a family of deformation curves $M_K(\xi)$ each of which corresponds to a fixed value of the pressure q. This case refers to a mobile-hinged cap with $\left|\xi\right| = 4$. The curve for $q = 0$, the "base" curve for this family, is a regular curve of a reciprocal system (in the sense discussed in Sections 2.2-2.6) of a type similar to a cubic parabola. When the values of $|q|$ increase, the difference between the upper and lower limit points (in absolute values) decreases, and for $|q| \geq 30$, the limit points disappear. In that case, all equilibrium states are stable. The same physical features characterize the family $q(\xi)$ when $M = \text{const.}$, as is shown in Figure 2.11, and the more complicated case (with the addition of a third loading factor, the edge membrane force $N = N_r^*(1)a^2 / Eh^3 = -5$), represented by the graphs in Figure 2.12. Thus, the additional factor, either q or M, independently from their sign, can be used to change the stability characteristics of the structures response, completely eradicating the loss of stability in the class of axisymmetric equilibrium states. (This is an interesting and important fact for a shell, where the change of sign of the load leads to different behavior.)

In Figures 4.11, 4.12, and 4.13, three cases are presented of a movable-hinged, rotationally orthotropic, circular plate subjected to a uniformly distributed edge load N and bending moment M. In the case presented in Figure 4.11, we have $M = 0$, while in Figure 4.12, $M =$ const. $\neq 0$. In Figure 4.13, $M = eN$ is generated by the edge load N applied eccentrically with eccentricity e. The latter is considered positive when the moment is acting in the clockwise direction. In all these three figures, the unstable states are marked by crosses.

The stability of these states was tested by the method of small oscillations. The corresponding dependencies of the first eigenvalue k_I^2 on N are given for the case $M = 0$ in Figure 6.49 for three different values of the coefficient of the orthotropy coefficient k (see (4.4)).

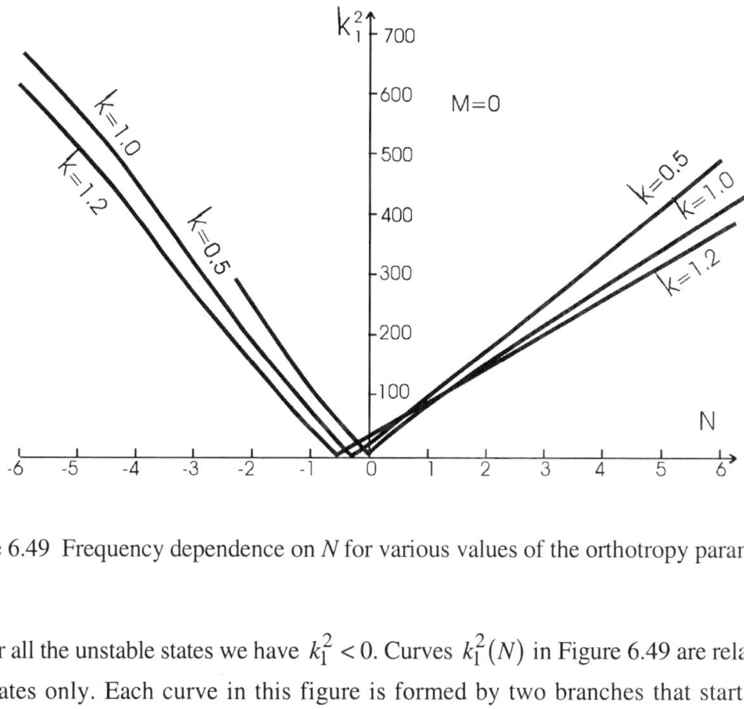

Figure 6.49 Frequency dependence on N for various values of the orthotropy parameter k.

For all the unstable states we have $k_I^2 < 0$. Curves $k_I^2(N)$ in Figure 6.49 are related to the stable states only. Each curve in this figure is formed by two branches that start from the corresponding bifurcation point ($k_I^2 = 0$). The right-hand branches correspond to the flat, prebuckling equilibrium states, and the left-hand ones characterize the post-critical stable states. Along the pre-critical states, k_I^2 increases linearly with N. In the post-critical region, these relationships are nonlinear. Thus, after bifurcation, the plate's response follows either one of the two branches of the graph of $N(\xi)$ (see Fig. 4.11) starting from N_1, but this assertion is only valid for axisymmetric deformations, to which the discussions in this section are restricted. In general, secondary bifurcation phenomena may occur, which are linked with the transition from the axisymmetric (post-critical) state to the nonsymmetric post-post-critical state. This effect was, in part, investigated by Cheo and Reiss [1973] for isotropic plates (see Section 3.3).

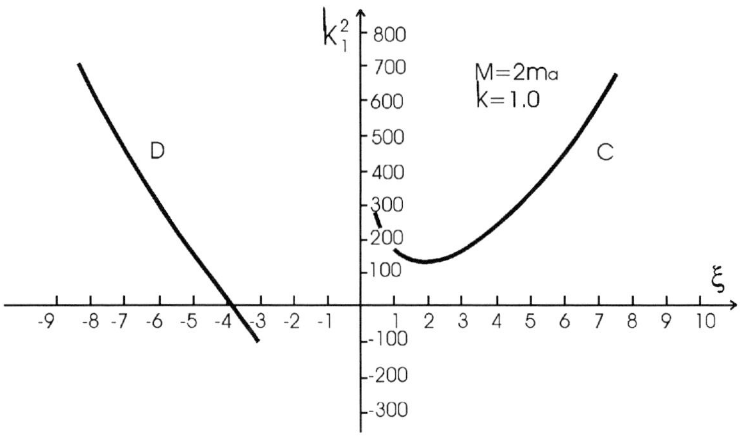

Figure 6.50 $k_{\text{I}}^2 = k_{\text{I}}^2(\xi)$ for the stable branch in Figure 4.12.

The graph $k_{\text{I}}^2 = k_{\text{I}}^2(\xi)$ for the stable, rather complicated, equilibrium branch in Figure 4.12 is presented in Figure 6.50 for $k = 1$. This curve consists of parts C and D. Part C corresponds to the right stable branch of the deformation curve shown in Figure 4.12, and D is related to the left stable branch of the curve in Figure 4.12. The minimum on part C (Fig. 6.50) corresponds to the inflection point I shown in Figure 4.12. The case of load eccentricity is shown in Figures 4.13, and in 6.51 and 6.52.

Consider now a three-parameter load system in a case that concerns the axisymmetric thermo-elastic deformation of a circular plate (see Fig. 4.29). The stability of the equilibrium states for $\gamma_1 = -3$ is determined by the graph of $k_{\text{I}}^2 = k_{\text{I}}^2(\xi)$ shown in Figure 6.53, where the curves O and L correspond to the equilibrium branches given in Figure 4.29. The same picture is available for other sections of the map, as, for example, depicted in Figure 4.30. It will be remembered that, in the case of immovable clamped, shallow, spherical segments under uniform external pressure, the solution set contains isolas (beginning from some value of $|\xi|$) (see Section 2.1, Fig. 2.6), but these closed loops are unstable (Section 6.3, part b).

Many other examples similar to these can be presented. They all seemed to confirm the existence of an important characteristic of the behavior of geometrically nonlinear shells and plates; i.e., there are no more than two, and not less than one, *stable* equilibrium states at each specific value of the external load. Quite recently, however, three stable equilibrium states were detected [Shilkrut, 1992a] in a case with three control parameters like the one pictured in Figure 4.29, which depicts a deformation curve in a different section ($\gamma_1 = -3.5$), and it is drawn in Figure 4.31. The three stable states are found in the area close to $M/m_a \approx 2$. This surprising result prompted us to conduct a more thorough investigation.

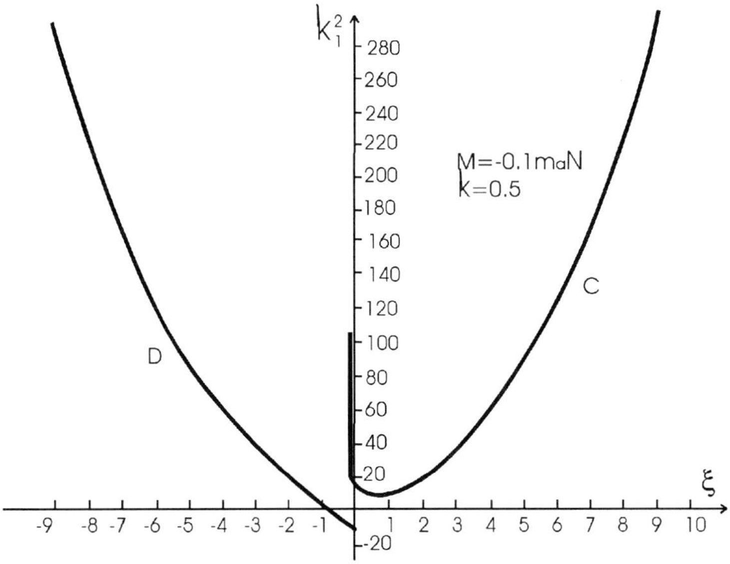

Figure 6.51 $k_1^2 = k_1^2(\xi)$ for the solutions in Figure 4.13, $k = .5$.

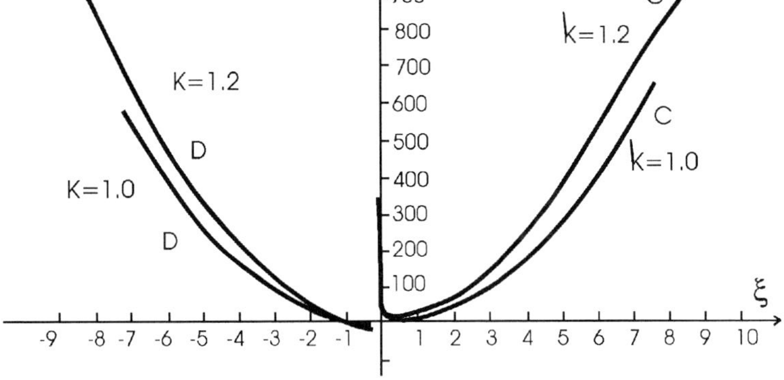

Figure 6.52 $k_1^2 = k_1^2(\xi)$ for the solutions in Figure 4.13, $k = 1$.

In Figure 6.54, section $\gamma_1 = -7.48$ of the deformation map for $q = 50$ (Fig. 4.36) is plotted, while all the other parameters are the same as in the case described in Figures 4.29, 4.30, and 4.31 [see, Ashkenazi, 1994; Shilkrut, Ashkenazi & Klimovitsky, 1996]. As can be seen from the solution curve in Figure 6.54, there are in this case four limit points (*A*, *B*, *C*, and *D*) that separate stable and unstable branches of the deformation curve in the section in question. The stable branches are marked by symbols I, II, and III (from the left to the right).

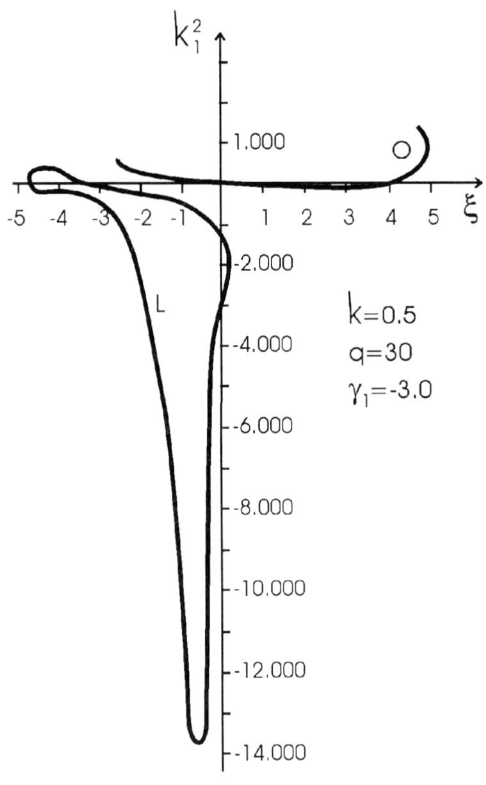

Figure 6.53 $k_1^2 = k_1^2(\xi)$ for the solutions of a three-parameter load system in Figure 4.29.

The jumping processes that start at the limit points *A* and *D* (Fig. 6.54) do not present anything new, because, at the critical values of loading parameters, there are only two stable equilibrium states. A different situation arises at point *B*, from which the system can jump into one of two possible stable equilibrium state (points *E*, *F* in Fig. 6.54). To which point, *E* or *F*, the system will jump cannot be answered in the framework of the existing theory of the static response of geometrically nonlinear systems. This question must be considered in the context of a time-dependent formulation, because, as we noted before, the jumping process is essentially a transient phenomenon. At this point of the discussion, we note that, there is apparently a need for the development of a new branch of nonlinear mechanics — the theory of dynamic systems described by partial differential equations in space and time — which should lead to the generalization of the theories of Poincaré. In particular, it is necessary to establish methods by which the areas of attraction with their boundaries of the (stable) equilibrium states in the infinite dimensional phase space can be determined.

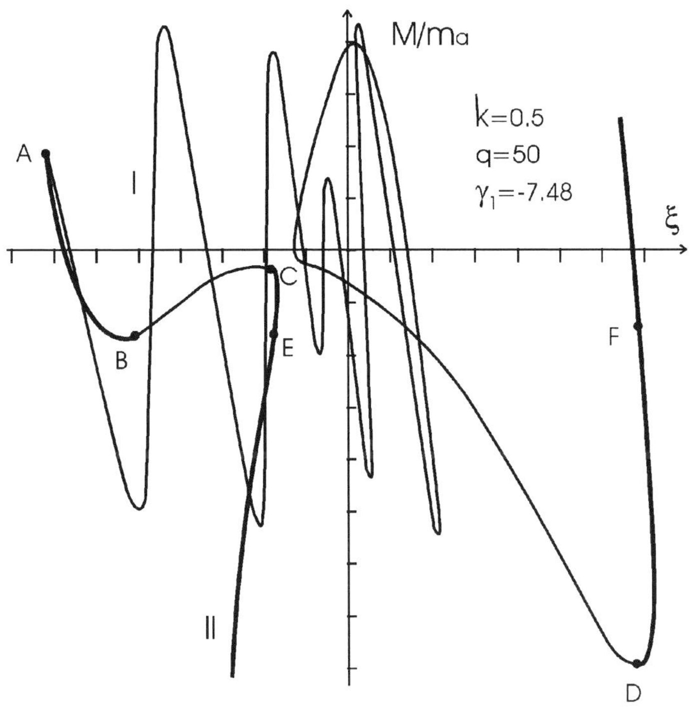

Figure 6.54 Section $\gamma_1 = -7.48$ of the deformation map for $q = 50$ in Figure 4.36.

An attempt to establish the influence factors that are responsible for the appearance of the three stable equilibrium states was reported in by Shilkrut, Ashkenazi and Klimovitsky [1996]. The case in question is related to the anisotropic plate with the coefficient of anisotropy k = 0.5. In order to study the role of k, the case k = 1 (isotropic) was also investigated, but this did not lead to any qualitative change. It was possible to arrive at the conclusion that the new phenomenon is probably caused by the thermostresses in combination with the two other factors, the parameters of the external load.

6.9 About stability regions and their boundaries.

Stability regions are the domain of parameter values where only stable equilibrium states exist. Thus, if a given combination of external loads generates one or more equilibrium states that are unstable, this combination does not belong to a stability region, but to an area of instability. It is clear that the search of the boundaries that separate the stable from the unstable zones is of fundamental importance, since this is, *de facto*, the search for the full global solution of the corresponding stability problem. The dimension of the region of stability coincides with the number of independent parameters of the external load system. For instance, in the case of one load parameter q, we have the one-dimensional area of stability $-\infty < q < q^-$ (where q^- and q^+ are the lower and upper critical loads, respectively), and $q^+ < q < \infty$. See, for example, the region of stability (Fig. 6.55) of a simply supported square plate that under-goes bi-axial compression by edge forces q_x and q_y (positive in compression). The buckling process in this case corresponds to the classical bifurcation buckling of the initially flat compressed plate, whereby the flat state of the plate is transformed into a buckled state. The dimensionless parameters $\bar{q}_x$ and $\bar{q}_y$, load intensities scaled to $(q_x)_{cr}$ and $(q_y)_{cr}$, are plotted along the horizontal and vertical axes, respectively, while $(q_x)_{cr}$ is the bifurcation load when $q_y = 0$ and $(q_y)_{cr}$ is the bifurcation load when $q_x = 0$. The boundary of the region of stability for this case is the heavy line in Figure 6.55, while the stability region itself is the semi-infinite shaded area; and n and m are the numbers of sinusoidal half-waves in the x and y directions, respectively.

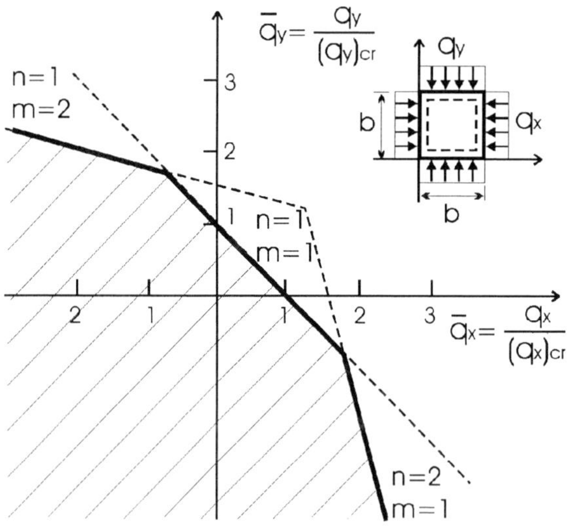

Figure 6.55 Stability zone for a bi-axially loaded plate.

A theorem exists that states that, for linear prebuckling equilibrium states, the stability boundary is always convex [Papkovitch, 1963; Volmir, 1963] (see the case depicted in Figure 6.55). For geometrically nonlinear systems such as shells, however, this theorem is no longer valid, as will be shown below.

As a characteristic example, we will consider the stability boundary of a geometrically nonlinear, isotropic, elastic, circular plate with uniform rigidity, the edges of which are movably hinged, which is subjected to a two-parameter axisymmetric load system. The stability region of this plate can be deduced from the deformation map depicted in Figure 4.9. On the vertical and horizontal axes, the moment M/m and dimensionless edge load N are plotted ($m = 12(1-\nu^2)$, where ν is Poisson's ratio).

It will be recalled that the deformation map contains the complete synopsis of the response of the plate to the influence of the two-parameter load system N and M, and that, in Figure 4.9, the points A, B, C... along the N axis represent the classical bifurcation points for the compressive load N [Timoshenko & Gere, 1961]. From each of these singular points emerge the branches $\left(S_A, S'_A\right)$; $\left(S_B, S'_B\right)$..., and so on. Outside the area enclosed by the "angle" formed by S_A and S'_A with vertex A, the nonlinear boundary problem has only one solution for each combination of M, N. Between S_A, S'_A and S_B, S'_B, there are already three equilibrium states at each pair of values of M and N, and so on. Each new pair of branches $\left(S_K, S'_K\right)$ adds two additional solutions. Keeping in mind that all the other sectors where multiple solutions exist are enclosed in the area inside the rays $\left(S_A, S'_A\right)$, it can be deduced that the branches $\left(S_A, S'_A\right)$ are the boundaries of the loading parameters (map) that separate the stable from the unstable areas. All points inside the area bounded by $\left(S_A, S'_A\right)$ are unstable equilibrium states, and the equilibrium states that correspond to points along the boundaries are the upper critical and lower critical loads, respectively. Outside of $\left(S_A, S'_A\right)$ (at the right) is the stability region. Once more, we must stress that, so far, all of the features described here are restricted to axisymmetric deformations of the plate.

A different situation arises when the symmetrically loaded circular plate bifurcates into a nonsymmetric equilibrium state; that is, when bifurcation in tension occurs (see Sections 2.1, 3.2, and 3.7). First of all, along the N axis ($M = 0$), there is the point α_1, which is the first classical bifurcation point of the hinged plate where bifurcation into a nonsymmetric mode occurs [see, e.g., Courant, 1938; Ponomarev, Biderman, Licharev, Makushin, Malinin & Feodosjev, 1959]. We have $\alpha_1 = 1.22$ for $\nu = 0.3$. (The corresponding actual physical value is $N_1 = -13.36\ Eh^3/12(1-\nu^2)\ a^2$, where a is the radius of the plate. This magnitude of $|N|$ is 3.18 times larger than the critical value of $|N|$, which corresponds to the first buckling load at which axisymmetric buckling occurs). Point α_1 lies between A and B (Fig. 4.9). There is, of course, a full discrete spectrum of α_i, ($i = 1,2,3$...). The second load, for example, is given by: $\alpha_2 \cong 2.32$ [Ponomarev *et al.*, 1959] (the corresponding value $N_2 \cong -25.08\ Eh^3/12(1-\nu^2)\ a^2$; $\nu = 0.3$).

The latter point lies, as does α_1, between the A and B. Thus, distributions of critical values for axisymmetric and for nonsymmetric buckling are not of a regular structure when each critical point for axisymmetric buckling is followed by a (corresponding) nonsymmetric critical point. It is worth noting that the member α_n of the discrete spectrum of bifurcation loads corresponds to a nonsymmetric buckling pattern with n waves in the circumferential direction, which has no nodes in the meridional direction (except for the edge of the plate). On the other hand, the magnitude of the lower critical value $M = M^* 12(1 - \nu^2)\, a^2/Eh^4$ ($N = 0$) — when bifurcation into a nonsymmetric form occurs for the first time (the loading is still axisymmetric) — is equal to ±318, where $n = 4$ and 7 [see Shilkrut, 1980; Shilkrut & Gerlaku, 1969]. At this value of M, bifurcation in tension occurs; that is, in this case, the meridian of the plate is stretched everywhere.

The corresponding points on the map are $N = 0$ and $M/m = \pm 29.121$ (see points 1 and 2 in Fig. 6.56).

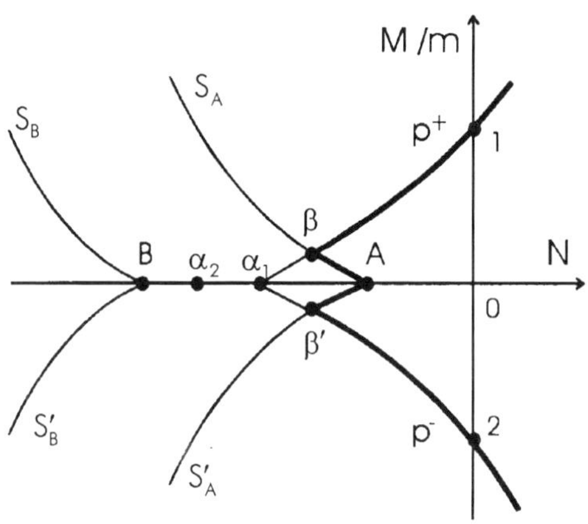

Figure 6.56 Stability boundary for a circular plate, loaded by N, M.

It must be clear that this lowest value of the bifurcation moment M will increase when N increases. It is thus expected that there will be two rays, P^+ and P^-, which emerge from point α_1 and which are the graphs of the variation of the lower bifurcation load M as a function of the edge force N (see qualitative picture in Fig. 6.56). The boundaries of the zone of stability are then the rays P^+ and P^-, which emerge from points β and β′, respectively, and the lines $A\beta$ and $A\beta'$ (see Fig. 6.56, where β, β′ are the intersections of S_A with P^+ and S'_A with P^-). This boundary is depicted in Figure 6.56 by a heavy line. The zone of stability will be inside the field bounded by the "polygon" $P^+\beta A\beta' P^-$; that is, at the right-hand side of this boundary. In this area, only one curve of the map passes through each of its points, and, as a consequence, there is here only a single equilibrium state for the corresponding load combination. Through each point of the area enclosed by the cusp $S_A \beta P^+$ and $S'_A \beta' P^-$, one curve represents an axisymmetric equilibrium state; the others that pass here through this point are related to nonsymmetric equilibrium states. The curves P^+ and P^- are probably envelopes of the latter for different numbers of circumferential waves n. Both types of curves pass through each point of the zone enclosed by the angle $S_A A S'_A$, representing, as before, an axisymmetric equilibrium state and "new" ones generated by a nonsymmetric equilibrium state (except in the quadrangle $A\beta\alpha_1\beta'$, where only axisymmetric states are represented).

To our best knowledge, these investigations are the first step to the solution of the problem of the stability boundaries of geometrically nonlinear systems. It can be expected that the main character traits of the stability boundary investigated above will be preserved for more complicated cases of geometrically nonlinear plate and shells of revolution loaded by axisymmetric loading systems. In any case, the boundary shown here indicates that the theorem concerning the convexity of the stability zone, which is valid for linear prebuckling states (see the beginning of the present section), is no longer valid in the nonlinear case.

FIGURES

Chapter 1

Chapter 2

Chapter 3

Chapter 4

Chapter 5

Chapter 6

REFERENCES

A.D. Adey and C.A. Brebbia. (1983). *Basic Computational Methods for Engineers.* New York: John Wiley & Sons.

L.M. Afanasjeva. (1970). On the stability of flexible, pressured shallow shells of revolution in a temperature field. *Mechanika Tverdogo Tela,* N2, 191-194. (In Russian).

L.M. Afanasyeva. (1975). The influence of the degree of orthotropy on the critical pressure of the orthotropic cap. *Mechanika Tverdogo Tela*, N2, 160-163. (In Russian).

L.M. Afanasjeva and B.J. Kantor. (1969). The direct computation of critical states of flexible shallow shells of revolution. *Mechanika Tverdogo Tela*, N5, 164-166. (In Russian).

N. Akkas and N.R. Bauld Jr. (1971). Buckling and postbuckling behavior of clamped shallow spherical shells under axisymmetric ring loads. *J. Appl. Mech.*, E38(4), 996-1002.

A.V. Aleksandrov, B.J. Ljastchenikov, N.I. Shaposhnikov and V.A. Smirnov. (1976). *Methods of Computing of Systems of Bars, Plates and Shells Using Computers, Part II.* Moscow: Strojizdat. (In Russian)

A.A. Andronov, E.A. Vitt and S.E. Khaikin. (1966). *Theory of Oscillators.* Oxford: Pergamon.

S.S. Antman, R.S. Marlow and C. Vlahacos. (Forthcoming). The complicated dynamics of heavy rigid bodies attached to deformable rods. *Quart. Appl. Math.*

P. Appel. (1932). *Traité de mécanique rationelle, v.4.* Paris: Gauthier-Villars. (In French)

R.R. Archer. (1962). On the numerical solution of the nonlinear equations of shells of revolution. *J. Math. Phys.*, 41, 165-178.

R.R. Archer and J. Famili. (1965). On the vibration and stability of finitely deformed shallow spherical shells. *J. Appl. Mech.,* E32(1), 116-120.

R.R. Archer and C.G. Lange. (1965). Nonlinear dynamic behavior of shallow spherical shells. *AIAA J.,* 3(12), 2313-2317.

V.I. Arnold. (1988). Bifurcations and singularities in mathematics and mechanics. Opening lecture, XVIIth International Congress of Theoretical and Applied Mechanics. IUTAM - Grenoble.

L. Ascione and D. Bruno. (1987). On the delamination problem of two-layer plate. In G. Del Piero and F. Macer, eds, *Unilateral Problems in Structural Analysis*, 2nd edn. Vienna, New York: Springer-Verlag, pp. 1-15.

R. Ashkenazi. (1994). Stability of nonlinear deformed systems. Master's Thesis: D.I. Shilkrut, supervisor). Beer-Sheva, Israel: Ben-Gurion University of the Negev, Dept. of Mechanical Engineering.

D.G. Aswell. (1959). On the large deflection of a spherical shell under inward point load. In *IUTAM Symp. on the Theory of Thin Elastic Shells*, Delft, The Netherlands.

V.L. Bajanov, I.I. Goldenblat, N.A. Nikolaenko and A.M. Sinyukov. (1969). *Computation of Structures Under Thermal Excitations.* Moscow: Mashinostrojenie. (In Russian).

P.K. Banerjee and R. Butterfield. (1981). *Boundary Element Methods in Engineering Sciences.* London: McGraw-Hill.

Yu. A. Barladeanu, P.M. Vyrlan, I.D. Gerlaku, N.V. Shevandronov and D.I. Shilkrut. (1973). Some results of solutions on electronic calculating machines of nonlinear problems of the theory of shells and plates. In *The Theory of Shells and Plates.* Moscow: Nauka, pp. 236-241. (In Russian).

J.L. Batoz, G. Dhatt and Y.P. Prost. (1975). Buckling behaviour of shells: An axisymmetrical element and a triangular element. In *3rd Intern. Conf. on Structural Mechanics in Reactor Technology*, M. 4/3, pp. 1-13.

L. Bauer, E.L. Reiss and H.B. Keller. (1970). Axisymmetric buckling of hollow spheres and hemispheres. *Communications on Pure and Applied Math.*, 23(4), 523-568.

L. Bauer, E.L. Reiss and H.B. Keller. (1973). Axisymmetric buckling of rigidly clamped hemispherical shells. *Int. J. Non-Linear Mech.*, 8, 31-39, 1973.

N.N. Benditch and V.M. Kornev. (1971). About density of eigenvalues in stability problems of elastic shells. *Prikladnaja Mathematika i Mechanika (PMM)*, 35(2). (In Russian) [An English translation exists.]

C.B. Biezeno. (1935). Uber die Bestimmung der "Durchschlagkraft" einer schwach-gekrümmth kriesformigen platte. *Z. angew. Math. und Mech.*, 15(1-2), 10-22. (In German)

C.B. Biezeno and R. Grammel. (1939). *Technishe Dynamic* (Berlin, 1939). (In German) [Pub. in English: *Engineering Dynamics, v.II.* London: Blackie and Sons, Ltd.]

D.P. Billington (1967). Note on finite symmetrical deflections in thin shells of revolution. *J. Appl. Mech.*, N3, 763-764.

S.R. Bodner. (1955). The postbuckling behavior of a clamped circular plate. *Quart. Appl. Math.,* 12(4), 397-401.

V.V. Bolotin. (1965). About the concept of stability in the theory of structures. In *Problems of Stability in the Theory of Structures.* Moscow: Izd. Lit. po Stroitelstvu, pp. 6-27. (In Russian)

V.V. Bolotin. (1966). The density of eigenvalues in the problems of vibrations of elastic plates and shells. In *Proc. VI All-Union Conference on the Theory of Shells and Plates.* Moscow: Nauka, pp. 161-168. (In Russian)

M. Born. (1906). Untersuchungen über die Stabilität der elastischen Linie in Ebene und Raum unter vershiedenen Grenzbedingungen [Investigation on the Stability of the Elastica in Plane and Space]. Dissertation. Göttingen: Dieterichsche Univ. Buchdruckerei

Göttingen. (In German)

C.A. Brebbia, J.C.F. Telles and L.C. Wrobel. (1984). *Boundary Element Techniques.* Berlin-Heidelberg: Springer-Verlag.

D.O. Brush, D.O. and B.O. Almroth. (1975). *Buckling of Bars, Plates and Shells.* New York: McGraw-Hill.

G.H. Bryan. (1891). On the stability of a plane plate under thrusts in its own plane with applications to the 'buckling' of the sides of a ship. *Proc. London Math. Soc.*, 22, 54-67.

B. Budiansky. (1960). Buckling of shallow spherical shells. In W.T. Koiter (ed), *Proc. Symp. on the Theory of Thin Elastic Shells,* IUTAM, Delft, 1959. Amsterdam: North-Holland Publ. Co., pp. 64-94.

B. Budiansky. (1965). Dynamic buckling of elastic structures: Criteria and estimates. In *Proc. Int. Conf. Dynamic Stability Structures.* New York: Pergamon Press, pp. 83-108.

B. Budiansky. (1974). Theory of bucklng and post-buckling of elastic structures. In Chia-Shun Yih (ed), *Advances in Applied Mechanics*, v.14. New York: Academic Press, pp. 2-65.

B. Budiansky and J.W. Hutchinson. (1966). Dynamic buckling of imperfection sensitive structures. In *Proc. 11th Int. Congr. Appl. Mech,* Munich, 1964. Berlin, New York: Springer-Verlag, pp. 636-651.

Bürgermeister. (1957). *Steup, Stabilitätstheorie*, v.1. Berlin. (In German)

D. Bushnell. (1967a). Bifurcation phenomena in spherical shells under concentrated and ring loads. *AIAA J.,* 5(11), 2034-2040.

D. Bushnell. (1967b). Buckling of spherical shells ring-supported at the edges. *AIAA J.*, 5(11), 2041-2046.

D. Bushnell. (1967c). Nonlinear axisymmetric behavior of shells of revolutions. *AIAA J.*, 5(3), 432-439.

D. Bushnell. (1967d). Symmetric and nonsymmetric buckling of finitely deformed eccentrically stiffened shells of revolution. *AIAA J.*, 5(8), 1455-1462.

D. Bushnell. (1989). *Computerized Buckling Analysis of Shells.* Dordrecht, Boston, London: Kluwer.

C.R. Calladine. *Theory of Shell Structures* (Cambridge (UK) & New Rochelle, NY: Cambridge University Press, 1983)

L.S. Cheo and E.L. Reiss. (1973). Unsymmetric wrinkling of circular plates. *Quart. Appl. Math.,* 31, 75-91.

Chia-Shun Yih. (1967). Finite deflections of thin shallow spherical shells under axisymmetrical line load and uniform pressure. *J. Appl. Mech.*, E34(4), 1022-1024.

Chia-Shun Yih. (1971). Buckling of thin spherical shells. *Ingenieur-Archiv,* 40, 227-237.

Chia-Shun Yih. (1976). Large deflections of axisymmetrically loaded spherical caps.

Aeronaut. Quarterly, 27(Pt 1), 29-39.

W.Z. Chien. (1947). Large deflection of a circular clamped plate under uniform pressure. *Chinese J. Phys.*, 7(2), 102-113.

L. Collatz. (1963). *Problems on Eigenvalues and Their Technical Applications.* Leipzig: Akad. Verlagsgesellschaft. (In German)

R. Courant. (1938). *Differential and Integral Calculus, v.II* [2nd English edn.]. New York/Glasgow: Backley.

R. Courant and D. Hilbert. (1953). *Methods of Mathematical Physics.* New York: Interscience Publ..

J.G.A. Croll and A.C. Walker. (1972). *Elements of Structural Stability.* London: Macmillan.

D.F. Davidenko. (1953). About approximative solution of systems of nonlinear equations. *Ukrainian Mathematical J.*, 5(2). (In Russian).

M.S. El Naschie. (1990). *Stress, Stability and Chaos in Structural Engineering: An Energy Approach.* London, Toronto: McGraw-Hill.

H.A. Evensen and R.M. Evan-Iwanowski. (1967). Dynamic response and stability of shallow spherical shells subject to time-dependent loading. *J. AIAA,* 5(5), 969-975.

J. Famili and R.R. Archer. (1965). Finite asymmetric deformation of shallow spherical shells. *AIAA J.,* 3(3), 506-510.

V.I. Feodosyev. (1945). About the calculation of a slamming membrane. *Prikladnaja Math. i Mech. (PMM)*, 9(5), 389-412. (In Russian).

V.I. Feodosjev.(1953). *Selected Problems and Questions on Strength of Materials.* Moscow: Gostechizdat. (In Russian) [There probably is an English translation. Two other corrected and expanded versions appear in Russian (Moscow, Fizmatgiz, 1967, 1973).]

V.I. Feodosjev. (1963). About one method of solving of stability of nonlinear problems of deformable systems. *Prikladnaja Mathematika i Mechanika (PMM)*, 27(2), 265-274. (In Russian)

V.I. Feodosjev. (1969). Axisymmetric elastic properties of a spherical shell. *Prikladnaja Mathematika i Mechanika (PMM)*, 33(2), 280-286. (In Russian)

M.M. Filonenko-Boroditch. (1965). *Theory of Elasticity.* New York: Dover.

J.R. Fitch. (1968). The buckling and postbuckling behavior of spherical caps under concentrated loads. *Int. J. Solids and Struct.*, 4(4), 421-446.

J.A. Fitch and B. Budiansky. (1970). Buckling and postbuckling behavior of spherical caps under axisymmetric load. *AIAA J.*, 8(4), 686-693.

W. Flugge. (1962). *Statik und Dynamik der Schallen*, 3rd edn. Berlin: J. Springer. (In German).

G. Freskasis and N. Morris. (1972). Asymmetric buckling of imperfect spherical shells. *J. of the Eng. Mech. Div. Proc. ASCE,* EM5, 1115-1131.

K.O. Friedrichs and J.J. Stoker. (1941). The non-linear boundary value problem of the buckled plate. *Amer. J. Math.*, 63(4), 839-888.

K.O. Friedrichs and J.J. Stoker. (1942). Buckling of the circular plate beyond the critical thrust. *J. Appl. Mech.*, 9(1), A7-A14.

Y.C. Fung and E.E. Sechler. (1960). Instability of thin elastic shells. In Y.N. Goodier and N.J. Hoff (eds), *Structural Mechanics. Proc. of the First Symp. on Naval Structural Mechanics*. Oxford, London, New York, Paris: Pergamon Press.

F.R. Gantmacher. (1960). *Lectures on Analytical Mechanics*. Moscow: Izd. Fizikomathematicheskoy Literatury. (In Russian)

I.D. Gerlaku, V.P. Morar and D.I. Shilkrut. (1970). Axisymmetric nonlinear elastic deformation of thin elastic shells of revolution. The influence of nonsymmetric deformations on the critical loading of symmetrically loaded shells of revolution. In *Proc. of the VIIth All-Union Conf. on the Theory of Shells and Plates*. Moscow: Nauka, pp. 172-176. (In Russian).

I.D. Gerlaku and D.I. Shilkrut. (1970). Determination of bifurcation values of loading for axisymmetrically loaded shells of revolution with allowance for nonsymmetric deformations. Izv. Akad. Nauk SSSR, *Mechanika Tverdogo, Tela,* 5(2), 184-190. (In Russian) [English transl. *Mech. Solids (USSR)*, 1971, pp. 166-172.]

A. Gjelsvik and S.R. Bodner. (1962). Nonsymmetrical snap buckling of spherical caps. *J. Eng. Mech. Div. Proc. ASCE v.88*, EM5(Pt 1), 135-167.

P.B. Goncalves and J.G.A. Croll. (1992). Axisymmetric buckling of pressure-loaded spherical caps. *J. of Structural Engineering*, 11(4), 970-985.

N.S. Gontkevitch. (1964). *Proper Vibrations of Plates and Shells.* Kiev: Naukova Dumka. (In Russian)

A.P. Gospodarikov and V.F. Terentjev. (1977). About a general algorithm of calculations of postcritical solutions of two points nonlinear boundary problems of the axisymmetrical deformations of shells of revolution. In *Questions of Mechanics and Control Processes, v.1.* Leningrad: Leningrad University, pp. 147-155. (In Russian).

E.A. Gotsuljak, V.I. Guliajev and G.I. Melnitchenko. (1974). Stability of nonshallow spherical panel under uniform distributed loads. *Strength of Materials and Theory of Structures*, 22, 109-115. (In Russian).

E.I. Grigolyuk. (1955). Nonlinear vibrations and stability of rods and shallow shells. *Izvestija AN SSSR. Otdelenje Technitcheskich Nauk*, N3, 35-68. (In Russian)

E.I. Grigolyuk. (1960). On the unsymmetrical snapping of shells of revolution. In *Proc. IUTAM Symp. Theory of Thin Elastic Shells,* Delft, 1959. Amsterdam: North Holland, pp. 112-121.

M. Gräff, R. Scheidl, H. Tröger and E. Weinmüller. (1985). An investigation of the complete post-buckling behavior of axisymmetric spherical shells. J. App. Math. and Phys. (ZAMP), 36, 803-821.

P.L. Grossman, B. Koplik and Yi-Yuan Yu. (1969). Nonlinear vibrations of shallow spherical shells. *J. Appl. Mech.,* 36(3), 451-458.

N.C. Huang. (1964). Unsymmetrical buckling of thin shallow spherical shells. *J. Appl. Mech.,* E31(3), 447-457.

G.W. Hunt, H.M. Bolt and J.M.T. Thompson. (1989). Structural localization phenomena and the dynamical phase-space analogy. *Proc. Royal Soc. London,* A425, 245-264.

J.W. Hutchinson. (1967). Imperfections sensitivity of externally pressurized spherical shells. *J. Appl. Mech.*, E34(1), 49-55.

14. J.W. Hutchinson and W.T. Koiter. (1970). Postbuckling theory. *Appl. Mech. Rev* 23(12), 1353-1366.

K. Huseyin. (1971). The post-buckling behaviour of structures under combined loading. *Z. Angew Math. Mech.*, 51, 177-182.

K. Huseyin. (1975). *Nonlinear Theory of Elastic Stability.* Gröningen: Noordhoff: .

K. Huseyin. (1978). *Vibration and Stability of Multiple Parameter Systems.* Gröningen: Noordhoff.

K. Huseyin. (1986). *Multiple-Parameter Stability Theory and Its Applications.* Oxford: Clarendon Press

J.W. Hutchinson and W.T. Koiter. (1970). Postbuckling theory. *Appl. Mech. Rev* 23(12), 1353-1366.

E. Itzhak and M. Baruch. (1986). Analogy for postbuckling structural resistance capability. *AIAA J.*, 24(10), 1719-1721.

A. Kalnins. (1979). Vibration and stability of prestressed shells. *Nuclear Eng. & Design,* 20(1).

B.J. Kantor. (1971). *Nonlinear Problems of the Theory of Nonhomogeneous Shallow Shells.* Kiev: Naukova Dumka. (In Russian)

L.V. Kantorovich and V.I. Krylov. (1964). *Approximate Methods of Higher Analysis.* New York: Interscience.

R. Kao. (1972). A note on buckling of spherical caps with initial asymmetric imperfections. *J. Appl. Mech.*, E39(3), 842-844.

R. Kao and N. Perrone. (1971). Asymmetric buckling of spherical caps with asymmetrical imperfections. *J. Appl. Mech.*, E38(1), 172-184.

A. Kaplan. (1974). Buckling of spherical shells. In Y.C. Fung and E.E. Sechler (eds)., *Thin Shell Structures, Experiment and Design.* Englewood Cliffs, NJ: Prentice-Hall, pp. 247-288.

T. von Karman and A.D. Kerr. (1965). Instability of spherical shells subjected to external pressure. *Topics of Applied Mechanics.* Amsterdam, London, New York.

T. von Karman and H.S. Tsien. (1939). The buckling of spherical shells by external pressure. *J. Aeronaut. Sci.*, 7(2), 43-50.

J.F. Kayuk. (1974). About improvement of the convergence of the method of simple iterations in nonlinear problems of plates and shallow shells. *Prikladnaja Mechanika,* 11(9). (In Russian).

H.B. Keller. (1968). *Numerical Methods for Two-Point Boundary Value Problems.* Waltham, MA: Blaisdell.

H.B. Keller and E.L. Reiss. (1958). Nonlinear bending and buckling of circular plates. In *Proc. 3rd U.S. Nat. Congr. Appl. Mech.*, Providence, RI. New York.

H.B. Keller and A.W. Wolfe. (1967). On the nonunique equilibrium states and buckling mechanism of spherical shells. *J. Soc. Ind. Appl. Math. (SIAM),* 13(3), 674-705.

H.B. Keller, J.B. Keller and E.L. Reiss. (1962). Buckled states of circular plates. *Quart. of Appl. Math.*, 20(1), 55-65.

P. Klemm and Cz. Wozniak. (1967). Perforated circular plates under large deflections. *Arch. Mech. Stosowanej*, 19(1), 45-56.

R.J. Knops and E.W. Wilkes. (1973). In *Handbuch der Physik, v.VI a/3.* Amsterdam: Springer-Verlag, pp. 125-302.

S.A. Kobrits and V.F. Terentjev. (1977). About the calculations of the curves loading — Displacement in one-dimensional nonlinear problems of the theory of rods and shells. In *Questions of Mechanics and Control Processes, v.1.* Leningrad: Leningrad University, pp. 155-172. (In Russian)

E. Kochavi and D. Shilkrut. (1984). Axisymmetric buckling of spherical caps with plane flanges. A model of the influence of edge imperfection on critical loading. *Mechanics Research Com.*, 11(3), 167-175.

W.T. Koiter. (1945). Over de Stabiliteit van het Elastisch Evenwicht. Doctoral thesis. Delft, Amsterdam. (In Dutch with English Summary). [English transl. Techn. Rep. AFFDL-TR-70-25. Air Force, USA, Wright-Patterson AFB, Ohio, 1970.]

W.T. Koiter. (1963a). On the concept of stability for continuous bodies. *Proc. Konikl. Nederl. Akad. Wet.*, B66(4), 173-177.

W.T. Koiter. (1963b) Elastic stability and post-buckling behavior. In R.E. Langer (ed), *Nonlinear Problems.* Proc. Symp. Univ. of Wisconsin. Madison, WI: University of Wisconsin Press, pp. 257-275.

W.T. Koiter. (1965). The energy criterion of stability for continuous elastic bodies. *Proc. Konikl. Nederl. Akad. Wet.*, B68(3), 107-113.

W.T. Koiter. (1967). In D. Muster (ed), *Proc. Symp. on the Theory of Shells*, in honor of L.H. Donnell. Houston, TX: University of Houston, pp. 187-277.

W.T. Koiter. (1969a). On the thermodynamic background of elastic stability theory. In *Problems of Hydrodynamics and Mechanics of Continuous Medium*, Anniv. Volume of Academician L.I. Sedov. Moscow: Nauka, pp. 277-285. (In English)

W.T. Koiter. (1969b). The nonlinear buckling problem of a complete spherical shell under uniform external pressure (I, II). *Proc. Kön. ned. Akad., Series B.*, 72(1), 40-79, and 72(2 - Pts III, IV), 81-123.

W.T. Koiter. (1971). *The Thermodynamics of Elastic Stability, Mechanika*, N.4. Moscow: Inostranaya Literatura. (In Russian)

W.T. Koiter. (1982). The application of the initial post-buckling analysis to shells. In *Buckling of Shells. A State of the Art Colloquium, v.1,* Stuttgart, May 1982. Stuttgart: Inst. für Baustatik, Univ. Stuttgart, 1.1-1.15.

B.G. Korenev (ed). (1972). *Reference Book on the Dynamics of Structures.* Moscow: Strojizdat. (In Russian)

M.S. Kornishin. (1964). *Nonlinear Problems of the Theory of Plates and Shells and Methods of Their Solutions.* Moscow: Nauka. (In Russian)

M.S. Kornishin and F.S. Isanbaeva. (1968). *Flexible Plates and Panels.* Moscow: Nauka. (In Russian)

V.M. Kornev. (1972). About approximations in the problems of stability and vibrations of elastic shells when condensation of eigenvalues takes place. *Mechanika Tverdogo Tela (MTT),* No. 2. (In Russian) [An English translation exists]

N.I. Krivosheev. (1959). On the stability of shallow spherical segments under the action of constant external pressure. *Izvestja vuzov. Stroitelstvo i Architektura,* 5(11), 48-54. (In Russian)

A.A. Kurdyumov. (1961). Concerning to the physical and geometrical nonlinear problems of bending and stability of plates and shells. In *The Works of the Shipbuilding Institute of Leningrad*, N34, 55-62. (In Russian).

C. Lanczos. (1949) *The Variational Principle of Mechanics.* Toronto: University of Toronto Press.

H.L. Langhaar. (1962). *Energy Methods in Applied Mechanics.* New York, London: John Wiley and Sons.

L.S. Leibenzon. (1917). About the application of harmonic functions to the problem of stability of spherical and cylindrical shells. *Proc. of the Yurjev University.* Tartu, 1917. Also in *Coll. of Works of the Academician L.S. Leibenson, v.1* (Moscow: AN SSSR, 1951) 50-85. (In Russian).

L.S. Leibenzon. (1943). *Variational Methods of Solving Problems of the Theory of Elasticity.* Moscow: Gostexizdat. (In Russian). [See also *Works of L.S. Leibenzon, v.1.* Moscow: Izd. AN SSSR, 1951.]

S.G. Lekhnitskii. (1963). *Theory of Elasticity of an Anisotropic Elastic Body.* San Francisco, CA: Holden-Day. (Translated from Russian)

Ta-Cheng Loo and R.M. Evan-Iwanowski. (1966). Interaction of critical pressures and critical concentrated loads acting on shallow spherical shells. *J. Appl. Mech.,* E33(3), 612-616.

A.E.H. Love. (1927). *The Mathematical Theory of Elasticity.* Cambridge (UK): Cambridge

University Press.

J. Makowski and L.-P. Nolte. (1987). Simple equations in terms of displacements for finite axisymmetric deflections of shells of revolution. *Int. J. of Non-Linear Mech.*, 22(1), 1-13.

I.G. Malkin. (1966). *The Theory of the Stability of Motion.* Moscow: Nauka. (In Russian)

L.I. Mandelshtam. (1959). *The Complete Works; v.IV: Lectures on Vibrations 1930-1932.* Moscow-Leningrad: Izd. AN SSSR. (In Russian)

D.S. Margolias and V.I. Weingarten. (1971). Free vibrations of pressure loaded paraboloidal shells. *AIAA J.,* 9(12), 2339-2345.

K. Marguerre. (1938). Zur theorie der gekrümmten platte grosser formänderung. In *Proceedings of the Fifth International Congress on Applied Mechanics,* Cambridge, MA, pp. 93-101. (In German) [English version: New York: John Wiley and Son, 1939)

L. Meirovitch. (1967). *Analytical Methods in Vibrations.* New York: Macmillan.

G.I. Melnitchenko. (1974). The stability of spherical shell under loadings of a local type. In *Strength of Materials and Theory of Structures,* 23, 56-61. (In Russian).

J.F. Mescall. (1965). Large deflections of spherical shells under concentrated loads. *J. Appl. Mech.,* E32(4), 936-938.

J.F. Mescall. (1967). Numerical solutions of non-linear equations for shells of revolution. *AIAA J.*, 4(11), 2041-2043.

J.F. Mescall. (1968). Large deflections of spherical shells under concentrated, distributed and ring loadings. In *Developments in Theoretical and Applied Mathematics. Proceedings of the Fourth Southwestern Conference on Theoretical Mechanics.* London: Pergamon, pp. 183-197.

S.G. Mikhlin. (1964). *Variational Methods in Mathematical Physics.* New York: Macmillan.

S.G. Mikhlin. (1971). *The Numerical Performance of Variational Methods.* Gröningen: Walters-Noordhoff.

F.C. Moon. (1987). *Chaotic Vibrations.* New York, Chichester, Toronto: John Wiley and Sons.

N.F. Morozov. (1958). The uniqueness of symmetric solutions of the problem of big deflection of symmetrically loaded circular plate. *Doklady AN SSSR,* 123(3), 417-419. (In Russian)

N.F. Morozov. (1961). To the question of the existence nonsymmetric solutions of the problems of big deflections of a circular plate loaded by symmetric loading. *Izvestija, vuzov. Mathematika*, 21(2), 126-129. (In Russian).

N.F. Morozov. (1963)/ Investigation of circular symmetrically compressed plate at big critical loadings. *Izv. vuzov. Mathematika*, N34, 95-97. (In Russian).

Kh.M. Mushtari and K.Z. Galimov. (1957). *Nonlinear Theory of Thin Elastic Shells.* Kazan: Tatknigoizgat. (In Russian) [There was an English translation made in 1961 by the Israel

Program for Scientific Translation.]

Yu.V. Nemirovsky. (1976). Stability and buckling constructive anisotropical heterogeneous shells and plates. In *Mechanics of Solid Deformable Bodies: v.9, Summaries of Science and Technology*, Moscow: VINITI. (In Russian).

V.M. Nenastjeva. (1970). About the influence of thermal conditions on the stability of the spherical cap. In *Proceedings of the VII All-Union Conference on the Theory of Shells and Plates.* Moscow: Nauka, pp. 448-450. (In Russian).

A.K. Noor. (1994). Recent advances and applications of reduction methods. *Appl. Mech. Rev.*, 47(5), 125-146.

V.V. Novozhilov. (1953). *Foundations of the Nonlinear Theory of Elasticity.* Rockesfer, NY: Greylock Press. (Translated from Russian)

V.V. Novozhilov. (1961). *Theory of Elasticity.* Oxford: Pergamon Press.

V.V. Novozhilov. (1964). *The Theory of Thin Shells.* Gröningen: Noordhoff. (Translated from Russian)

J.T. Oden. (1972). *Finite Elements of Nonlinear Continua.* New York: McGraw-Hill.

S. Okubo and J.S. Whittier. (1967). A note on buckling and vibrations of an externally pressurized shallow spherical shell. *J. Appl. Mech.*, 34(4), 1032-1034.

O.D. Oniashvili. (1957). *Some Dynamical Problems of the Theory of Shells.* Tbilisi: Izd. AN GSSR. (In Russian)

J.R. Otter. (1965). Computations for prestressed concrete reactor pressure vessels using dynamic relaxation. *Nuclear Structural Engineering*, 1, 61.

P.O. Panagiotopoulos. (1985). *Inequality Problems in Mechanics and Applications. Convex and Nonconvex Energy Functions.* Boston, Basel, Stuttgart: Birkhaüser.

D. Yu. Panov. (1939a). On big deflections of circular plate. *Transactions of the Central Aero-Hydronimcal Inst. (TzAGI),* No. 450. (In Russian)

D. Yu. Panov. (1939b). Application of the method of Academician B.G. Galerkin for solutions of some nonlinear problems of the theory of elasticity. *Prikladnaja Math. i Mech. (PMM)*, 3(2), 139. (In Russian).

D. Yu. Panov and V.I. Feodosiev. (1948). On equilibrium and stability loss of shallow shells with big deflections. *Prikladnaja Math. i Mech.,* 12(4), 334-406. [also *Prikladnaja Math. i Mech.,* 13(1), 116, 1949.] (In Russian)

J.G. Panovko and I.I. Gubanova. (1965). *Stability and Oscillations of Elastic Systems (Paradoxes, Fallacies and New Concepts).* New York: Consultants Bureau Larrick. (Translated from Russian)

P.F. Papkovitch. (1963). *Works on Navy Structural Mechanics (selected works), v.4.* Leningrad: Sudpromgiz. (In Russian)

N. Perrone and R. Kao. (1970). Large deflection response and buckling of partially and fully loaded spherical caps. *AIAA J.*, 8(12), 2130-2136.

V.V. Petrov. (1975). *The Method of Successive Loadings in the Nonlinear Theory of Plates and Shells.* Saratov: University of Saratov. (In Russian).

T.H.P. Pian and P. Tong. (1971). Variational formulation of finite-displacement analysis. In B.F. de Veubeke, ed, *High Speed Computing of Elastic Structures*, Proc. of the Symp. of IUTAM, Liege, 1970. Liege: University of Liege, pp. 43-63.

A.V. Pogorelov. (1966). *The Geometrical Theory of Shell Stability.* Moscow: Nauka. (In Russian)

A.V. Pogorelov. (1967). *Geometrical Methods in the Nonlinear Theory of Elastic Shells.* Moscow: Nauka. (In Russian)

A.V. Pogorelov. (1986). *Bending of Surfaces and Stability of Shells.* Moscow: Nauka. (In Russian)

H. Poincaré. (1885). Sur l'équilibre d'une masse fluide animée d'un mouvement de rotation. *Acta Mathematica,* 7, 259-283. (In French)

H. Poincaré. (1902). *Figures d'équilibre d'une masse fluide.* Paris: Gauthier-Villars. (In French)

S.D. Ponomarev, V.L. Biderman, K.K. Licharev, V.M. Makushin, N.N. Malinin and V.I. Feodosjev. (1959). *Calculations of Strength in Machine Building, v.3.* Moscow: Maschgiz. (In Russian)

R.I. Rabinowitch. (1968). About free vibrations of flexible shallow shells. *Stroitelinaja Mechanika i Rastchet Soorujeni*, N4. (In Russian)

E. Ramm. (1981). Strategies for tracing the nonlinear response near limit points. Nonlinear FEM analysis in structural mechanics. In *Proc. of the Europe-USA Workshop,* Ruhr University, Bochum. Berlin, Heidelberg, New York: Springer-Verlag.

E. Reissner. (1946). Stresses and small displacements of shallow spherical shells. *J. Math. and Phys.,* 25, 80-85.

E. Reissner. (1950). On axisymmetrical deformations of thin shells of revolution. In *Proc. of Symposia in Appl. Math., v.3.* New York: McGraw-Hill, pp. 27-52.

E. Reissner. (1958a). On variational principles in elasticity. *Proc. Symposia Appl. Math.*, 8, 1-6.

E. Reissner. (1958b). Symmetrical bending of shallow shells of revolution. *J. Math. and Mech.*, 7(2), 121-140.

E. Reissner. (1969). On finite symmetrical deflections of thin shells of revolution. [trans. ASME, Series E, v.91, 267-270].

E. Riks. (1971). An incremental approach to the solution of snapping and buckling problems. *Int. J. of Solids and Structures*, 7, 1581-1599.

E. Riks. (1973). *The Incremental Solution of Some Basic Problems in Elastic Stability.* Rep. NLRTR 74005 U of The National Aerospace Laboratory NLR of the Netherlands.

E. Riks. (1998). Buckling analysis of elastic structures: A computational approach. In E. van der Giessen and Th. Y. Sih (eds), *Advances of Applied Mechanics, v.34*. Cambridge, MA: Academic Press, pp. 1-81.

E. Riks and C.C. Rankin. (1997). Computer simulation of the buckling behavior of thin shells under quasi-static loads. *Arch. Computatl. Meth. Appl. Mech.*, December Issue, 1-16.

E. Riks, C.C. Rankin and F.A. Brogan. (1996). On the solution of mode jumping phenomena in thin-walled shell structures. *Int. J. Comp. Meth. Appl. Mech. and Eng.*, 4(4), 325-351.

S.M. Roberts and J.S. Shipman. (1972). *Two-point Boundary Value Problems: Shooting Methods.* New York, London, Amsterdam: Elsevier.

J.L. Sanders Jr. (1963). Nonlinear theories for thin shells. *Quarterly Appl. Math.*, 21, 21-36.

E. Schwerin. (1922). Zur Stabulität der dünnwandingen Hohlkugel unter gleichmässigem Ausendruck. Zangew. *Math und Mech.*, 2(2), 81-91. (In German).

P. Seide. (1974). A reexamination of Koiter's theory of initial postbuckling behaviour and imperfection sensitivity of structures. In Y.C. Fung and E.E. Sechler (eds), *Proc. Symp. on Thin Shell Structures,* Pasadena (1972), pp. 59-82.

R. Seydel. (1988). *From Equilibrium to Chaos.* New York, London: Elsevier.

V.E. Shamansky. (1966). *Methods of Numerical Solving of Boundary Value Problems on the Digital Computer, v.2.* Kiev: Naukova Dumka. (In Russian).

I.H. Shames and C.L. Dym. (1985). *Energy and Finite Element Methods in Structural Mechanics.* New York: McGraw Hill.

D.I. Shilkrut. (1965). About solution of nonlinear problems of the theory of plates and shells by the method of small loadings. In *The Works of the Polytechnical Institute of Kishinev, v.4.* Kishinev: Kartea Moldoveneaska. (In Russian).

D. Shilkrut. (1967). *Some Problems of the Nonlinear Theory of Shells and Plates.* Kishinev: Izd.-vo AN Mold. SSR. (In Russian)

D.I. Shilkrut. (1969a). Theorem of symmetry for shallow forms of equilibrium of geometrically nonlinear shallow and nonshallow shells. *Dok. Akad. Nauk SSSR,* 187(1), 64-67. (In Russian) [English translation in *Sov. Phys. Dokl.* 14(7), 724-727, 1970].

D.I. Shilkrut. (1969b). Some questions of the theory of nonshallow nonlinear elastic shells. In D.I. Shilkrut, ed, *Questions of the Nonlinear Theory of Shells and Rods.* Kishinev: Polytechn. Inst. Kishinev, pp. 3-29. (In Russian).

D.I. Shilkrut. (1974). *Problems in the Qualitative Theory of Nonlinear Shells.* Kishinev: Izd.-vo Stinjntza Kishinev. (In Russian)

D.I. Shilkrut. (1980). Solutions of some stability problems in the theory of geometrically nonlinear shells. *Israel Journal of Technology*, 18, 76-83.

D. Shilkrut. (1982). The deformation map as a means for investigating the behavior of deformable bodies. *Mech. Research. Com.,* 9(2), 77-84.

D. Shilkrut. (1983a). Bifurcation in tension of nonlinear spherical caps. *J. of Eng. Mechanics, ASCE, Eng. Mech. Division,* 109(1), 289-296.

D. Shilkrut. (1983b). Investigations of axisymmetric deformation of geometrically nonlinear, rotationally orthotropic circular plates. *Int. J. Non-Linear Mech.*, 18(2), 95-118.

D. Shilkrut. (1983c). Axisymmetrical deformation of geometrically nonlinear cylindrically orthotropic circular plates on elastic foundation. *Isr. J. of Technology,* 21, 85-90.

D. Shilkrut. (1984). Stability and vibration of geometrically nonlinear, cylindrically orthotropic circular plates. *J. Appl. Mech.*, E51, 354-360.

D. Shilkrut. (1987). Unilateral contact between a geometrically nonlinear plate and elastic foundation. A model of delamination phenomena in a two-layered composite. In AG. Del Piero and F. Maceri, eds, *Unilateral Problems in Structural Analysis*, 2nd edn. Vienna, New York: Springer-Verlag, pp. 298-312.

D. Shilkrut. (1988). The influence of the paths of multiparametrical conservative loading on the behavior of a geometrically nonlinear deformable elastic body in buckling of structures. In I. Elishakoff, J. Arbocz, C.D. Babcock, Jr. and A. Libai, eds, *Y. Singer Anniversary Volume*. Amsterdam: Elsevier, pp. 381-407.

D. Shilkrut. (1992a). Stability of equilibrium states of nonlinear structures and chaos phenomenon. Chaos in statics? *Int. J. of Bifurcation and Chaos,* 2(2), 271-283.

D. Shilkrut. (1992b). On the Influence of the paths of conservative multiparametrical loading on equilibrium states of geometrically nonlinear structures. *Computers and Struct.,* 44(1/2), 137-146.

D.I. Shilkrut and I.D. Gerlaku. (1969). Determination of bifurcation critical loads of symmetrically loaded shallow geomerically nonlinear shells of revolution taken into account nonsymmetric states of equilibrium.. In D.I. Shilkrut, ed, *Voprosy Nelinoynoi Teorij Obolotchek i Sterzney* [Questions on Nonlinear Theory of Shells and Rods]. Kishinev: Polytechn. Inst. Kishinev, pp. 37-89. (In Russian)

D. Shilkrut and E. Kochavi. (1983). Stability loss phenomena in geometrically nonlinear spherical caps under multiparamtrical loading. In *Proceedings of the 25th Israel Annual Conf. Aviation and Astronautics*, Israel, pp. 89-95.

D. Shilkrut and E. Kochavi. (1986). Asymmetrical bifurcation phenomenon in geometrically nonlinear spherical caps subjected to multiparametrical loading. *Thin-Walled Struct.,* N4, 83-99.

D.I. Shilkrut and P.M. Vyrlan. (1975a). On the stability of geometrically nonlinear shells. *Doklady AN SSSR,* 225(4), 782-785. (In Russian). [Translated into English: *Soviet Physics - Doklady*, 20(12), 1976.]

D.I. Shilkrut and P.M. Vyrlan. (1975b). Dynamic investigations of the stability of the whole multitude of the forms of equilibrium of the geometrically non-linear shallow spherical shells. *Stroitel 'naja mechanika i rastchet sooruzenij,* 5, 54-58. (In Russian)

D.I. Shilkrut and P.M. Vyrlan. (1976a). The stability of the symmetric forms of equilibrium of the geometrically non-linear shallow and non-shallow shells of revolution. *Trudy X*

Vsesoiuznoij konf. po teorij obolotchek i plastin., Misnierebe v.1 (Tbilisi) 562-568. [Report to the X All Union Conf. on the Theory of Shells and Plates, USSR, Tbilisi, 1975]. (In Russian)

D.I. Shilkrut and P.M. Vyrlan. (1976b). On the stability of the symmetric forms of equilibrium of the geometrically nonlinear shells of revolution. In B.F. Korenev, I.M. Rabinovitch and A.F. Smirnov, eds, *Scientific Annual Issledovanja po teorij sooruzenij, v.XXII*. Moscow. (In Russian)

D.I. Shilkrut and P.M. Vyrlan. (1977). *Stability of Nonlinear Shells.* Shtiyntza, Kishinev: Polytechn. Inst. of Kishinev. (In Russian). [Translation into English by US Air Force, Report FTD-ID (RS) T-0246-79, 1979, Wright-Patterson AFB. OHIO.]

D. Shilkrut, R. Ashkenazi and I. Klimovsky. (1996). About the global postbuckling behavior of geometrically nonlinear deformable systems. *Proceedings of the 36th Annual Conference on Aerospace Science.* Haifa, Israel: Technion, pp. 336-342.

D.I. Shilkrut, V.P. Morar and I.D. Gerlaku. (1970). Symmetric nonlinear elastic deformations of thin nonshallow shells of revolution. The influence of the nonsymmetric deformations on the critical loads of the symmetrically loaded shallow shells. In *The Theory of Shells and Plates, Proc. of VII All-Union Conference on the Theory of Shells and Plates*, SSSR, 1969. Moscow: Nauka, pp. 172-176. (In Russian)

D.I. Shilkrut, N.V. Shevandronov, V.P. Morar and Yu.A Maksimov. (1969). *Analog Computer Solution of Problems of Nonlinear Shell Theory by Direct Simulation and Their Digital Computation by Direct Reduction to a Cauchy Problem.* Kishinev: Izd. -vo AN Mold. SSR. (In Russian)

D.I. Shilkrut, Yu. A. Barladeanu, P.M. Vyrlan, I.D. Gerlaku, G.B. Fishberg and V.P. Tcherkezov. (1978). Some results of the theory of geometrically non-linear shells. In D.I. Shilkrut, ed., *Voprosy Mechaniki deformiruemyh sistem* [Questions of the Mechanics of Deformable Systems]. Kishinev: Polytechn. Inst. Kishinev, pp. 44-83. (In Russian)

J.K. Silverman and J.R. Mays. (1973). A collocation solution of the nonlinear equations for axisymmetric bending of shallow spherical shells. *J. of the Franklin Inst.*, 294(3), 181-192.

G.I. Simitses and C. Blackmon. (1975). Snap-through buckling of eccentrically stiffened shallow spherical caps. *Int. J. Solids Structures*, 11, 1035-1049.

G.I. Simitses and R.T. Cole. (1968). Effect of load behavior on the buckling of thin spherical shells under pressure. *J. Appl. Mech.*, E37(4), 1165-1168.

G.I. Simitses and R.T. Cole. (1970). General instability of eccentrically stiffened spherical shells under pressure. *J. Appl Mech.*, 4, 1165-1168.

A.F. Smirnov, A.V. Aleksandrov, N.N. Shaposhnikov and B.J. Ljastchennikov. (1964). *Computation of Structures by Computers.* Moscow: Strojizdat. (In Russian).

L.S. Srubtchik. (1964). About asymptotic integration of the system of nonlinear equation of the theory of plates. *Prikladnaja Mathematika i mechanika (PMM),* 30(2), 335-349. (In Russian)

L.S. Srubtchik and V.I. Yudovitch. (1962). Asymptotic integration of the system equations of great displacements of symmetrically loaded shells of revolution. *Prikladnaja Mathematika i Mechanika (PMM),* 26(5), 913-922. (In Russian)

Y. Stavsky. (1970). Non-linear theory of axisymmetric deformations of heterogeneous shells of revolution. In D. Abir, ed, *Contribution to Mechanics*. Oxford, NY: Pergamon Press, pp. 181-194.

V.P. Stegniy. (1969). About the stability of shallow spherical cap when ring forces are acting. *Izv. vuzov, Machinostroemie*, N12, 34-36. (In Russian).

E. Stein, W. Wagner and K.H. Lambertz. (1984). Geometrically nonlinear theory and incremental analysis of thin shells. In E.L. Axelrod and F.A. Emmerling, eds, *Flexible Shells, Theory and Applications*. Berlin, Heidelberg, New York: Springer-Verlag.

W.B. Stephens and R.E. Fulton. (1969). Axisymmetric static and dynamic buckling spherical caps due to centrally distributed pressures. *AIAA Pap. N69-89*, 1-14.

J.J. Stoker. (1950). *Nonlinear Vibrations in Mechanical and Electrical Systems.* New York.

N.H. Suyunshkakov. (1970). Flexible circular plate in a stationary temperature field. In *Proceedings of the VII All-Union Conference on the Theory of Shells and Plates.* Moscow: Nauka, pp. 574-577. (In Russian).

I.V. Svirsky. (1968). Methods of Bubnov-Galerkin and successive approximation types. Moscow: Nauka. (In Russian).

Yu.A. Tchernucha. (1967). The influence of the nonuniform heating on the stability of spherical cap which undergoes finite deflections and nonsymmetric deformations. In *Thermal Stresses in Elements of Structures, N7.* Kiev: Haukova Dumka, 234-242. (In Russian).

J.M.T. Thompson. (1979). Stability predictions through a succession of folds. *Phil. Trans. of the Royal Soc. of London. Ser. A*, 292(1386), 1-23.

J.M.T. Thompson and G.W. Hunt. (1973). *A General Theory of Elastic Stability.* London: John Wiley.

J.M.T. Thompson and G.W. Hunt. (1984). *Elastic Instability Phenomena.* Chichester, New York, Toronto: John Wiley and Sons.

J.M.T. Thompson and L.N. Virgin. (1988). Spatial chaos and localization in nonlinear elasticity. *Physics Letters,* A126(8,9), 491-496.

G.A. Thurston. (1961). A numerical solution of the nonlinear equations for axisymmetrical bending of shallow spherical shells. *J. Appl. Mech.*, 28(4), 557-562.

G.A. Thurston. (1964). Asymmetrical buckling of spherical caps under uniform pressure. *AIAA J.*, 2(10), 1832-1833.

G.A. Thurston. (1971). A new method for computing axisymmetrical buckling of spherical caps. *J. Appl. Mech.*, E38(1), 185-193.

G.A. Thurston and F.A. Penning. (1966). Effect of axisymmetric imperfections on the buckling of spherical caps under uniform pressure. *AIAA J.*, 4(2), 319-327.

S.C. Tillman. (1970). On the buckling behavior of shallow spherical caps under a uniform pressure load. *Int. J. Solids, Struct.*, 6(1), 37-52.

S. P. Timoshenko. (1915). About large displacements of circular plates. In *Collection of Works of the Institute of Engineers on Ways of Communications, v.89,* 1-10, Petrograd. (In Russian)

S. P. Timoshenko. (1955). *Vibration Problems in Engineering*, 3rd edn. Toronto, New York, London: D. van Nostrand Co.

S.P. Timoshenko and L.M. Gere. (1961).*Theory of Elastic Stability*. New York: McGraw-Hill.

S.P. Timoshenko and J.M. Gere. (1963). *Theory of Elastic Stability*, 2nd edn. New York: McGraw-Hill.

S.P. Timoshenko and J.N. Goodier. (1970). *Theory of Elasticity*, 3rd edn. New York: McGraw-Hill.

S. P. Timoshenko and S. Woinowsky-Krieger. (1959). *Theory of Plates and Shells*, 2nd edn. New York: McGraw-Hill.

E. Touti. (1967). Variational principles in elastostatics. *Meccanica*, 2(4), 201-208.

E. Trefftz. (1928). *Handbuch der Physik, v.6*, p. 130. (In German).

E. Trefftz. (1930). Über die Ableitung der Stabilitätskriterien des elastischen Gleichgewichts aus der Elastizitätstheorie endlicher Deformationen. *Proceedings, International Congress of Applied Mechanics, v.3*, Stockholm, p. 44. (In German)

E. Trefftz. (1933). Zur Theorie der Stabilität des elastischen Gleichgewichts. *Zeitschr. f. angew. Math. u Mech.*, Bd. 13, p. 160. (In German)

F.G. Tricomi. (1961). *Differential Equations.* London: Blackie and Son Ltd.

K. Uchiyama and M. Yamada. (1974). Buckling of clamped imperfect thin shallow spherical shells under external pressure. (i). the effects of geometrically symmetrical initial imperfections. *Techn. Reports, Tokyo University,* 39(1), 101-132.

K. Uchiyama and M. Yamada. (1975). Buckling of clamped imperfect thin shallow spherical shells under external pressure. (ii). The effects of geometrically asymmetrical initial imperfections. *Tech. Reports, Tokyo Univ.*, 40(1), 1-23.

D.P. Updike and A. Kalnins. (1970). Axisymmetrical behavior of an elastic spherical shell compressed between rigid plates. *J. Appl. Mech.*, Sept, 635-640.

D.P. Updike and A. Kalnins. (1972a). Contact pressure between an elastic spherical shell and a rigid plate. *J. Appl. Mech.*, Dec, 1110-1114.

D.P. Updike and A. Kalnins. (1972b). Axisymmetric postbuckling and nonsymmetric buckling of a spherical shell compressed between rigid plates. *J. Appl. Mech.*, March,

172-178.

D.B. Vainberg. (1973). *Reference Book on Strength, Stability and Vibrations of Plates.* Kiev: Budivelnik. (In Russian).

N.V. Valishvili. (1970). On stability of shallow spherical shells. *Proc. VII All-Union Conference on the Theory of Shells and Plates.* Moscow: Nauka, pp. 114-119. (In Russian)

N.V. Valishvili. (1976). *Methods of Computer Calculations of Shells of Revolution.* Moscow: Mashinostroenije. (In Russian)

N.V. Valishvili and V.P. Stegniy. (1970). About the stability of shallow spherical shells when concentrated forces are acting. *Izv. vuzov. Mashinostrojenie*, N9, 28-30. (In Russian).

N.V. Valishvili and V.B. Sylkin. (1975). Strength calculations of plates when big displacements take place. In *Computation of the Strength of Structures in Machine Design.* Moscow: Machine Design, pp. 145-156. (In Russian).

A. van der Neut. (1932). De elastische stabiliteit van de dunwandinge bol. Dissert. Delft. (In Dutch).

A.B. Vasyljeva and V.F. Butuzov. (1973). *Asymptotic Series of Solutions of Singular Perturbed Equations.* Moscow: Fizmatgiz. (In Russian)

M.I. Vishik and L.A. Lyusternik. (1958). About asymptotics of solution of boundary problems for quasilinear differential equations. *Doklady Akad. Nauka SSSR,* 121(5). (In Russian)

V.Z. Vlasov. (1958). *General Theory of Shells and Its Applications to Engineering.* [Transl. from Russian, 1949; NASA Techn. Translation TTF-99, 1964. Also, in German (Berlin: Akademic Verlag, 1958).]

A.S. Volmir. (1956) *Flexible Plates and Shells.* Moscow: Gostechnizdat. (In Russian). [Also in German translation: A.S. Wolmir, *Biegsamme platten und Shalen.* Berlin: VEB, 1961.]

A.S. Volmir. (1963). *Stability of Elastic Systems.* Moscow: Fizmatgiz. (In Russian) [English transl. by Air Force USA FTD 64-335, 1964.]

T. von Karman and H.S. Tsien. (1939). The buckling of spherical shells by external pressure. *J. Aeronaut. Sci.*, 7(2), 43-50.

I.I. Vorovitch. (1955). On existence of solutions in the nonlinear theory of shells. *Izvestija AN SSSR, Math Series,* 19(4), 173-186. (In Russian)

I.I. Vorovitch. (1957). About the existence of solutions in the nonlinear theory of shells. *Doklady AN SSSR,* 117(2), 203-206. (In Russian)

I.I. Vorovitch and N.I. Minakova. (1968a). Investigations of stability of nonshallow spherical cap. *Prikladnaja Mathematika i Mechanika (PMM)*, 32(2), 332-338. (In Russian).

I.I. Vorovitch and N.I. Minakova. (1968b). Investigations of stability of nonshallow spherical cap by high approximations algorithm. *Mechanika Tverdogo Tela*, N2, 121-128b. (In

Russian)

I.I. Vorovitch and N.I. Minakova. (1973). The problem of stability and numerical methods in the theory of spherical shells. In *Mechanics of Solid Deformable Bodies, v.7: Summaries of Science and Technology.* Moscow: VINITI, pp. 5-87. (In Russian)

I.I. Vorovitch and L.S. Srubtschik. (1970). Asymptotic analysis of basic equations of the nonlinear theory of shallow shells. In *Proc. of the VII All-Union Conf. on Theory of Shells and Plates.* Moscow: Nauka, pp. 156-159. (In Russian)

I.I. Vorovitch and V.F. Zipalova. (1965). About the solution of nonlinear boundary value problems of the theory of elasticity by the method of transition to the problem of Cauchy. *Prikladnaja Math. i Mech. (PMM)*, 29(5), 894-901. (In Russian).

P.M. Vyrlan and D. Shilkrut. (1978). On the stability of the forms of equilibrium of geometrically non-linear spherical shells. *Izv. AN SSSR, Mechanika Tverdogo Tela,* 13(4), 170-176. (In Russian) [Translated into English — *Journal of Mechanics of Solids,* 1979]

W. Wagner. (1985). Eine Geometrish Nichtlineare Theorie Schubelastischer Schalen mit Anwendeng auf Finite-Element-Berechnungen. Doctoral Dissertation, Engineering. Hanover: University of Hanover, March. (In German)

L.R.L. Wang. (1967). Discrepancy of experimental buckling pressures of spherical shells. *AIAA J.*, 5, 357-359.

K. Washizu. (1968). *Variational Methods in Elasticity and Plasticity.* New York: Pergamon Press. [Later editions exist.]

S. Way. (1934). Bending of circular plates with large deflection. *Trans. ASME*, 56(8), 627-638.

H.J. Weinitschke. (1962). Asymmetric buckling of clamped shallow spherical shells under uniform pressure. NASA Techn. Note, No. D1510.

H.J. Weinitschke. (1965). On asymmetric buckling of shallow spherical caps. *J. Math. and Phys.*, 44(2), 141-163.

H.J. Weinitschke. (1988). On uniqueness of axisymmetric deformations of elastic plates and shells. *SIAM J. Math. Anal.*, 19(3), 580-592.

P.E. Wilson and E.E. Spier. (1965). Numerical analysis of large axisymmetric deformations of thin spherical shells. *AIAA J.*, 3(9), 1716-1725.

A.E.R. Woodcock and T. Poston. (1970). *A Geometrical Study of the Elementary Catastrophes.* Berlin, Heidelberg, New York: Springer-Verlag.

M.T. Wu and Shun Cheng. (1970). Nonlinear asymmetric buckling of truncated shells. *J. Appl. Mech.*, Sept, 651-660.

S. Yaghmi and E.P. Popov. (1971). Incremental analysis of large deflections of shells of revolution. *Int. J. Solids Structures*, 7, 1375-1393.

T.Y. Yang and H.W. Kim. (1973). Vibration and buckling of shells under initial stresses.

AIAA J., 11(11), 1525-1531.

M. Yanowitch. (1956). Non-linear buckling of circular elastic plates. *Communications in Pure and Appl. Math.*, 9(4), 633-659.

J.C. Yao. (1963). Buckling of a truncated hemisphere under axial tension. *AIAA J.*, 1(10), 2316-2319.

Y. Yoshimura. (1951). On the mechanism of buckling of a circular cylindrical shell under axial compression. *Rep. of the Inst. of Sci. and Techn.*, Univ. of Tokyo, 5(5).

H. Ziegler. (1968). *Principles of Structural Stability*. Waltham, Toronto, London: Blaisdell.

O.C. Zienkiewicz. (1971). Incremental displacement in non-linear analysis. *Int. J. Num. Method. Engng.*, 3, 587-588.

O.C. Zienkiewicz. (1977). *The Finite Element Method.* New York: McGraw-Hill.

V.F. Zipalova. (1967). The stability of cap with hinged supports. *Mechanika Tverdogo Tela*, N1, 172-177. (In Russian).

V.F. Zipalova and V.M. Nenastjeva. (1966). Investigation in high approximations of the snapping of spherical shell subjected to ring loading. *Proc. of the VI All-Union Conf. on the Theory of Shells and Plates (Baku).* Moscow: Nauka, pp. 413-414. (In Russian).

V.F. Zipalova and V.M. Nenastjeva. (1971). About the application of the Galerkin method in problems of stability of shells subjected to temperature and forces excitations. *Mechanika Tverdogo Tela,* N1, 100-103. (In Russian).

V.F. Zipalova and V.M. Nenastjeva. (1973). Thermal stability of hinged shallow spherical shell. In *The Theory of Shells and Plates.* Moscow: Nauka, pp. 289-293. (In Russian).

R. Zoelly. (1915). Uber ein Knickungsproblem an des Kügelshale. Dissert. Zurich 1915. (In German).

INDEX

C

D

M

N

O